# Progress on Porous Media Research - Fundamentals and Applications

*Edited by Huijin Xu, Ruming Pan, Hanbing Ke and Liu Yang*

Published in London, United Kingdom

Progress on Porous Media Research - Fundamentals and Applications
http://dx.doi.org/10.5772/intechopen.1007994
Edited by Huijin Xu, Ruming Pan, Hanbing Ke and Liu Yang

Contributors
Ana Arenillas de la Puente, Ana Letícia Pilz de Castro, Ana Luísa Santos Martins, Antonio Montes, Antonio Santos Sánchez, Anum Shafiq, Ashwini Sharan, Balasubramanian Karuppiah, Celia Toyos Rodríguez, Chen-Min Zheng, Chuang-Yao Zhao, Clara Pereyra, Daniela Antunes Lessa, Diego Valor, Ghulam Rasool, Guilherme Augusto Magalhães Gomes, Huijin Xu, Ignacio Cameán Martínez, Ignacio García-Casas, Kenneth S. Sorbie, Li-Yuan Yang, Ludisbel León-Marcos, Luis Henríquez-Vargas, Luiz Henrique Cardoso, Malla Reddy Perati, Mario Sánchez Suárez, Muhammad Ijaz Khan, Murugesan S.N., Natalia Rey Raap, Pablo Donoso-García, Rames C. Panda, Ruming Pan, Shreedevi Kalyan, Sobana Subramani, Sundararaman T.R., Venkatachalam Varadharajan, Yu-Ming Guo, Yuhan Jin



First published in London, United Kingdom, 2025 by IntechOpen
IntechOpen is the global imprint of INTECHOPEN LIMITED, registered in England and Wales, registration number: 11086078, 167-169 Great Portland Street, London, W1W 5PF, United Kingdom

For EU product safety concerns: IN TECH d.o.o., Prolaz Marije Krucifikse Kozulić 3, 51000 Rijeka, Croatia, info@intechopen.com or visit our website at intechopen.com.

British Library Cataloguing-in-Publication Data
A catalogue record for this book is available from the British Library

Progress on Porous Media Research - Fundamentals and Applications
Edited by Huijin Xu, Ruming Pan, Hanbing Ke and Liu Yang
p. cm.
Print ISBN 978-0-85466-866-3
Online ISBN 978-0-85466-865-6
eBook (PDF) ISBN 978-0-85466-867-0

# Meet the editors

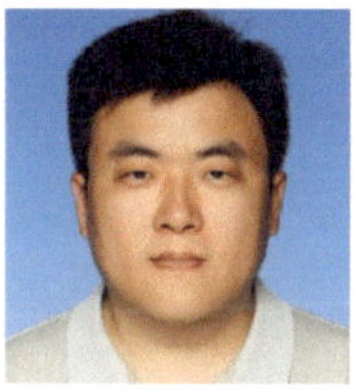

Huijin Xu is a Professor at Shanghai Maritime University, China. He graduated from Xi'an Jiaotong University, China. He has a strong record in research on porous media, carbon capture, ocean energy, and seawater desalination, as well as educating undergraduates and postgraduates, with notable achievements. He acts as the co-organizer for Interpore conferences (Interpore 2019, Interpore 2020, Interpore 2021; Interpore 2022, and Interpore 2023). He is the associate editor of STRPM and JPM in Begell House.

Ruming Pan is an Associate Professor at Harbin Institute of Technology, China. He earned his Ph.D. from Toulouse INP in France and focuses on reactive porous media, solid waste valorization, and solar thermochemical processes. An award-winning researcher, he has delivered keynote talks and serves as a Youth Editorial Board Member of Rare Metals. He has authored 34 journal papers and holds three Chinese patents.

Hanbing Ke is a Senior Researcher at the Wuhan Second Ship Design and Research Institute. He holds a Ph.D. from Xi'an Jiaotong University and works on thermodynamics, heat and mass transfer, and applications in porous media for shipping and energy systems. He has published 30 journal papers and led several technical research projects.

Liu Yang is an Associate Professor at China University of Mining and Technology, Beijing. He graduated from China University of Petroleum, Beijing, and researches soft rock mechanics, $CO_2$ geological storage, and multi-physics coupling in fractured rocks. He serves as a reviewer and editorial board member for over 20 journals and holds more than 30 invention patents.

# Contents

# Preface

Porous media research has become a unifying force across energy, environmental, and materials science, bridging fundamental physics with urgent societal needs. Enabled by advances in computation, diagnostics, and fabrication, we can now revisit long-standing transport, structure, and scaling challenges from new angles. This book comprises eleven chapters written by scientists, engineers, and technologists who share a common conviction: porous media are not an academic niche, but a decisive lever for generating clean energy, securing water, and manufacturing advanced materials. Together, these contributions offer a concise snapshot of the state of the art and a roadmap for the decade ahead.

The book is organized into two complementary sections.

Section I, "Theoretical Aspects", revisits, refines, and extends the conceptual pillars upon which all reliable engineering predictions must rest. Here, the reader will find rigorous analyses of solar-driven biomass valorization, magneto-hydrodynamic radiative heat transfer, and the enduring Rapoport scaling laws that govern multiphase flow in porous networks. We examine the elusive physics of micropolar fluids in complex geometries and revisit classical poroelasticity to shed new light on plane-stress phenomena. Even the seemingly applied problem of aquifer prospecting by airborne geophysics is treated from a foundational perspective: the chapter presents a methodology whose practical value is inseparable from the physical insight that underpins it.

Section II, "Engineering Applications", moves from insight to impact. Flow-boiling intensification strategies, porous-media-controlled combustion, supercritical processing of biodegradable polymers, and sol-gel electrocatalysts for sustainable energy conversion are presented as tangible demonstrations of how porous-media science translates into cleaner processes and novel products. A concluding chapter surveys the landscape of emerging development technologies, offering a state-of-the-art realization that integrates lessons learned throughout the volume.

We owe sincere gratitude to the contributors, who distilled years of research into concise yet comprehensive chapters, and to the reviewers, whose constructive scrutiny sharpened every argument and figure. We thank the editorial and production teams for transforming a collection of manuscripts into a coherent and visually appealing book. We also acknowledge the funding agencies and institutions that provided support, making this work possible.

Whether you are a student entering the field, a researcher seeking fresh insight, or an engineer tackling real-world designs, we hope this book serves as both a concise reference and lasting inspiration. Porous media lie beneath our feet, inside our devices, and within our bodies, yet their secrets are far from fully revealed. May these pages speed your own discoveries. We thank every contributor, reviewer, and the IntechOpen team for bringing this volume to life; above all, we thank you, the reader, whose curiosity will define the next chapter of porous-media science.

**Huijin Xu**
Shanghai Maritime University,
Shanghai, China

**Ruming Pan**
Harbin Institute of Technology,
Harbin, China

**Hanbing Ke**
Wuhan Second Ship Design and Resource Institute,
Wuhan, China

**Liu Yang**
China University of Mining and Technology,
Beijing, China

Section 1

# Theoretical Aspects

Chapter 1

# Theoretical Insights on Solar-Driven Biomass Valorization

*Yuhan Jin, Huijin Xu and Ruming Pan*


## Abstract

The growing global energy demand and increasing environmental challenges have accelerated research into renewable energy utilization technologies. Biomass and solar energy, as promising renewable resources, can be combined through solar-driven pyrolysis/gasification for producing biofuels and biochar. This study reviews comprehensive numerical simulation methods for solar-driven biomass thermochemical conversion, focusing on solar heat source treatment methods, the equivalent porous media model of heat and mass transfer, and common reaction kinetics models. Three principal solar heat source modeling approaches are examined: temperature boundary, heat flux boundary, and Monte Carlo ray tracing (MCRT), with MCRT providing superior accuracy in capturing directional radiation in complex reactor geometries. The thermal analysis incorporates local thermal equilibrium (LTE) and local thermal non-equilibrium (LTNE) approaches within the equivalent porous media model, along with the radiative transfer models including the Rosseland approximation, P1 approximation, and Discrete Ordinates (DO) method. LTNE is crucial for systems with pronounced solid-fluid temperature gradients, while the Rosseland and P1 approximation provide the efficient computational methods for media with high optical thickness. Three fundamental reaction kinetics models are presented for biomass pyrolysis: single-component single-step, single-component multistep, and multicomponent multistep reaction kinetic models. Multistep reaction kinetics models are shown to capture intermediate pathways and product distributions more realistically than single-step approaches. The numerical model coupling multi-physical fields can provide a solid theoretical basis for optimizing reactor design, regulating target products, and enhancing energy conversion efficiency. Future research should integrate high-fidelity simulations with experimental validation to enhance predictive accuracy and support scale-up of solar-driven biomass conversion systems.


**Keywords:** solar energy, biomass, numerical simulation, equivalent media model, reaction kinetics models

## 1. Introduction

The global energy demand escalation driven by population expansion and improved living standards has led to unprecedented pressure on energy systems. Over the past decades, conventional fossil energy sources, including petroleum, natural gas, and coal, have dominated the energy mix, accounting for approximately 80% of global consumption [1–3]. This heavy reliance has resulted in severe environmental consequences, including environmental pollution and accelerated climate change due to carbon emissions. Renewable energy development emerges as an essential strategy to address these challenges. Renewable energy systems provide a sustainable solution by diversifying energy sources and reducing fossil fuel dependence, thereby stabilizing energy supply, mitigating environmental impacts, and achieving carbon neutrality targets.

Among various renewable energy sources, biomass, which is widely available globally and considered a clean energy due to its low nitrogen and sulfur content, has garnered considerable attention in the recent decades for its potential contributions to energy security and sustainable development. Biomass is a carbonaceous material consisting principally of carbon (C), hydrogen (H), and oxygen (O) in its elemental composition, which stores chemical energy that could be liberated through direct combustion. Furthermore, biomass could be transformed through thermochemical conversion into higher energy-density liquid, solid, and gaseous products. These conversion methods enable biomass to yield solid, liquid, and gaseous fuels while generating minimal greenhouse gas emissions and exhibiting negligible contribution to the greenhouse effect [4]. Studies have demonstrated that biomass utilization can achieve carbon neutrality with appropriate process design. Pyrolysis and gasification represent the predominant thermochemical conversion pathways for transforming biomass into sustainable biofuels, value-added chemicals, and functional biochar materials.

The thermochemical conversion of biomass pyrolysis has been industrially practiced since the nineteenth century, primarily for charcoal production from woody biomass [5–7]. During pyrolysis, biomass undergoes gradual thermal decomposition reactions in an anaerobic or oxygen-deficient environment to produce biochar, bio-oil, and pyrolysis gas. Biochar is a porous medium containing carbon that can be used for carbon sequestration. Furthermore, biochar application can enhance soil water retention capacity and improve soil fertility, thereby achieving economic and environmental benefits in agricultural systems [8, 9]. Bio-oil represents a promising chemical feedstock that can be further upgraded to high-quality fuels or utilized directly for electricity and heat generation [9]. Similarly, pyrolysis gas containing syngas as its major component can be employed for thermal applications or power production [9]. Notably, pyrolysis product distribution is significantly affected by operational parameters including temperature, residence time, and heating rate. Pyrolysis temperature has a significant impact on product distribution: Lower temperatures favor biochar formation, while higher temperatures shift product selectivity toward gaseous species. Furthermore, an optimal intermediate temperature range exists that maximizes bio-oil production. The gasification technology was first demonstrated in 1659 through the thermal conversion of coal to methane-rich syngas [10]. Gasification and pyrolysis are similar thermochemical processes that convert carbon-based materials through controlled heating. However, gasification occurs at relatively higher temperatures (>700°C) and produces mainly gas and ash as residues. The syngas can be directly utilized in boilers/engines/turbines or catalytically processed into liquid fuels and chemicals [11]. The gasification process yields ash as a byproduct, consisting of inorganic residues, which exhibits potential applications in soil enhancement and construction materials [12, 13].

Simultaneously, solar energy represents a widely distributed, abundant, and pollution-free renewable resource, with total potential power output significantly exceeding global annual energy consumption [14]. However, a significant portion of solar energy cannot currently be fully utilized and converted into useful energy, due to technological limitations and thermodynamic constraints, leaving substantial potential for research and development to enhance conversion efficiency [15]. Nowadays, solar energy utilization can be classified into three primary approaches: photochemical, photovoltaic, and photothermal conversion. Photochemical processes employ solar radiation to drive chemical reactions [15, 16]. Photovoltaic systems directly transform solar irradiance into electrical energy through the photovoltaic effect in semiconductor materials [17]. Photothermal technologies transform solar energy into thermal energy for applications including power generation [18]. Solar photochemical technologies can serve as a heat source to drive chemical reactions and integrate with biomass thermochemical conversion processes for targeted chemical production.

Solar-driven biomass thermochemical conversion represents an energy transformation and storage approach that utilizes concentrated solar radiation as the thermal energy source to initiate biomass chemical reactions, thereby storing solar energy in high-density chemical form. Solar-driven pyrolysis/gasification constitutes a significant thermochemical conversion technology that synergistically integrates two renewable resources, solar energy and biomass, to generate high-value products including bio-oil, pyrolysis gas, and biochar. The utilization of solar energy to provide heat for pyrolysis reactors presents an attractive solution for meeting the demands of highly endothermic conversion reactions when adequate solar radiation is accessible. The approach represents an emerging technology that integrates concentrated solar energy with biomass thermochemical conversion within the reactor system, referred to as solar-driven biomass thermochemical conversion. Additionally, certain biomass types can be characterized as homogeneous porous media exhibiting the anisotropic physical properties attributable to their pore structure. These porous media comprise a solid matrix interconnected with pores containing gases or liquids, which significantly influence heat and mass transfer mechanisms as well as chemical reaction kinetics during the pyrolysis and gasification processes.

Currently, researchers are focused on investigating the pyrolysis reaction mechanisms and biomass decomposition pathways to optimize the product distribution. Experiments with multiphysics-coupled numerical simulations have been integrated to investigate the temperature distribution and mass transport in biomass. Their goal is to produce high-value pyrolysis oil at low temperatures and high-calorific-value gas at high temperatures. However, there is a lack of research on coke preparation and its applications. Therefore, this study aims to comprehensively summarize the numerical models of reactive porous media during solar-driven biomass pyrolysis/gasification, further providing a solid theoretical foundation for optimizing the design of solar pyrolysis/gasification reactors to regulate the target products and improve energy conversion efficiency.

## 2. Solar heat source treatment

### 2.1 Fixed temperature boundary

The fixed temperature boundary refers to the direct use of experimental data (e.g., temperatures measured by thermocouples) as simulation inputs without the need for theoretical derivations or complex physical equations. The temperature of the reactor wall is obtained through infrared temperature measurement or thermocouples, and it is directly set as a boundary condition for the CFD simulation. Zeng et al. [19, 20]

simulated the solar pyrolysis reactor of beech wood porous pellets, incorporating experimentally measured temperatures as thermal boundary conditions. The two-dimensional model results demonstrated good agreement with experimental data across a comprehensive parameter space, spanning final temperatures of 600–2000°C and heating rates of 10–50°C/s. Zuo et al. [21] established a new numerical model applying the experimentally measured temperature input as the solar radiation condition. The model describes the multi-dimensional shrinkage behavior of the packed bed in the solar-driven gasification reactors.

## 2.2 Fixed heat flux boundary

The focused sunlight direct irradiation model simulates a concentrator system to focus sunlight onto the reactor surface, ignoring scattering, assuming parallel sunlight, and considering only direct radiation. The focused spot is simplified to a heat flow boundary condition (e.g., Gaussian distribution) on the reactor surface. The advantage of this radiation model is that it is computationally efficient and suitable for optical design optimization. The disadvantage is that the radiation absorption and scattering of the medium inside the reactor are ignored. The concentrated solar flux distribution is mathematically represented using the Gaussian function to model the incident radiation in the solar thermal reactor system [22]:

$$q_G = I\mathrm{e}^{\left(-\frac{r^2}{2\sigma^2}\right)} \tag{1}$$

The coefficient $\sigma$ is directly associated with the diameter of the light spot [22]. To maximize solar power utilization, it is crucial to maintain the light spot diameter smaller than that of the reactor. Following the three-sigma principle of normal distributions, the criterion $3\sigma < R$ (reactor radius) ensures that the process remains confined within the boundaries of the reactor. The radiative power from the solar concentrator is calculated using the following expression:

$$Q_G = \int_{-\infty}^{+\infty} q_G \cdot 2\pi r dr \tag{2}$$

Zhang et al. [22] developed a numerical model using the Gaussian distribution to represent the spatial distribution of incident solar radiative flux at the aperture surface, aiming to determine the optimal structural parameters of the ceramic foam. The simulated results demonstrate that ceramic foam with higher porosity and larger pore size achieves optimal thermochemical performance. Liu et al. [23] studied the capability of silicon carbide (SiC) foam to improve reaction rates in the solar-driven packed-bed gasification reactor, employing the Gaussian distribution to model the spatial solar flux profile at the cavity aperture surface. The results showed that incorporating SiC foam could obviously improve the thermal conductivity of the filled bed.

## 2.3 Monte-Carlo ray-tracing

For modeling radiation transport processes with directional and spectral properties, the most widely used method is the Monte-Carlo Ray-Tracing method for calculating the concentrated irradiation intensity on the irradiated surfaces [24, 25]. The MCRT method has proven to be an effective computational approach for determining heat flux distributions in solar collectors and receiver systems, as demonstrated in numerous previous studies [26–28]. The MCRT method is a stochastic computational simulation approach based on probabilistic statistics. The method is recognized for its simplicity, scalability, accuracy, and capacity to handle complex geometries, and has found broad application in concentrated solar studies. This method is particularly valued in concentrated solar power research due to its inherent advantages: algorithmic simplicity, excellent scalability, high computational accuracy, and powerful capability for modeling complex geometries. The MCRT method solves radiative transfer problems by statistically simulating numerous energy-carrying rays interacting with materials and interfaces. Each ray carrying the same energy is tracked and judged, and its propagation direction is randomly determined by the probability density function. A series of independent sub-processes (emission, reflection, absorption, and scattering behavior) are represented based on a specific probabilistic model during transmission. The reflective direction of each ray adheres to Fresnel's laws of reflection.

For solar radiation simulations, the incident solar irradiance $q_{\text{sun}}$ is adopted as the direct normal irradiance. The beam number density, $n_s$, is defined as the quantity of sampled photon bundles emitted per unit area. Consequently, each ray carries an energy of $q_{\text{sun}}/n_s$. The receiver surface concentrated heat flux $q_{\text{r,i}}$, based on energy conservation principles, can be formulated as [29]:

$$q_{\text{r,i}} = \frac{q_{\text{sun}}}{A_{\text{i}} n_{\text{s}}} \sum_{j=1}^{N} M_i^j \tag{3}$$

where $q_{\text{r,i}}$ is the concentrated radiation flux on the $i$th surface element of porous media receiver, $A_{\text{i}}$ is the area of the $i$th element, and $N$ is the total number of rays incident on the aperture plane. If the $j$th ray arrives at the $i$th surface element, $M_i^j = 1$, otherwise $M_i^j = 0$.

Chen et al. [30] adopted a Monte-Carlo Ray-Tracing simulation to investigate the spatial distribution of absorbed radiative energy within the porous absorber and the optical performance of the entire system. They used the volumetric receiver with a solar dish concentrator system. Incident sunlight strikes the concentrator as a solar cone, with the solar shape effect represented by a columnar distribution, implying uniform radiation intensity within the solar cone. Solar radiative transfer is computationally modeled using a custom MCRT code, tracing sunlight rays from the concentrator to the porous absorber. Following Monte Carlo principles, the incident ray's zenith and azimuth angles are determined through the following probabilistic model [30]:

$$\theta = \arcsin \sqrt{R_\theta \sin^2 \theta_s} \tag{4}$$

$$\varphi = 2\pi R_\varphi \tag{5}$$

where $\theta_s$ is the half solar cone angle of 4.65 mrad, and $R_\theta$ and $R_\varphi$ are uniformly distributed random variables on the interval [0,1].

Based on the above, the MCRT method is selected due to its superior capability in accurately modeling the directional and spectral characteristics of radiation transport processes. Unlike the simplified boundary conditions, MCRT effectively captures complex interactions such as emission, reflection, absorption, and scattering within materials and at interfaces, adhering to physical laws like Fresnel's reflection. By statistically simulating numerous energy-carrying rays, the method provides high-precision predictions of heat flux distributions in solar concentrators and receiver systems, accommodating complex geometries and heterogeneous media. Furthermore, the MCRT approach inherently satisfies energy conservation principles, ensuring physically consistent and numerically stable results. Its comprehensive treatment of radiative transfer phenomena makes it particularly suitable for optimizing the design and performance evaluation of solar-driven thermochemical reactors, where accurate representation of concentrated solar irradiation is critical.

## 3. Heat and mass transfer models

Biomass particles undergo morphological evolution during pyrolysis, including shrinkage, abrasion, and pore structure transformation [31]. These microscale changes directly alter critical properties (e.g., particle size, density, porosity), thereby significantly impacting heat/mass transfer and reaction kinetics. However, elucidating microstructural evolution remains challenging due to inherent complexity and experimental limitations [32]. DiBlasi [33] introduced a shrinkage factor to model the biomass particle shrinkage phenomenon. This model assumes a linear relationship between particle microstructure collapse and pyrolytic conversion. However, such models solely consider the effect of the conversion degree, neglecting other critical factors. Consequently, progress in microscale biomass morphology models has stagnated, primarily due to the enormous challenges behind the problem.

However, the equivalent medium model for porous media can transform the microscopic complex pore structure into macroscopic equivalent parameters by homogenization, avoiding the need to directly resolve the details of each pore or particle. This model can significantly reduce the computational effort and is suitable for reactor-scale simulation of biomass pyrolysis over long periods. Currently, the modeling of heat transport in porous media employs two approaches: local thermal equilibrium (LTE) and local thermal non-equilibrium (LTNE), differentiated by interfacial temperature treatment. The LTE model assumes identical fluid-solid phase temperatures, whereas LTNE resolves separate energy equations for each phase. Notably, LTNE requires higher computational costs than LTE, necessitating method selection based on geometric and thermo-hydraulic conditions.

### 3.1 Continuity equation and momentum equation

Continuity equation:

$$\frac{\partial\left(\varepsilon\rho_f\right)}{\partial t}+\nabla.\left(\rho_f\vec{v}\right)=R_m \tag{6}$$

where$\varepsilon$ is the porosity, $\rho_f$ is the density, $\vec{v}$ is the velocity, and $R_m$ is the quality source.

Momentum equation:

$$\frac{\rho_f}{\varepsilon}\frac{\partial\vec{v}}{\partial t}+\frac{\rho_f}{\varepsilon^2}\left(\vec{v}\cdot\nabla\right)\vec{v}=-\nabla p+\frac{\mu}{\varepsilon}\nabla^2\vec{v}-\left(\frac{\mu}{K}+\frac{c_F\rho_f}{\sqrt{K}}|\vec{v}|+\frac{R_m}{\varepsilon^2}\right)\vec{v} \tag{7}$$

where $p$ is the pressure, $\mu$ is the dynamic viscosity, $K$ is the effective permeability, and $c_F$ is the Forchheimer parameter.

### 3.2 Energy equation

#### *3.2.1 LTE*

In the Local Thermal Equilibrium (LTE) model, it is assumed that the solid matrix and the fluid phase share the same temperature at any local position. This assumption is valid when the interphase heat transfer rate is sufficiently high to eliminate any significant temperature difference between phases, leading to a single, unified temperature field for the porous medium. Consequently, only one energy equation is required to describe the thermal behavior of the system, making the LTE approach suitable for scenarios with low fluid velocities or highly efficient heat exchange between phases.

Energy equation:

$$(c_p\rho)_e\frac{\partial T}{\partial t} + (c_p\rho)_f\vec{v}\cdot\nabla T = \nabla\cdot\left(k_{eff}\nabla T\right) + Q \tag{8}$$

$$(\rho c_P)_e = \varepsilon(\rho c_P)_f + (1-\varepsilon)(\rho c_P)_s \tag{9}$$

$$k_{eff} = \varepsilon k_f + (1-\varepsilon)k_s \tag{10}$$

where $(c_p\rho)_e$ is the effective material property of the porous media, which is calculated as the volume-fraction-weighted average of the solid and fluid phase properties. $k_{eff}$ is the effective thermal conductivity. $Q$ is the reaction heat.

#### *3.2.2 LTNE*

In the Local Thermal Non-Equilibrium (LTNE) model, the temperatures of the solid and fluid phases are allowed to differ locally. This approach accounts for finite interphase heat transfer rates and the resulting temperature gradients between phases by solving separate energy equations for the solid and fluid domains, coupled through an interphase heat exchange term. The LTNE model is particularly appropriate for conditions involving high fluid velocities, low heat transfer coefficients, or significant local temperature differences within the porous structure, where the LTE assumption may no longer hold.

Energy equation:

for fluid phase

$$\varepsilon(\rho c_P)_f\frac{\partial T_f}{\partial t} + (\rho c_P)_f\vec{v}.\nabla\left(T_f\right) = \varepsilon\nabla.\left(k_f\nabla T_f\right) + h_{sf}a_{sf}\left(T_s - T_f\right) + \varepsilon S_f \tag{11}$$

for solid phase

$$(1-\varepsilon)(\rho c_P)_s\frac{\partial T_s}{\partial t} = (1-\varepsilon)\nabla.(k_s\nabla T_s) + h_{sf}a_{sf}\left(T_f - T_s\right) + (1-\varepsilon)S_s \tag{12}$$

where $c_P$, $k_f$, $k_s$, $h_{sf}$, and$a_{sf}$ are the heat capacity, fluid conductivity, solid conductivity, interfacial heat transfer coefficient, and specific surface area. $S_f$ and $S_s$ are heat sources in the fluid phase and solid phase.

The LTE model, while widely adopted for predicting thermo-hydraulic properties in porous media, exhibits significant limitations. It remains valid only when solid-fluid temperature differences are negligible and interfacial temperatures are constant.

However, in complex systems (e.g., nuclear reactors, high-speed flows), such temperature gradients become critical, rendering LTE's equilibrium assumption invalid. Therefore, the LTNE model is essential, as it explicitly resolves conjugate heat transfer by accounting for internal solid-fluid phase heat exchange.

## 3.3 Radiation heat transfer equation

The radiative transfer equation (RTE) in the numerical simulation is as follows [34]:

$$\frac{dI\left(\vec{r},\vec{s}\right)}{ds} = -\kappa_e I\left(\vec{r},\vec{s}\right) + \kappa_a I_b(s) + \frac{\kappa_s}{4\pi}\int_{4\pi} I\left(\vec{r},\vec{s}'\right)\Phi\left(\vec{s},\vec{s}'\right)d\Omega' \tag{13}$$

where the symbol $I$ is the local radiation intensity in the porous media, which depends on the position vector $\vec{r}$ and direction vector$\vec{s}$, $I_b$ is the blackbody radiative intensity, the phase function $\Phi$ varies with direction, $\vec{s}'$ is the scattering direction vector, and $\Omega'$ is the solid angle. $\kappa_e$, $\kappa_a$, and$\kappa_s$ are the extinction, absorption, and scattering coefficient of porous media, $\kappa_e = \kappa_a + \kappa_s$.

The Rosseland approximation, the P1 approximation, or the discrete coordinate (DO) method are commonly used for radiative transfer processes involving photosensitive media. In high-temperature environments, radiative heat transfer becomes the predominant mode of thermal exchange in porous media. Due to strong solar absorption, these media typically exhibit large optical thicknesses ($\tau > 5$) and short radiative mean free paths. The Rosseland and P1 approximation are applicable for problems with large optical thickness, whereas the Discrete Ordinates (DO) model is suitable for a wider range of optical thicknesses [35].

### *3.3.1 Rosseland approximation*

The Rosseland approximation is applicable to participating media with high optical thickness, where radiative transfer behaves similar to a diffusion process. It assumes that the radiation intensity is nearly isotropic and that the radiative heat flux can be modeled using an effective radiative thermal conductivity. This simplification transforms the radiative transfer equation into a conduction-type equation, significantly reducing the computational complexity. Boundary conditions for the Rosseland approximation typically involve specifying temperature or heat flux at the domain boundaries, with radiation boundary conditions treated as equivalent conduction conditions. Convective and surface radiative heat exchange can also be incorporated at boundaries to reflect realistic thermal environments.

The Rosseland approximation applies to systems with greater optical thicknesses than the P1 approximation, offering better computational efficiency. The Rosseland approximation provides fairly good predictions compared to experimental results for investigating the detailed mechanisms of radiation transfer in porous media [36]. With the Roseland approximation applied, the radiative transfer equation simplifies to the following equation [34]:

$$q_r = \frac{4\sigma n^2}{3\kappa_e}\frac{dT^4}{dx} \tag{14}$$

The extinction $\kappa_e$, absorption $\kappa_a$ and scattering $\kappa_s$ coefficients are calculated based on the assumption of geometrical optics approximation:

$$\kappa_\alpha = \frac{3e(1-\varepsilon)}{2d_p} \quad (15)$$

$$\kappa_s = \frac{3(2-e)(1-\varepsilon)}{2d_p} \quad (16)$$

$$\kappa_e = \kappa_\alpha + \kappa_s = \frac{3(1-\varepsilon)}{d_p} \quad (17)$$

where $d_p$ is the pore size, $\varepsilon$ is the porosity, and $e$ is the emissivity of the bulk surface.

As evident from Eq. (14), the Rosseland approximation offers significant computational advantages. By defining irradiative thermal conductivity as $k_r$:

$$k_r = \frac{16\sigma n^2 T^3}{3\kappa_e} \quad (18)$$

So that

$$q_r = k_r \frac{dT}{dx} \quad (19)$$

The radiative heat transfer problem is simplified to the thermal conduction problem featuring the highly temperature-dependent effective thermal conductivity. Wang et al. [29] adopted the LTNE model coupled with the Rosseland approximation to analyze the temperature field evolution in porous media receivers.

### 3.3.2 P1 approximation

The P1 approximation is a first-order spherical harmonics method suitable for media with moderate-to-high optical thickness. It approximates the radiative intensity as the sum of an isotropic component and a linearly anisotropic term, enabling the modeling of absorption, scattering, and emission processes within the participating medium. This approach assumes a relatively diffuse radiation field with the limited directional variation. Boundary conditions generally require specification of incident radiation intensity or radiative flux, while reflection can be modeled as either diffuse or specular at surfaces. The P1 approximation can be coupled with convective and surface radiation boundary conditions to accurately represent the combined heat transfer mechanisms.

In the porous media region of solar receiver, concentrated solar radiation penetrates through the aperture surface into the participating medium, where it undergoes absorption, scattering, and heat conversion. For porous media with large optical thickness, the P1 approximation (an approximate spherical harmonics method) is adopted. The radiative transfer equation simplifies to [37]:

$$-\nabla \cdot \left[\frac{1}{3(\kappa_a + \kappa_s)} \nabla G_d\right] = \kappa_a \left(4\sigma T_s^4 - G_d\right) \quad (20)$$

where $G_d$ is the diffused integrated intensity, and σ is the Stefan-Boltzmann constant.

Moreover, the modified P1 approximation [38] could better model radiative transfer, treating the irradiation intensity as a superposition of collimated and diffuse components. This approach effectively accounts for the highly directional nature of the solar radiation in the system. The radiative heat transfer equation applying the modified P1 approximation to account for both the diffuse integrated intensity $G_s$ and coherent radiative intensity $G_c$ is as follows [38]:

$$\nabla\left[\left(\frac{1}{3 - A_1\omega}\right)\nabla G_s\right] + (1-\omega)(4\pi I_b - G_s) + \omega G_c = 0 \tag{21}$$

$$G_c = q_s e^{-\tau} \tag{22}$$

The divergence of irradiative heat transfer is

$$\nabla q_s = (1-\omega)(4\pi I_b - G_s - G_c) \tag{23}$$

where $A_1$ designates the first-order coefficient of the modified P1 approximation and $\tau$ is the optical thickness. The albedo ω could be described as [38]:

$$\omega = \frac{\kappa_s}{\kappa_\alpha + \kappa_s} \tag{24}$$

The extinction, absorption, and scattering coefficients $\kappa_e$, $\kappa_a$, $\kappa_s$in the P1 approximation, are consistent with those in the Rosseland approximation.

The P1 approximation and Rosseland approximation are among the most frequently adopted approaches for radiative transfer problems in porous media. Wang et al. [38] systematically evaluated the effect of radiative transfer models on the thermal performance of porous media receiver, comparing the Rosseland approximation with the modified P1 approximation. The results show that solid-fluid temperature differences between the modified P1 and Rosseland approximations disappear at high optical thickness with the higher fluid velocities. Consistent temperature distributions diminish with increasing solar irradiance and mean cell size, while lower fluid inlet velocity yields greater relative temperature differences at the outlet surface. The simulation shows that the difference in peak temperatures predicted by the modified P1 approximation and the Rosseland approximation is minimal for the porous media receiver. However, the thickness of the thermal non-equilibrium region is smaller with the Rosseland approximation compared to the modified P1 approximation. The external concentrator system uses ray tracing and the reactor interior uses Rosseland or P1 approximation modeling. This model can better simulate reactive porous media reactions (e.g., biomass particles) within a solar reactor. Wu et al. [39] developed the equivalent media model to describe the radiation transfer in participating porous media and established the radiation transfer model integrating the ray tracing method. The model offered pioneering theoretical insights that can guide the further optimization of system-level photothermal catalysis processes.

### *3.3.3 Discrete Ordinates Model*

The Discrete Ordinates Model (DOM) solves the irradiative transfer equation (RTE) *via* directional and spatial discretization, generating coupled linear equations for radiative intensity at each point [40]. This approach requires solving as many transport equations as discrete directions and introduces no simplifying assumptions except for partial differential equations with discretization [40], ensuring high accuracy. The model discretizes the computational domain of polar/azimuthal directions into solid angles and solves the RTE for each direction.

$$\nabla\left(I\left(\vec{r},\vec{s}\right)\vec{s}\right)+(\alpha+\sigma_s)I\left(\vec{r},\vec{s}\right)=\alpha n^2\frac{\sigma T^4}{\pi}+\frac{\sigma_s}{4\pi}\int_0^{4\pi}I\left(\vec{r},\vec{s}'\right)\Phi\left(\vec{s}\cdot\vec{s}'\right)d\Omega' \tag{25}$$

where the symbol $I$ represents the local radiation intensity dependent on the position vector $\vec{r}$ and direction vector $\vec{s}$. $a$ and $\sigma_s$ are the absorption and scattering coefficients. $n$ is the refractive index of the media. $\sigma$ is the Stefan-Boltzmann constant. The phase function $\Phi$ varies with direction. $\vec{s}'$ is the scattering direction vector, and $\Omega'$ is the solid angle.

Over the years, the Discrete Ordinate Method (DOM) has been widely utilized to simulate various types of photochemical and photocatalytic reactors, owing to the high accuracy. Pareek and Adesina [40] employed the MC method to estimate the light intensity distribution in an annular photocatalytic reactor while utilizing the DO model to assess the effect of wall reflectivity, catalyst loading, and the phase function parameters on the light intensity distribution. Bellouard et al. [41] developed a novel solar spouted-bed reactor incorporating continuous biomass feeding, with reactor design optimized using the Discrete Ordinates radiation model. The gasification rates and gas yields were enhanced through continuous stirring of the biomass particles, enabling steady operation. Zhang et al. [42] established a simulated model with the local thermal equilibrium model, Discrete Ordinates Model, and dry reforming of methane reaction kinetics to analyze the thermochemical performance of a catalytic porous material-filled solar reactor. The results indicate that the peak heat flux of the incident solar radiation correlates linearly with reaction temperature, enhancing $CO_2$ conversion and CO yield but increasing temperature differences and carbon deposition. The reaction temperature varies quadratically with the inlet flow rate (IFR), with low IFR improving syngas yield and limiting carbon formation, while excessive IFR reduces quality and $H_2$/CO ratio.

In summary, the P1 approximation and Rosseland approximation are widely adopted for modeling radiative transfer in porous media due to their suitability for systems with large optical thickness and their ability to significantly simplify the radiative transfer equation (RTE) while maintaining adequate accuracy. Both methods have been validated against experimental data, showing minimal differences in predicted peak temperatures for porous media receivers, and can be readily coupled with external ray-tracing for concentrated solar input as well as reactive porous media heat transfer models such as LTE or LTNE. This balance of computational efficiency, physical fidelity, and integration capability makes these methods well-suited for designing and optimizing solar-driven porous media reactors.

### 3.4 Mass equation

Species equation

$$\frac{\partial\varepsilon\rho_f Y_i}{\partial t}+\nabla\cdot\left(\rho_f Y_i\vec{v}\right)=-\nabla\cdot\left(\vec{J}_i\right)+R_i \tag{26}$$

where $\rho_f$ and $\vec{v}$ are the density and velocity. $Y_i$, $J_i$, and $R_i$ are the mass fraction, diffusive mass flux, and mass generated from reactions for species $i$.

Consequently, the equivalent porous medium model more accurately simulates the pyrolysis reaction process in reactive porous media (e.g., biomass particles) within solar reactors. Zeng et al. [19, 20] established a two-dimensional transient CFD model for single-particle porous media to simulate beech wood pyrolysis under solar irradiation. The simulated results exhibited strong concordance with experimental data, validating the model's ability to predict biomass pyrolysis product yields and compositions. Furthermore, the model demonstrates broad applicability to pyrolysis processes and can be readily integrated with reactor-scale simulations. Sangaré et al. [43] implemented a numerical model analyzing species transport and flow in porous media to study reactivity and kinetics during slow pyrolysis of lignocellulosic biomass. The model predicted non-condensable gas yields with mean relative errors of 6.90% for α-cellulose and 11.59% for avocado stone. Qi et al. [44] developed a pyrolysis and secondary cracking kinetics model incorporating porous media characteristics (porosity, permeability, conductivity) and transient gas-phase processes governed by Darcy's law, convective heat transfer, diffusion, and radiation. The results demonstrate that porous media flow characteristics affect volatile residence time, thereby influencing secondary cracking reaction products.

In conclusion, the porous media approach incorporating particulate heat transfer represents the most comprehensive CFD model for biomass pyrolysis simulation. The particle-scale simulation approach could simulate the entire reactor and obtain detailed and accurate characterizations of biomass pyrolysis. However, the current computational level confines the model to the validation of laboratory-scale systems, precluding reliable application to industrial-scale processes.

## 4. Kinetics models

The most commonly accepted simplification for approximating complex chemical reaction chains is the simple decomposition of biomass feedstocks into tar, gas, and char. There are many ways to categorize chemical reaction kinetics models for biomass pyrolysis. The most common method is to classify existing kinetics models into single-component single-step, single-component multistep, and multicomponent multistep reaction kinetics models.

### 4.1 Single-component single-step reaction kinetics

The single-component single-step reaction kinetics model is a simplified reaction kinetics modeling approach. The kinetics model's central feature is the generalization of complex biomass pyrolysis reaction networks into a single macroscopic reaction step. Biomass pyrolysis under constant temperature conditions is described, and typically biomass materials such as wood, straw, and waste are converted into biochar, bio-oil, and syngas products. The pyrolysis products can also be described in detail as char, tar, and the gas mixture of $H_2O$, $H_2$, $CH_4$, CO, and $CO_2$. The single-component single-step reaction equation is expressed in Eq. (27). The significant drawback of this kinetics model is the constant composition of the process products, independent of the reaction conditions. Muktar Bashir et al. [18] modeled the pyrolysis biomass in a solar thermal reactor incorporating single-component single-step reaction kinetics models. The reactor reached a total energy efficiency of 67.8%, with the product distribution of 51.5% bio-oil, 43.7% char, and 4.8% non-condensable gases. Yu et al. [45] proposed an Eulerian-Eulerian CFD model coupled with a single-component single-step pyrolysis reaction kinetics to study the biomass fast pyrolysis featuring a novel gas-solid separation mechanism. The final product composition consisted of 56.85% bio-oil,

37.87% bio-char, and 5.28% non-condensable gas, and the model demonstrated robustness and high efficiency in computational time.

$$\text{Biomass} \rightarrow \alpha_1\text{char} + \alpha_2\text{oil} + \alpha_3\text{H}_2\text{O} + \alpha_4\text{H}_2 + \alpha_5\text{CO} + \alpha_6\text{CO}_2 + \alpha_7\text{CH}_4 \tag{27}$$

where $\alpha_1$-$\alpha_7$ are the stoichiometric coefficients of the pyrolysis reaction.

### 4.2 Single-component multistep reaction kinetics

The single-component multistep reaction kinetics model is important in biomass pyrolysis kinetics. The foundational framework assumes that there are multiple parallel reaction paths in the biomass pyrolysis process, which compete with each other to determine the final product distribution (e.g., char, gas, tar). This type of model is closer to the actual reaction mechanism than the single-component single-step reaction kinetics model, but still maintains a relatively simple mathematical form. The competing model considers the temperature dependence of the reaction rate. The model describes the entire biomass pyrolysis as completely independent sub-reactions occurring at different intensities depending on the given conditions. In addition, the single-component multistep reaction kinetics model can characterize the intermediate and final products in the system in detail. The model can also capture the different levels of detail of secondary reactions. The reaction rate for each sub-reaction follows first-order Arrhenius kinetics, as shown:

$$k_j = A_j \exp\left(\frac{E_{aj}}{RT}\right) \tag{28}$$

where $j$ is the index of each reaction, $A$ is the pre-exponential factor, $E_a$ is the activation energy, and $R$ is the ideal gas constant.

Papadikis et al. [46] simulated the fluid-particle interaction and the effect of shrinkage on biomass pyrolysis within a 150 g/h fluidized bed reactor using the two-step competing model shown in **Figure 1**. The results indicate that particle shrinkage had a negligible effect on both product yields and pyrolysis duration for thermally thin biomass particles. Hooshdaran et al. [47] developed a hydrodynamic model integrated with the pyrolysis kinetics model consisting of three parallel reactions and a secondary reaction converting bio-oil into gas in **Figure 2**. The simulation results showed excellent agreement with experimental data, accurately predicting both the flow characteristics and the yields and production rates of pyrolysis products.

### 4.3 Multicomponent multistep reaction kinetics

The multicomponent multistep reaction kinetics model considers biomass as a mixture of three main components: cellulose, hemicellulose, and lignin. This model is

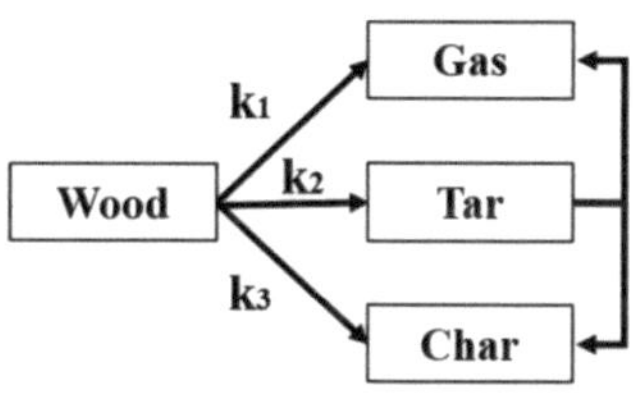


**Figure 1.**
*Two-stage, single-component kinetics model from the study by Papadikis et al. [46].*

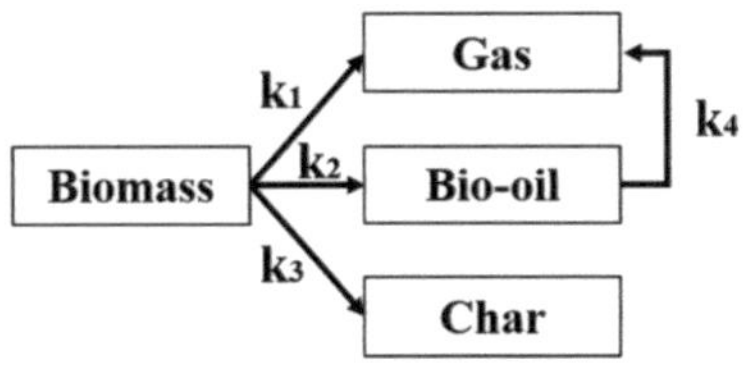


**Figure 2.**
*The biomass fast pyrolysis kinetics model from the study by Hooshdaran et al. [47].*

more complex and closer to the real reaction mechanism in biomass pyrolysis kinetics models. Moreover, the model combines the characteristics of multicomponent decomposition and multistep competing reaction and can more accurately describe the independent decomposition paths of different components (cellulose, hemicellulose, and lignin) and their interactions during biomass pyrolysis. The model can predict the product distribution with high accuracy, distinguishing between the sources of gas, tar, and coke (e.g., cellulose contributes mainly to tar and lignin to coke). By explicitly resolving biomass constituents and mechanistically representing reaction pathways, the multicomponent multistep reaction kinetics model demonstrates superior performance to simplified approaches in both predictive accuracy of product distributions and mechanistic interpretability of pyrolysis processes.

Therefore, the multicomponent multistep reaction kinetics model is particularly applicable to mechanistic studies of pyrolysis processes or applications requiring precise product control. Xue et al. [48] established an Euler-Euler multiphase CFD model that integrates the biomass pyrolysis kinetics model, incorporating simultaneous hemicellulose, cellulose, and lignin decomposition reactions in **Figure 3**. The simulated product yields revealed strong agreement with the experimental data for pure cellulose and red oak pyrolysis. Liu et al. [49] established an Eulerian-Eulerian multiphase CFD model incorporating the multicomponent multistep chemical reaction kinetics mechanism to simulate biomass fast pyrolysis. The model was further utilized to investigate the evolution of biomass particle size under different particle shrinkage patterns within the fluidized bed system.

Therefore, multistep reaction kinetics models are preferred for biomass pyrolysis because they more accurately represent the actual reaction mechanisms than simplified single-step approaches. By describing biomass decomposition through multiple parallel or competing reactions, these models capture the temperature dependence of reaction rates and the effects of operating conditions on product distribution. The models also provide detailed characterization of intermediate and secondary reactions, improving predictive accuracy and mechanistic interpretability. The ability to represent processes at varying levels of complexity, from single-component competing schemes to multicomponent models resolving cellulose, hemicellulose, and lignin decomposition. These features make multistep kinetics well suited for mechanistic studies, precise product control, and integration into CFD simulations for optimizing biomass pyrolysis systems.

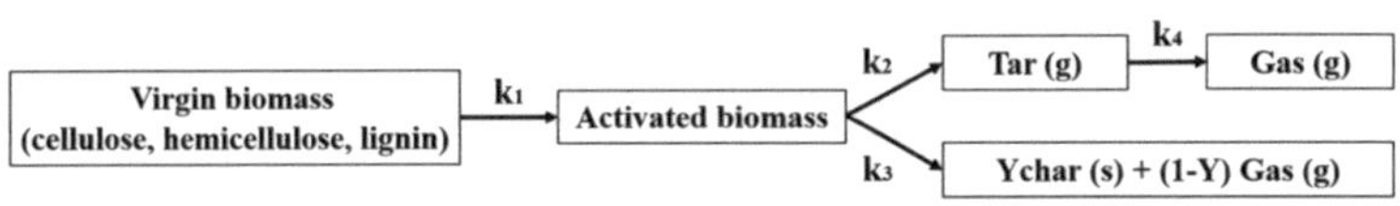


**Figure 3.**
*Kinetic scheme for cellulose, hemicellulose, and lignin from the study by Xue et al. [48].*

## 5. Conclusions

Solar-driven biomass thermochemical conversion technology provides a sustainable pathway for producing high-value fuels and carbon-rich materials. This study presents a systematic analysis of the simulation theoretical mechanism for solar-driven biomass pyrolysis/gasification, with particular emphasis on reactive porous media modeling approaches. This study highlights the importance of accurately simulating the solar heat source, heat and mass transfer, radiative transfer, and chemical kinetics within the porous media biomass for improving reactor performance and product yield. This study discusses common heat source treatments, the equivalent media models of heat and mass transfer, and reaction kinetics models. The Monte Carlo Ray Tracing (MCRT) provides higher accuracy for directional radiation in complex geometries. The Equivalent media models effectively simplify the microscale complexity and achieve efficient reactor-scale simulations. The Local Thermal Non-Equilibrium (LTNE) model is essential for systems with significant temperature gradients between the solid and liquid phases. For radiative transfer, the Rosseland and P1 approximations provide computationally efficient solutions for optically thick media. The multistep reaction kinetics models more accurately represent actual reaction mechanisms compared to single-component single-step models, facilitating a comprehensive description of the intermediate and final products. Therefore, a comprehensive numerical model of solar-driven biomass pyrolysis/gasification could be established through these coupled simulation methods, which could achieve the goals of optimizing the solar pyrolysis/gasification reactor design and regulating the product yields, as well as enhancing the energy conversion efficiency. This study innovatively combines advanced multiphysics modeling approaches, establishing a foundation for subsequent research that integrates high-fidelity numerical simulations with the experimental validation. This synergistic methodology is crucial for enhancing predictive precision and facilitating the scaling and industrial deployment of solar-driven biomass conversion systems, ultimately contributing to the advancement of sustainable energy solutions.

## Acknowledgements

This work was supported by National Key R&D Program of China (No. 2024YFE0116800) and Natural Science Foundation of Heilongjiang Province, China (No. LH2024E040).

## Conflict of interest

The authors declare no conflict of interest.

## Author details

Yuhan Jin[1], Huijin Xu[2] and Ruming Pan[1]*

1 School of Energy Science and Engineering, Harbin Institute of Technology, Harbin, China

2 Merchant Marine College, Shanghai Maritime University, Shanghai, China

*Address all correspondence to: ruming.pan@hit.edu.cn

## References

[1] Wang Z, Song H, Liu H, et al. Coupling of solar energy and thermal energy for carbon dioxide reduction: Status and prospects. Angewandte Chemie International Edition. 2020;**59**: 8016-8035. DOI: 10.1002/anie.201907443

[2] Gustavsson L, Nguyen T, Sathre R, et al. Climate effects of forestry and substitution of concrete buildings and fossil energy. Renewable and Sustainable Energy Reviews. 2021;**136**:110435. DOI: 10.1016/j.rser.2020.110435

[3] Yolcan OO. World energy outlook and state of renewable energy: 10-Year evaluation. Innovation and Green Development. 2023;**2**(4):100070. DOI: 10.1016/j.igd.2023.100070

[4] Sulaiman C, Abdul-Rahim AS, Ofozor CA. Does wood biomass energy use reduce $CO_2$ emissions in European Union member countries? Evidence from 27 members. Journal of Cleaner Production. 2020;**253**:119996. DOI: 10.1016/j.jclepro.2020.119996

[5] Demirbas A. Biomass resource facilities and biomass conversion processing for fuels and chemicals. Energy Conversion and Management. 2001;**42**(11):1357-1378. DOI: 10.1016/S0196-8904(00)00137-0

[6] Demirbas MF, Balat M. Biomass pyrolysis for liquid fuels and chemicals: A review. Journal of Scientific & Industrial Research. 2007;**66**(10): 797-804

[7] Jahirul MI, Rasul MG, Chowdhury AA, et al. Biofuels production through biomass pyrolysis-a technological review. Energies. 2012;**5**(12):4952-5001. DOI: 10.3390/en5124952

[8] Sánchez M, Clifford B, Nixon JD. Modelling and evaluating a solar pyrolysis system. Renewable Energy. 2018;**116**:630-638. DOI: 10.1016/j.renene.2017.10.023

[9] Anca-Couce A. Reaction mechanisms and multi-scale modeling of lignocellulosic biomass pyrolysis. Progress in Energy and Combustion Science. 2016;**53**:41-79. DOI: 10.1016/j.pecs.2015.10.002

[10] Sikarwar VS, Zhao M, Clough P, et al. An overview of advances in biomass gasification. Energy & Environmental Science. 2016;**9**(10): 2939-2977. DOI: 10.1039/c6ee 00935b

[11] Molino A, Larocca V, Chianese S, et al. Biofuels production by biomass gasification: A review. Energies. 2018; **11**(4):811. DOI: 10.3390/en11040811

[12] Skousen J, Yang JE, Lee JS, et al. Review of fly ash as a soil amendment. Geosystem Engineering. 2013;**16**(3): 249-256. DOI: 10.1080/12269328.2013.832403

[13] Azevedo P, Costa P. Electrical driven pyrolysis reactor retrofit for indirect concentrated solar heat. Thermal Science and Engineering Progress. 2024;**50**: 102524. DOI: 10.1016/j.tsep.2024.102524

[14] Anon. Statistical Review of World Energy. London, UK: BP p.l.c; 2022. p. 2022

[15] Spasiano D, Marotta R, Malato S, et al. Solar photocatalysis: Materials, reactors, some commercial, and pre-industrialized applications. A comprehensive approach. Applied Catalysis B: Environmental. 2015; **170-171**:90-123. DOI: 10.1016/j.apcatb.2014.12.050

[16] Cambié D, Dobbelaar J, Riente P, et al. Energy-efficient solar

photochemistry with luminescent solar concentrator based photomicroreactors. Angewandte Chemie. 2019;**58**: 14374-14378. DOI: 10.1002/anie.201908553

[17] Sobri S, Koohi-Kamali S, Rahim NA. Solar photovoltaic generation forecasting methods: A review. Energy Conversion and Management. 2018;**156**:459-497. DOI: 10.1016/j.enconman.2017.11.019

[18] Bashir M, Yu X, Hassan M, et al. Modeling and performance analysis of biomass fast pyrolysis in a solar-thermal reactor. ACS Sustainable Chemistry & Engineering. 2017;**5**(5):3795-3807. DOI: 10.1021/acssuschemeng.6b02806

[19] Zeng K, Soria J, Gauthier D, et al. Modeling of beech wood pellet pyrolysis under concentrated solar radiation. Renewable Energy. 2016;**99**:721-729. DOI: 10.1016/j. renene.2016.07.051

[20] Soria J, Zeng K, Asensio D, et al. Comprehensive CFD modelling of solar fast pyrolysis of beech wood pellets. Fuel Processing Technology. 2017;**158**: 226-237. DOI: 10.1016/j.fuproc.2017.01.006

[21] Zuo H, Zeng K, Zhong D, et al. Multi-dimensional shrinkage models developed by phase field method for gasification of carbonaceous feedstock in packed-bed solar reactor. Fuel. 2023;**331**: 125749. DOI: 10.1016/j.fuel.2022.125749

[22] Zhang H, Shuai Y, Lougou BG, et al. Effects of foam structure on thermochemical characteristics of porous-filled solar reactor. Energy. 2020; **239**:122219. DOI: 10.1016/j.energy.2021.122219

[23] Liu Y, Pan R, Ansart R, et al. Numerical simulation of solar-driven biomass gasification by using ceramic foam. Process Safety and Environmental Protection. 2024;**184**:300-313. DOI: 10.1016/j.psep.2024.02.008

[24] Jin J, Wei X, Liu M, et al. A solar methane reforming reactor design with enhanced efficiency. Applied Energy. 2018;**226**:797-807. DOI: 10.1016/j.apenergy.2018.04.098

[25] Tong K, Chen L, Yang L, et al. Energy transport of photocatalytic carbon dioxide reduction in optical fiber honeycomb reactor coupled with trough concentrated solar power. Catalysts. 2021;**11**(7):829. DOI: 10.3390/catal11070829

[26] Li ZG, Tang DW, Du LT. Study on the radiation flux and temperature distributions of the concentrator-receiver system in a solar dish/stirling power facility. Applied Thermal Engineering. 2011;**31**:1780-1789. DOI: 10.1016/j.applthermaleng.2011.02.023

[27] Nedea SV, Markvoort AJ, Steenhoven AV, et al. Heat transfer predictions for micro-/nanochannels at the atomistic level using combined molecular dynamics and Monte Carlo techniques. Journal of Heat Transfer. 2009;**131**(3): 033104. DOI: 10.1115/1.3056592

[28] He YL, Cheng ZD, Cui FQ, et al. Numerical investigations on a pressurized volumetric receiver: Solar concentrating and collecting modelling. Renewable Energy. 2012;**44**:368-379. DOI: 10.1016/j.renene.2012.02.001

[29] Wang F, Shuai Y, Tan H, et al. Thermal performance analysis of porous media receiver with concentrated solar irradiation. International Journal of Heat and Mass Transfer. 2013;**62**:247-254. DOI: 10.1016/j.ijheatmasstransfer.2013.03.003

[30] Chen X, Xia X, Dong X, et al. Integrated analysis on the volumetric

absorption characteristics and optical performance for a porous media receiver. Energy Conversion and Management. 2015;**105**:562-569. DOI: 10.1016/j.enconman.2015.08.028

[31] Kumar RR, Kolar AK, Leckner B. Shrinkage characteristics of Casuarina wood during devolatilization in a fluidized bed combustor. Biomass and Bioenergy. 2006;**30**:153-165. DOI: 10.1016/j.biombioe.2005.11.005

[32] Xiong Q, Xu F, Pan Y, et al. Major trends and roadblocks in CFD-aided process intensification of biomass pyrolysis. Chemical Engineering & Processing: Process Intensification. 2018; **127**:206-212. DOI: 10.1016/j.cep.2018.04.005

[33] DiBlasi C. Heat, momentum and mass transport through a shrinking biomass particle exposed to thermal radiation. Chemical Engineering Science. 1996;**51**:1121-1132. DOI: 10.1016/0009-2509(95)00356-8

[34] Modest M. Radiative Heat Transfer. 3rd ed. San Diego: Academic Press; 2013

[35] Barbosa ISO, Santos RJ, Dias MM, et al. Radiation models for computational fluid dynamics simulations of photocatalytic reactors. Chemical Engineering and Technology. 2023;**46**(6):1059-1077. DOI: 10.1002/ceat.202200551

[36] Zhao CY. Review on thermal transport in high porosity cellular metal foams with open cells. International Journal of Heat and Mass Transfer. 2012; **55**:3618-3632. DOI: 10.1016/j.ijheatmass transfer. 2012.03.017

[37] Chen X, Wang FQ, Yan X, et al. Thermal and chemical analysis of methane dry reforming in a volumetric reactor under highly concentrated solar radiation. Solar Energy. 2018;**162**: 187-195. DOI: 10.1016/j.solener.2018.01.032

[38] Wang FQ, Tan JY, Shuai Y, et al. Thermal performance analyses of porous media solar receiver with different irradiative transfer models. International Journal of Heat and Mass Transfer. 2014; **78**:7-16. DOI: 10.1016/j.ijheatmasstransfer.2014.06.035

[39] Wu LX, Lougou BG, Jiang B, et al. Experimentally validated numerical model of multi-spectral bands radiative transport in solar receiver/reactor with photo-active porous absorber reacting media. Energy Conversion and Management. 2023;**278**:116740. DOI: 10.1016/j.enconman.2023.116740

[40] Pareek VK, Adesina AA. Light intensity distribution in a photocatalytic reactor using finite volume. American Institute of Chemical Engineers. 2004; **50**:1273-1288. DOI: 10.1002/aic.10107

[41] Bellouard Q, Rodat S, Abanades S, et al. Design, simulation and experimental study of a directly-irradiated solar chemical reactor for hydrogen and syngas production from continuous solar-driven wood biomass gasification. International Journal of Hydrogen Energy. 2019;**44**:19193-19205. DOI: 10.1016/j.ijhydene.2018.04.147

[42] Zhang H, Shuai Y, Yuan Y, et al. Thermal-chemical reaction characteristics of $Ni/Al_2O_3$ catalytic porous material filled solar reactor for dry reforming of methane process. Applied Thermal Engineering. 2020;**180**: 115901. DOI: 10.1016/j.applthermaleng.2020.115901

[43] Sangaré D, Moscosa-Santillan M, Bostyn S, et al. Multi-step kinetic mechanism coupled with CFD modeling of slow pyrolysis of biomass at different

heating rates. Chemical Engineering Journal. 2024;**479**:147791. DOI: 10.1016/j.cej.2023.147791

[44] Qi GL, Wang ZW, Zhang S, et al. Numerical simulation on biomass-pyrolysis and thermal cracking of condensable volatile component. International Journal of Hydrogen Energy. 2020;**45**:12283-12297. DOI: 10.1016/j.ijhydene.2020.02.199

[45] Yu X, Hassan M, Ocone R, et al. A CFD study of biomass pyrolysis in a downer reactor equipped with a novel gas-solid separator-II thermochemical performance and products. Fuel Processing Technology. 2015;**133**:51-63. DOI: 10.1016/j.fuproc.2015.01.002

[46] Papadikisa K, Gub S, Bridgwater AV. CFD modelling of the fast pyrolysis of biomass in fluidised bed reactors: Modelling the impact of biomass shrinkage. Chemical Engineering Journal. 2009;**149**:417-427. DOI: 10.1016/j.cej.2009.01.036

[47] Hooshdaran B, Haghshenasfard M, Hosseini SH, et al. CFD modeling and experimental validation of biomass fast pyrolysis in a conical spouted bed reactor. Journal of Analytical and Applied Pyrolysis. 2021;**154**:105011. DOI: 10.1016/j.jaap.2020.105011

[48] Xue Q, Dalluge D, Heindel TJ, et al. Experimental validation and CFD modeling study of biomass fast pyrolysis in fluidized-bed reactors. Fuel. 2012;**97**:757-769. DOI: 10.1016/j.fuel.2012.02.065

[49] Liu B, Papadikis K, Gu S, et al. CFD modelling of particle shrinkage in a fluidized bed for biomass fast pyrolysis with quadrature method of moment. Fuel Processing Technology. 2017;**164**:51-68. DOI: 10.1016/j.fuproc.2017.04.012

Chapter 2

# Recent Theoretical Approaches to MHD Radiative Heat Transfer in Chemically Reactive Powell-Eyring Nanofluid Flow through Porous Media

*Ghulam Rasool, Anum Shafiq and Muhammad Ijaz Khan*

## Abstract

The chapter establishes a comprehensive theoretical basis for analyzing magnetohydrodynamic (MHD) radiative heat and mass transport in chemically reactive Powell-Eyring nanofluid flow through porous media. The model includes nonlinear thermal radiation, heat sink/source terms, Brownian motion, thermophoresis, and convective boundary conditions. The nanofluid's electromagnetic properties benefit from an angled magnetic field, but the low Reynolds number assumption allows us to disregard magnetic induction effects. The conversion of Cartesian boundary-layer equations into nonlinear ODEs through similarity transformations results in nonlinear ODEs. The numerical solution of these ODEs reveals how dimensionless factors influence velocity, temperature, and concentration values. We evaluate and discuss the engineering quantities such as the skin-friction coefficient, Nusselt number, and Sherwood number. The results demonstrate that thermal radiation and chemical reactions, together with porous media properties and magnetic field effects, substantially influence non-Newtonian nanofluid heat and mass transmission. The obtained results can be used in complex thermal management systems, propulsion devices, gas turbines, nuclear energy systems, and aeronautical engineering technologies.

**Keywords:** Powell-Eyring nanofluid flow, magnetohydrodynamics (MHD), nonlinear thermal radiation, chemically reactive flows, porous media heat, mass transfer

## 1. Introduction

Nanofluids, formed by dispersing nanoparticles into base fluids, have emerged as efficient heat transfer agents due to their enhanced thermal conductivity and energy transport capabilities. Among various rheological models, the Powell-Eyring fluid model is particularly useful for describing non-Newtonian behavior in complex industrial systems such as polymer processing, geothermal plants, and thermal management systems. The flow behavior of nanofluids becomes even more intricate when subject to multiple interacting phenomena such as magnetic fields, nonlinear radiation, porous structures, and chemical reactions. These coupled effects play a pivotal role in applications including nuclear reactor cooling, gas turbines, and biomedical heat exchangers. Previous studies have examined nanofluid flow over stretching or shrinking surfaces under various thermal and chemical influences. Despite advancements, limited attention has been given to models that simultaneously incorporate Powell-Eyring rheology, inclined MHD forces, nonlinear radiative heat transfer, binary chemical reactions, and convective boundary conditions in a porous medium. This chapter aims to fill that gap by developing a comprehensive theoretical model that integrates all these mechanisms and by analyzing the resulting thermophysical behavior using numerical simulations.

The dispersion of nanoparticles throughout base fluids resulted in heat transfer nanofluids. Nanofluids with the Powell-Eyring model represent complex rheological behaviors encountered during industrial processes and require additional analysis. The system becomes complicated when these fluids with these features undergo chemical reactions within porous medium flow when exposed to magnetic fields and heat radiation. These phenomena combine their effects for industrial applications, including polymer production, cooling systems, geothermal power generation, and biomedical technologies. Chemical reactivity produces exothermic/endothermic characteristics of nanofluids by modifying liquid dynamics as well as thermal interactions, which affect temperature gradients and concentration distributions, and particle adherence. Example: computer chips, wafers, nuclear reactors. The formulation of the nanofluid directly defines its concept. A base fluid containing large amounts of nanoparticles constitutes a nanofluid. The fluid displays better heat conductivity. Nanofluids developed by Choi [1] have received significant attention in fluid research because they enhance thermal properties. Numerous investigations involving different conditions like MHD, heat source/sink, and chemical reaction exist for linear stretching surfaces and sheets. The nanofluidic flow across a linearly stretched surface with nonlinear internal thermal radiation was analytically studied by Hayat et al. [2], who examined the influence of an inclined magnetic field in combination with heat source/sink terms. Their results demonstrated how radiative heat transport and electromagnetic forces affect the thermal boundary layer in nanofluid systems. Sandeep et al. [3] presented a theoretical investigation of unsteady magnetohydrodynamic (MHD) radiative flow of a dusty nanofluid over an exponentially stretching surface. Their work highlighted how the presence of dust particles modifies thermal and momentum transfer, particularly under transient flow conditions. The influence of porous substrates on nanofluid behavior was evaluated by Ziaei-Rad et al. [4], who employed artificial neural networks to simulate MHD dissipative nanofluid flow over a permeable stretching sheet. Their study provided predictive capabilities for flow and heat transfer in porous and magnetically influenced environments. Mohammadpour et al. [5] investigated forced convection heat transfer of nanofluid flow over a stretching surface in the presence of a heat source. Using similarity transformations, the study analyzed how nanoparticle concentration and heat generation affect the thermal boundary layer, revealing enhanced heat transfer performance due to nanoparticle inclusion. Hsiao [6] explored stagnation-point mixed convection flow of an electrically conducting nanofluid over a stretching surface, incorporating the effects of slip boundary conditions. The study showed that the presence of slip significantly alters the velocity and thermal profiles, which is critical in microscale heat exchanger designs. Zeeshan et al. [7] focused on ferrofluid flow affected by a magnetic dipole over a stretching sheet, including the effects of linear thermal radiation. Their results provided important insights into the heat transfer characteristics of magnetic nanofluids used in advanced thermal systems. Further work on stagnation-point flows was conducted by Nandy and Mahapatra [8], who studied the combined effects of slip and internal heat generation/absorption on MHD nanofluid stagnation flow over a stretching or shrinking surface. This study is important for applications in thermal processing and extrusion. Parsa et al. [9] extended this analysis to MHD boundary-layer flow with internal heat generation or absorption. They examined the interplay between electromagnetic and thermal energy transport in nonuniform heating environments. In a related study, Pavithra and Gireesha [10] investigated the impact of internal heat generation/absorption and viscous dissipation on dusty fluid flow over exponentially stretching surfaces. Their findings contribute to the understanding of dusty plasma and aerosol flows in thermal systems. Mukhopadhyay and Vajravelu [11] analyzed unsteady Maxwell fluid flow over a stretching sheet, incorporating effects of transpiration and internal heat generation/absorption. This work is foundational for modeling complex non-Newtonian flows with energy exchange mechanisms. A more recent contribution by Rasool et al. [12] presented a detailed analysis of Darcy-Forchheimer nanofluid flow over a nonlinearly stretching sheet under the influence of a magnetic field. Their work investigated the interplay of drag forces and convective heat transfer in porous media, revealing complex behavior due to nonlinear momentum interactions. Sheikh and Abbas [13] studied MHD nanofluid flow with thermophoresis and heat generation/absorption in the presence of chemically reactive species. Their findings highlighted how mass transport and chemical reactions can alter thermal performance in MHD systems. In a broader meta-analysis, Animasaun et al. [14] discussed the random motion of nano-sized particles and their impact on physical fluid properties. They consolidated findings from numerous studies, offering generalizations relevant to fluid instability and dispersion behavior. Lund et al. [15] explored the flow of Cu- and Ag-based nanofluids over nonlinear shrinking surfaces, revealing conditions for multiple solutions and flow bifurcations. Their analysis aids in understanding solution stability in highly conductive fluids. Beyond flow modeling, Wang et al. [16] conducted thermodynamic and optical analyses of a hybrid solar CPV/T system employing spectral beam splitting for enhanced concentration uniformity. Their study exemplifies how nanofluid technologies are integrated into renewable energy applications. In environmental engineering, Yu et al. [17] evaluated the degradation performance of nano-$TiO_2$ coatings for asphalt pavements, demonstrating their potential in reducing NOx emissions under solar irradiation. Guo et al. [18] developed composite nanostructures combining zeolitic imidazolate frameworks with carbon nanofibers to sense $\alpha$-synuclein oligomers. Their work is significant for nanobiosensing in biomedical applications. Liu et al. [19] provided a comprehensive review of cement and geopolymer materials modified with graphene and graphene oxide. They detailed enhancements in thermal conductivity, mechanical strength, and durability—vital for green construction. Finally, Yan et al. [20] investigated sodium ion intercalation in a polypyrrole-decorated $NaTi_2(PO_4)_3$ nanocomposite, revealing excellent electrochemical performance. This contributes to the development of advanced energy storage systems like Na-ion batteries.

Fluid flow and heat and mass transfer for a conductive nanofluid across a convectively heated surface have been of interest to geophysical, physical, and industrial nanofluid researchers. Many industrial processes stretch sheets and filaments across quiescent fluids to heat and cool them. Thus, cooling speed and method impact product quality. This cooling process speed may be modified in fluids exposed to magnetohydrodynamics (MHD), allowing regulated product manufacture. Metallurgy is another key MHD use. Using MHD, heat and fluid flow have been thoroughly explored. The literature shows that flows with linear stretching and linear thermal radiation have been well studied, but nonlinearity in these parameters has not been thoroughly investigated, and the study is limited to non-Newtonian and incompressible fluids. Internal thermal radiation plays a crucial role in the design and performance of high-temperature systems such as heating chambers, propulsion devices, gas turbines, nuclear reactors, spacecraft, and satellites, where radiative heat transfer dominates conduction and convection processes. Cortell [21] analyzed the influence of thermal radiation on nonlinear heat transfer in boundary-layer flow over a stretching surface. His study laid the groundwork for understanding how radiative flux modifies thermal gradients and boundary-layer thickness in stretched-sheet configurations, which are common in manufacturing and material processing applications. Shehzad et al. [22] extended this concept to three-dimensional flow of Jeffrey nanoliquids, incorporating nonlinear thermal radiation effects. Their work demonstrated how radiation significantly alters the thermal boundary layer and heat transfer rate in nanofluids with viscoelastic properties, contributing valuable insights into solar energy and cooling technologies. Wakif [23] developed a novel numerical scheme to simulate steady magnetohydrodynamic (MHD) convective flow of Casson fluids over a horizontally stretching sheet. The model included temperature-dependent thermal conductivity and viscosity as well as nonlinear radiative heat flux. This multifactorial approach allowed for a more realistic assessment of nanofluid behavior under combined physical effects, applicable to electronic cooling and biomedical devices. Animasaun et al. [24] investigated the effects of unequal diffusivities in homogeneous and heterogeneous chemical reactions within a viscoelastic fluid subjected to an external magnetic field and nonlinear radiation. Their findings emphasized the need to consider species-dependent diffusion rates in reacting flows, especially when radiation and MHD forces are present, which is relevant in catalytic reactors and plasma applications. Hayat et al. [25] focused on three-dimensional MHD flow with slip boundary conditions under the influence of nonlinear radiative heat transfer. Their study addressed the interaction between velocity slip, stretching surfaces, and radiation absorption, providing a realistic model for applications involving microfluidic and high-speed aerodynamic systems, although their treatment remained limited to simplified geometries and boundary conditions.

In general, the classical Navier-Stokes equations do not adequately represent the complex behavior of non-Newtonian nanofluids. These fluids, including greases, paints, blood, oils, and polymer-saturated solutions, exhibit a nonlinear relationship between stress and strain rate, which significantly complicates their flow characteristics. Because of this complexity, no universal model can fully characterize all types of non-Newtonian fluids, especially those involving nanoparticles. To address this, various rheological models have been proposed, among which the Powell-Eyring fluid model stands out for its ability to capture the shear-thinning behavior of complex fluids derived from kinetic theory rather than empirical fitting. This model has gained wide attention in recent years for its applicability to nanofluid systems under thermal, magnetic, and chemically reactive conditions. Rasool et al. [26] investigated the chemically reactive Powell-Eyring nanofluid flow over a radiative Riga plate under convective boundary conditions. Their study integrated chemical reactions, thermal radiation, and Lorentz forces and provided valuable insights into the thermophysical behavior of nanofluids in electrically actuated systems such as MHD pumps and sensors. Hayat et al. [27] explored magnetohydrodynamic (MHD) Powell-Eyring nanofluid flow with linear thermal radiation and convective boundary conditions. The results showed that radiation and MHD parameters significantly influence the velocity and temperature fields, offering practical implications for heat exchangers and materials processing units operating under electromagnetic fields. The foundational theory for this fluid model was introduced by Powell and Eyring [28], who proposed a mechanism based on the relaxation behavior of polymeric fluids using gas kinetic principles. Their theoretical formulation laid the groundwork for later studies on non-Newtonian flow in industrial and biological systems. Jalil et al. [29] obtained self-similar solutions for the heat and flow characteristics of Powell-Eyring fluid over a moving surface in a free stream. Their analysis, based on similarity transformations, contributed to the understanding of shear-thinning behavior in boundary-layer flows. Mushtaq et al. [30] extended the study of Powell-Eyring fluids to exponentially stretching sheets. They provided both numerical and analytical (series) solutions to predict velocity and temperature distributions, revealing the sensitivity of flow behavior to stretching rates and fluid parameters. Poonia and Bhargava [31] performed a finite element analysis of Powell-Eyring fluid flow under convective boundary conditions. Their work offered numerical verification and a deeper understanding of how wall heating and fluid rheology interact in heat transfer applications. Hayat et al. [32] investigated radiation effects on Powell-Eyring fluid flow past an unsteady inclined stretching sheet, considering nonuniform heat generation or absorption. Their work showed how unsteadiness and spatial heating variations significantly affect thermal boundary layers and flow acceleration or deceleration. In a later study, Hayat et al. [33] explored the influence of homogeneous and heterogeneous chemical reactions in Powell-Eyring fluid flow. Their findings are important for modeling chemically reactive flows in porous and catalytic media, where diffusion and reaction interplay strongly with fluid deformation and heat transfer.

The goal of this chapter is to develop a general theoretical model of the behavior of Powell-Eyring nanofluid flow through porous media in the presence of MHD forces, radiative heat transfer, and chemical reactions.

## 2. Governing equations and assumptions

The flow is assumed to be:

- Two-dimensional, steady, and laminar.
- Incompressible and single-phase.
- Influenced by a uniform magnetic field perpendicular to the flow.
- Subject to thermal radiation and a first-order chemical reaction.
- Through a homogeneous porous medium.

### 2.1 Continuity equation

$$\nabla \cdot \vec{v} = 0 \tag{1}$$

### 2.2 Momentum equation (modified Navier-Stokes with MHD and porous medium effects)

$$\rho\left(u\frac{\partial u}{\partial x} + v\frac{\partial u}{\partial y}\right) = -\frac{\partial p}{\partial x} + \mu\frac{\partial}{\partial y}\left(\frac{1}{\beta}\sinh^{-1}\left(\beta\frac{\partial u}{\partial y}\right)\right) - \sigma B_0^2 u - \frac{\mu}{K}u \tag{2}$$

where:

- $\beta$ is the Powell-Eyring fluid parameter.
- $\sigma$ is the electrical conductivity.
- $B_0$ is the magnetic field strength.
- $K$ is the permeability of the porous medium.

### 2.3 Energy equation (including radiation and viscous dissipation)

$$\rho c_p\left(u\frac{\partial T}{\partial x}+v\frac{\partial T}{\partial y}\right)=k\frac{\partial^2 T}{\partial y^2}-\frac{\partial q_r}{\partial y}+\Phi \tag{3}$$

where $q_r$ is the radiative heat flux and $\Phi$ is the viscous dissipation term.
Using the Rosseland approximation:

$$q_r=-\frac{4\sigma^*}{3k^*}\frac{\partial T^4}{\partial y} \tag{4}$$

For small temperature differences:

$$T^4\approx 4T_\infty^3 T-3T_\infty^4 \tag{5}$$

### 2.4 Species concentration equation

$$u\frac{\partial C}{\partial x}+v\frac{\partial C}{\partial y}=D\frac{\partial^2 C}{\partial y^2}-k_r(C-C_\infty) \tag{6}$$

where $k_r$ is the reaction rate constant.

## 3. Boundary conditions

At the wall ($y=0$):

$$u=0,\quad v=0,\quad T=T_w,\quad C=C_w \tag{7}$$

As $y\to\infty$:

$$u\to u_\infty,\quad T\to T_\infty,\quad C\to C_\infty \tag{8}$$

## 4. Nondimensionalization

Introduce dimensionless variables:

$$\eta=\frac{y}{L},\quad f'(\eta)=\frac{u}{u_\infty},\quad \theta(\eta)=\frac{T-T_\infty}{T_w-T_\infty},\quad \phi(\eta)=\frac{C-C_\infty}{C_w-C_\infty} \tag{9}$$

Define dimensionless parameters:

- Magnetic parameter: $M = \frac{\sigma B_0^2 L^2}{\mu}$
- Radiation parameter: $R = \frac{4\sigma^* T_\infty^3}{k^* k}$
- Prandtl number: $Pr = \frac{\mu c_p}{k}$
- Schmidt number: $Sc = \frac{\mu}{\rho D}$
- Chemical reaction parameter: $\gamma = \frac{k_r L^2}{D}$
- Darcy number: $Da = \frac{K}{L^2}$

The dimensionless equations become:

### 4.1 Dimensionless momentum equation

$$\left(1+\beta\frac{d^2 f}{d\eta^2}\right)\frac{d^2 f}{d\eta^2} - \left(M+\frac{1}{Da}\right)f' = 0 \tag{10}$$

### 4.2 Dimensionless energy equation

$$\frac{1}{Pr}\frac{d^2\theta}{d\eta^2} + R\frac{d\theta}{d\eta} + Ec\left(\frac{df}{d\eta}\right)^2 = 0 \tag{11}$$

where $Ec$ is the Eckert number representing viscous dissipation.

### 4.3 Dimensionless species concentration equation

$$\frac{1}{Sc}\frac{d^2\phi}{d\eta^2} - \gamma\phi = 0 \tag{12}$$

### 4.4 Numerical setup

The system of coupled, nonlinear ordinary differential equations (ODEs) resulting from the similarity transformations was solved numerically using a combination of the shooting method and a fourth-order Runge-Kutta integration scheme. The computational steps are outlined as follows:

- The boundary value problem is first transformed into an initial value problem by estimating the missing initial conditions.
- The nonlinear algebraic system resulting from boundary mismatches is solved using the Newton-Raphson method.
- Numerical integration is carried out using a step size of $\Delta\eta = 0.01$ over the domain $0 \leq \eta \leq 10$.

- The convergence criterion is set to $10^{-6}$ for all functions and their derivatives.
- All simulations were implemented in MATLAB (R2023a) using built-in solvers and custom scripts for parameter sweeps.

This numerical approach ensures stability and accuracy across a broad range of thermophysical parameters, including those related to magnetic effects, porous resistance, radiation, and chemical reactivity.

## 5. Solution techniques

Given the nonlinear and coupled nature of the equations, various techniques can be employed:

- Semi-analytical: Homotopy Analysis Method (HAM), Optimal Homotopy Asymptotic Method (OHAM).
- Numerical: Runge-Kutta method with shooting technique, Finite Difference Method (FDM).

## 6. Engineering relevance: A practical case study

To demonstrate the practical utility of the developed theoretical framework, we consider a representative engineering scenario: a nanofluid-based thermal management system for a nuclear reactor cooling channel. In such systems, effective heat dissipation is critical for safe and efficient operation. Non-Newtonian nanofluids, particularly those governed by Powell-Eyring rheology, offer superior thermal performance and are increasingly being explored for use in advanced cooling loops.

In our model, we assume a reactor wall acts as a nonlinearly stretching surface, with chemically reactive nanofluid circulating over it. A transverse magnetic field is applied at an inclination to suppress turbulent fluctuations and enhance electromagnetic conductance. Heat generation due to radioactive decay and exothermic reactions is modeled as a volumetric heat source. The porous media represent engineered flow pathways in the coolant loop, while radiation effects simulate the high-temperature environment inside the reactor.

Using the specified boundary conditions and physical properties (as outlined in **Table 1**), we simulate the temperature and velocity distributions under typical operating conditions. The results demonstrate the ability of the model to predict thermal boundary-layer thickness, surface heat flux (Nusselt number), and solutal concentration gradients (Sherwood number). Such predictions are crucial for evaluating the effectiveness of nanofluid-based cooling systems in high-performance energy applications.

## 7. Example problem

Consider a boundary-layer problem incorporating a chemically reactive, radiative, Powell-Eyring type of nanofluid flow initiated by an inclined magnetic field under the

| Parameter | Value |
|---|---|
| Magnetic parameter ($M$) | 2.0 |
| Radiation parameter ($R_1$) | 1.5 |
| Prandtl number ($Pr$) | 6.2 |
| Schmidt number ($Sc$) | 0.9 |
| Reaction rate parameter ($\gamma$) | 0.5 |
| Eckert number ($Ec$) | 0.1 |
| Porosity factor ($Fr$) | 0.2 |

**Table 1.**
*Representative dimensionless parameters used in the nuclear cooling channel scenario.*

direct impact of nonlinear radiation, heat source/sink, and Brownian diffusion due to thermophoresis over a nonlinearly stretching surface [34]. The boundary-layer equations are:

*Continuity equation*:

$$\frac{\partial u}{\partial x}+\frac{\partial v}{\partial y}=0 \tag{13}$$

*Momentum equation*:

$$\begin{aligned} v\frac{\partial u}{\partial y}+u\frac{\partial u}{\partial x}=&\left(\nu+\frac{1}{\rho_{\mathrm{fl}}\delta C_1}\right)\frac{\partial^2 u}{\partial y^2}-\frac{1}{2\rho_{\mathrm{fl}}\delta C_3}x^{3n-1}\left(\frac{\partial u}{\partial y}\right)^2\frac{\partial^2 u}{\partial y^2}\\ &-\frac{\sigma B_0^2 x^{n-1}}{\rho_{\mathrm{fl}}}u\,\sin^2\alpha-C_b\frac{u^2}{x\sqrt{K}}-\frac{\nu}{Kx^{1-n}}u \end{aligned} \tag{14}$$

*Energy equation*:

$$\begin{aligned} v\frac{\partial T}{\partial y}+u\frac{\partial T}{\partial x}=&\,\alpha_{\mathrm{fl}}\frac{\partial^2 T}{\partial y^2}+\tau\frac{D_{\mathrm{Th}}}{T_\infty}\left(\frac{\partial T}{\partial y}\right)^2+\tau D_{\mathrm{Br}}\frac{\partial T}{\partial y}\frac{\partial C}{\partial y}\\ &-\frac{1}{(\rho c)_{\mathrm{fl}}}\frac{\partial q_r}{\partial y}+\frac{Q_0}{(\rho c_p)_{\mathrm{fl}}}(T-T_\infty) \end{aligned} \tag{15}$$

*Concentration equation*:

$$\begin{aligned} v\frac{\partial C}{\partial y}+u\frac{\partial C}{\partial x}=&\frac{D_{\mathrm{Th}}}{T_\infty}\frac{\partial^2 T}{\partial y^2}+D_{\mathrm{Br}}\frac{\partial^2 C}{\partial y^2}\\ &-K_r\left(\frac{T}{T_\infty}\right)^n\exp\left(-\frac{E_1}{kT}\right)(C-C_\infty) \end{aligned} \tag{16}$$

Radiative heat flux with nonlinear radiation is modeled using the Rosseland approximation.

### 7.1 Nondimensionalization

Introducing similarity transformations:

$$\eta = \sqrt{\frac{b(n+1)}{2\nu}}x^{\frac{n-1}{2}}y, \quad u = bx^n f'(\eta), \quad v = -\sqrt{\frac{b\nu(n+1)}{2}}x^{\frac{n-1}{2}}\left(f(\eta) + \frac{n-1}{n+1}\eta f'(\eta)\right) \tag{17}$$

$$\theta(\eta) = \frac{T - T_\infty}{T_{fl} - T_\infty}, \quad \phi(\eta) = \frac{C - C_\infty}{C_{fl} - C_\infty} \tag{18}$$

The dimensionless ordinary differential equations (ODEs) become:
*Dimensionless momentum equation*:

$$(1+P)f''' + ff'' - \frac{(n+1)}{2}PQ(f'')^2 f''' - \frac{2n}{n+1}(f')^2 - \frac{M_2^2}{2}f' \sin^2\alpha - \lambda f' - Fr\frac{2n}{n+1}(f')^2 = 0 \tag{19}$$

*Dimensionless energy equation*:

$$\begin{aligned}&3R_1(\theta_1-1)^3(\theta')^2\theta^2 + 6R_1(\theta_1-1)^2\theta(\theta')^2 + 3R_1(\theta_1-1)(\theta')^2\\&+R_1(\theta_1-1)^3\theta^3\theta'' + 3R_1(\theta_1-1)^2\theta^2\theta'' + 3R_1(\theta_1-1)\theta\theta'' + (1+R_1)\theta''\\&+Pr\left(N_2\theta'\phi' + f\theta' + N_1(\theta')^2 + \beta\theta\right) = 0\end{aligned} \tag{20}$$

*Dimensionless concentration equation*:

$$\phi'' + Sc\phi' f + \frac{N_1}{N_2}\theta'' - Sc\gamma N_2(1+\delta_1\theta)^n \exp\left(-\frac{E_1}{1+\delta_1\theta}\right)\phi = 0 \tag{21}$$

Boundary conditions:

$$f = 0, \quad f' = 1, \quad \theta' = -\alpha_1(1-\theta), \quad \phi' = -\alpha_2(1-\phi) \quad \text{at} \quad \eta = 0 \tag{22}$$

$$f' \to 0, \quad \theta, \phi \to 0 \quad \text{as} \quad \eta \to \infty \tag{23}$$

The following dimensionless parameters appear in the model:

- Pr—Prandtl number, representing the ratio of momentum diffusivity to thermal diffusivity.
- $M_2$—Magnetic parameter, quantifying the influence of the applied magnetic field on the flow.
- $R_1$—Nonlinear radiation parameter, accounting for radiative heat flux under the Rosseland approximation.
- $P$, $Q$—Fluid parameters associated with the Powell-Eyring rheological model.
- $N_1$—Thermophoresis parameter, describing nanoparticle motion induced by temperature gradients.
- $N_2$—Brownian motion parameter, associated with random particle diffusion.

- $Sc$—Schmidt number, indicating the relative effectiveness of momentum to mass diffusion.
- $\alpha_1$, $\alpha_2$—Biot numbers related to convective heat and mass transfer at the wall, respectively.
- $\beta$—Internal heat source/sink coefficient.
- $Fr$—Forchheimer number, characterizing inertial resistance within the porous medium.
- $\lambda$—Porosity-related permeability parameter.

Here, primes (e.g., $f'$, $\theta'$) denote differentiation with respect to the similarity variable $\eta$.

The nondimensional expressions are given as:

$$\left.\begin{array}{l} \Pr = \dfrac{\nu}{\alpha}, \quad M_2^2 = \dfrac{2\sigma B_0^2}{b\rho_{\text{fl}}(n+1)}, \quad R_1 = \dfrac{4\sigma^* T_\infty^3}{kk^*}, \quad P = \dfrac{1}{\mu\delta C_1}, \quad Q = \dfrac{u_w^3}{2\nu x C_1^2}, \\ N_1 = \dfrac{(\rho c)_p D_{\text{Th}}(T_{fl} - T_\infty)}{(\rho c)_{\text{fl}} \nu T_\infty}, \quad N_2 = \dfrac{(\rho c)_p D_{\text{Br}}(C_{fl} - C_\infty)}{(\rho c)_{\text{fl}} \nu}, \quad Sc = \dfrac{\nu}{D_{\text{Br}}}, \\ \alpha_1 = \dfrac{h_m}{k}\sqrt{\dfrac{2\nu}{b(n+1)}}, \quad \alpha_2 = \dfrac{h_t}{D_{\text{Br}}}\sqrt{\dfrac{2\nu}{b(n+1)}}, \quad Fr = \dfrac{C_b}{\sqrt{K}}, \quad \lambda = \dfrac{2\nu}{Ka(n+1)}. \end{array}\right\} \tag{24}$$

### 7.2 Solution methodology

The system of coupled nonlinear ODEs is solved numerically by:

- Using a shooting method combined with a Runge-Kutta fourth-order scheme.
- Appropriate initial guesses for unknown boundary conditions are iteratively refined using a Newton-Raphson technique.
- MATLAB, Fortran, or Python computational codes are typically used.

### 7.3 Key outcomes

- Enhanced porosity factor ($Fr$) reduces fluid velocity significantly.
- Inclined magnetic field ($\alpha$) and MHD effect ($M_2$) suppress the momentum profile.
- Nonlinear radiation ($R_1$) increases thermal boundary-layer thickness.
- Thermophoresis ($N_1$) elevates temperature distribution, while Brownian motion ($N_2$) has the opposite effect on concentration.
- Activation energy ($E_1$) enhances the nanoparticle concentration distribution; see **Figures 1–4** for generalized trends in some cases.

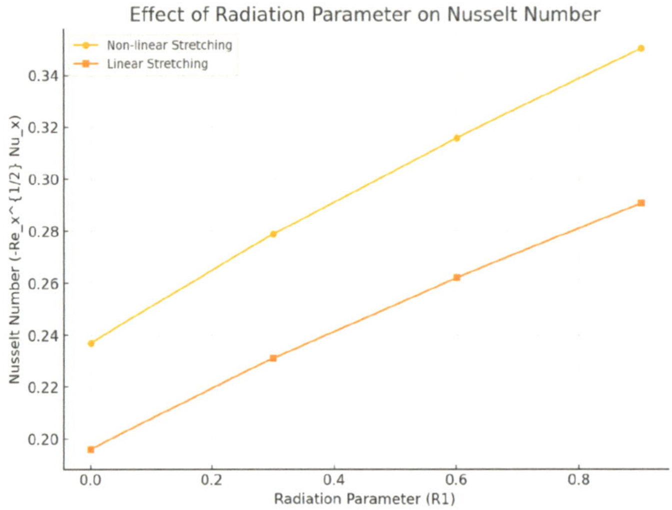


**Figure 1.**
*Effect of radiation parameter $R_1$ on the Nusselt number for nonlinear and linear stretching surfaces. Boundary conditions: convective thermal BC at surface, $\theta'(0) = -\alpha_1(1 - \theta(0))$. Fixed parameters: $Pr = 6.2$, $M = 2$, $Ec = 0.1$, $Fr = 0.2$.*

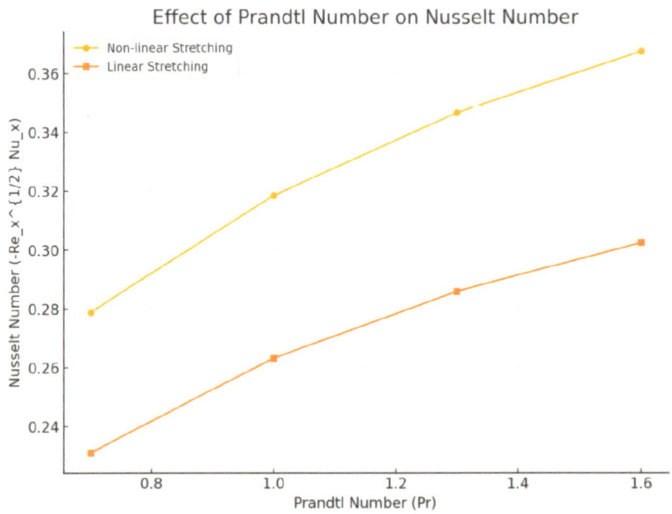


**Figure 2.**
*Variation of the Nusselt number with Prandtl number (Pr), showing a decrease in heat transfer as Pr increases. Boundary condition: nonlinear convective surface heating. Other parameters: $R_1 = 1.5$, $M = 2$, $N_1 = 0.3$, $N_2 = 0.2$.*

## 8. Recent advances

Recent studies have explored the synergistic effects of MHD, nonlinear radiation, heat sink/source, and chemical reactions on Powell-Eyring nanofluid flows. In particular, Rasool and Shafiq [34] conducted a detailed numerical study involving:

- *Nonlinear radiation*: Enhanced temperature distribution was observed when nonlinear radiative effects were incorporated, compared to linear radiation.

*DOI: http://dx.doi.org/10.5772/intechopen.1011310*

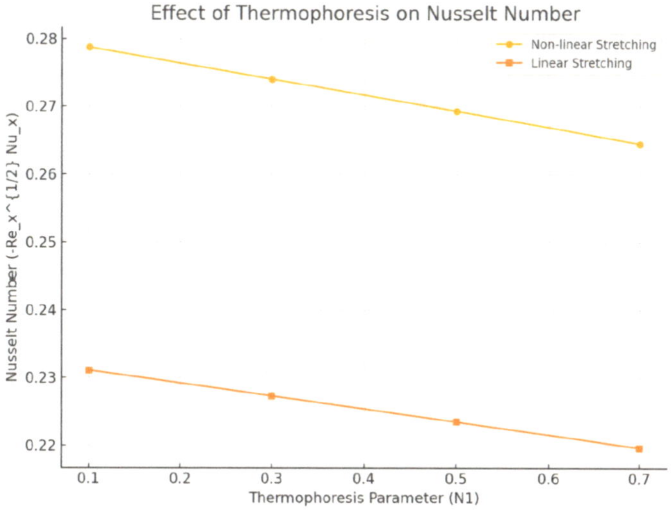


**Figure 3.**
*Influence of thermophoresis parameter ($N_1$) on the thermal boundary layer. Thermophoresis enhances temperature near the surface. Parameters: $Pr = 6.2$, $M = 2$, $Ec = 0.1$, $R_1 = 1.5$, $\alpha_1 = 0.3$.*

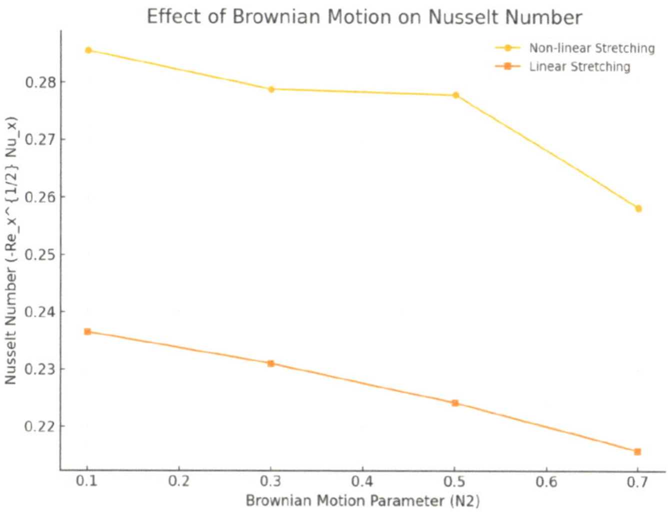


**Figure 4.**
*Effect of Brownian motion parameter ($N_2$) on temperature and mass transfer characteristics. Higher $N_2$ reduces temperature and enhances concentration diffusion. Fixed parameters: $Pr = 6.2$, $Sc = 0.9$, $Ec = 0.1$, $Fr = 0.2$.*

- *Binary chemical reactions*: Modeled using Arrhenius activation energy, significantly impacting nanoparticle concentration near the surface.
- *Inclined magnetic fields*: The inclination angle ($\alpha$) in the transverse magnetic field was found to decrease the fluid velocity profile due to enhanced Lorentz forces.
- *Convective boundary conditions*: Heat and mass fluxes were treated with convective heating/mass transfer, allowing realistic boundary interactions.

Their numerical findings revealed:

- Increased porosity factor leads to a decline in fluid velocity and an increase in drag force.
- Radiation parameter (nonlinear) strongly enhances temperature and thermal boundary-layer thickness.
- Prandtl number ($Pr$) increase reduces the heat transfer rate, while thermophoresis increases the temperature distribution.
- The Forchheimer number ($Fr$) influences both the momentum and thermal fields significantly due to porous resistance.

Quantitative data from their results include:

- Skin-friction coefficient ($C_f$), Nusselt number ($Nu_x$), and Sherwood number ($Sh_x$) variations were provided for different parameters.
- For instance, increasing the radiation parameter from ($R_1 = 0$) to (0.9) increased the Nusselt number (heat transfer rate) from 0.236963 to 0.350334 in the nonlinear stretching case.
- An increment in activation energy ($E_1$) led to higher nanoparticle concentration profiles, enhancing the Sherwood number (mass transfer rate).

These results align well with industrial needs such as controlling heat transfer in heating equipment, nuclear reactors, and propulsion systems, where non-Newtonian and radiative effects are dominant.

## 9. Conclusion

This chapter gave a detailed theoretical explanation of MHD radiative heat transfer in chemically reactive Powell-Eyring nanofluid flow across porous media. The governing equations were created and nondimensionalized to reveal heat and mass transport characteristics. The study enables improved energy systems and industrial application research. The study focuses on a stable, incompressible, radiative Powell-Eyring nanofluid that is chemically reactive as it flows over a nonlinearly expanding flat surface or sheet. Convection is used on the border. A nonlinear thermal radiation effect is also relevant. The velocity field and momentum layer increase fluid parameter $P$ as observed in the velocity field and momentum layer. Nonlinear radiation's thermal field and thermal layer thickness increase in both linear and nonlinear stretching scenarios. Thermophoresis enhances heat distribution and layering while the Prandtl number decreases. The Biot number coefficient for heat transport enhances the thermal layer. Similar but heightened results are seen for the Biot number coefficient for mass transfer in concentration distribution. High $Sc$ values lead to a significant decrease in concentration distribution and boundary layer. The concentration distribution increases with the activation energy for linear and nonlinear stretching. Skin friction increases with $P$ and $\alpha$, but decreases with $Q$. Heat and mass flux characteristics increase as the nonlinear radiation parameter increases. However, heat and mass flow properties show opposing trends for incremental change in *Pr*. Brownian diffusion and thermophoretic force reduce heat flow (Nusselt).

## 10. Applications and future directions

The advanced modeling of inclined MHD and nonlinear radiation in chemically reactive Powell-Eyring nanofluids opens up new avenues for smart thermal management systems.

### 10.1 Applications

- Satellite and spacecraft thermal control systems benefiting from radiative heat enhancement.
- Solar collectors and energy conversion devices where nonlinear radiation must be accounted for.
- Nanotechnology-enhanced heat exchangers in propulsion devices and gas turbines.

### 10.2 Future research directions

- Developing dynamic models considering time-varying magnetic fields.
- Machine learning integration for real-time control and optimization of nanofluidic systems.
- Multiphase nanofluid models incorporating particle agglomeration effects.

## Availability of data

The data that support the findings of this study are available within the article.

## Author's contribution

All authors contributed equally.

## Conflict of interest

The authors have no conflict of interest.

## Nomenclature

| | |
|---|---|
| $x, y$ | cartesian coordinates/m |
| $u_w(x)$ | stretching velocity/m·s$^{-1}$ |

| | |
|---|---|
| $u, v$ | horizontal and vertical velocity components/m·s$^{-1}$ |
| $\mu$ | dynamic viscosity/Pa·s |
| $\nu$ | kinematic viscosity/m$^2$·s$^{-1}$ |
| $b$ | stretching rate/s$^{-1}$ |
| $k$ | thermal conductivity/W·m$^{-1}$·K$^{-1}$ |
| $\rho_{fl}$ | fluid density/kg·m$^{-3}$ |
| $(\rho c)_{fl}$ | fluid heat capacity/J·m$^{-3}$·K$^{-1}$ |
| $(\rho c)_p$ | nanoparticle heat capacity/J·m$^{-3}$·K$^{-1}$ |
| $\sigma$ | electrical conductivity/(Ω m)$^{-1}$ |
| $B_0$ | magnetic field intensity/A·m$^{-1}$ |
| $T$ | temperature/K |
| $C$ | concentration/kg·m$^{-3}$ |
| $f$ | dimensionless stream function or velocity function |
| $\theta$ | dimensionless temperature |
| $\phi$ | dimensionless concentration |
| $\eta$ | similarity variable |
| $Sh_x$ | Sherwood number (mass flux parameter) |
| $Nu_x$ | Nusselt number (heat flux parameter) |
| $\tau$ | heat capacity ratio of fluid to nanoparticles |
| $M$ | magnetic parameter |
| $R$ | radiation parameter |
| $Pr$ | Prandtl number |
| $Sc$ | Schmidt number |
| $Da$ | Darcy number |
| $\gamma$ | chemical reaction parameter |
| $Ec$ | Eckert number (viscous dissipation) |
| $N_1$ | thermophoresis parameter |
| $N_2$ | Brownian motion parameter |
| $\beta$ | nonlinear stretching exponent or slip coefficient |
| $Fr$ | Forchheimer number (porous resistance) |
| $\delta_1$ | temperature ratio for Arrhenius activation energy |
| $E_1$ | activation energy |
| $\alpha_1$, $\alpha_2$ | biot numbers for thermal and mass convection |
| $\theta_1$ | nonlinear radiation temperature ratio |

## Abbreviations

| | |
|---|---|
| MHD | magnetohydrodynamics |
| ODE | ordinary differential equation |
| PDE | partial differential equation |

## Author details

Ghulam Rasool[1*†], Anum Shafiq[2,3†] and Muhammad Ijaz Khan[1†]

1 Department of Mechanical Engineering, College of Engineering, Prince Mohammad Bin Fahd University, Al-Khobar, KSA

2 IT4Innovations, VSB -Technical University of Ostrava, Ostrava, Czech Republic

3 School of Mathematics and Statistics, Nanjing University of Information Science and Technology, Nanjing, China

*Address all correspondence to: grasool@pmu.edu.sa

† These authors contributed equally.

## References

[1] Choi SUS. Enhancing thermal conductivity of fluids with nanoparticles, USA. ASME. 1995;**66**: 99-105. FED 231/MD

[2] Hayat T, Qayyum S, Alsaedi A, Shafiq A. Inclined magnetic field and heat source/sink aspects in flow of nanofluid with nonlinear thermal radiation. International Journal of Heat and Mass Transfer. 2016;**103**:99-107

[3] Sandeep N, Sulochana C, Kumar BR. Unsteady MHD radiative flow and heat transfer of a dusty nanofluid over an exponentially stretching surface. Engineering Science and Technology. 2016;**19**:227-240

[4] Ziaei-Rad M, Saeedan M, Afshari E. Simulation and prediction of MHD dissipative nanofluid flow on a permeable stretching surface using artificial neural network. Applied Thermal Engineering. 2016;**99**:373-382

[5] Mohammadpour M, Valipour P, Shambooli M, et al. Nanofluid flow and forced convection heat transfer over a stretching surface considering heat source. European Physical Journal. 2015;**130**:155. DOI: 10.1140/epjp/i2015-15155-8

[6] Hsiao KL. Stagnation electrical MHD nanofluid mixed convection with slip boundary on a stretching sheet. Applied Thermal Engineering. 2016;**98**:850-861

[7] Zeeshan A, Majeed A, Ellahi R. Effect of magnetic dipole on viscous Ferro-fluid past a stretching surface with thermal radiation. Journal of Molecular Liquids. 2016;**215**:549-554

[8] Nandy SK, Mahapatra TR. Effects of slip and heat generation/absorption on MHD stagnation flow of nanofluid past a stretching/shrinking surface with convective boundary conditions. International Journal of Heat and Mass Transfer. 2013;**64**:1091-1100

[9] Parsa AB, Rashidi MM, Hayat T. MHD boundary-layer flow over a stretching surface with internal heat generation or absorption. Heat Transfer Asian Research. 2013;**42**:500-514

[10] Pavithra GM, Gireesha BJ. Effect of internal heat generation/absorption on dusty fluid flow over an exponentially stretching sheet with viscous dissipation. Journal of Mathematics. 2013;**2013**:10. Article ID: 583615. DOI: 10.1155/2013/583615

[11] Mukhopadhyay S, Vajravelu K. Effects of transpiration and internal heat generation/absorption on the unsteady flow of a Maxwell fluid at a stretching surface. Journal of Applied Mechanics. 2012;**79**:044508

[12] Rasool G, Shafiq A, Khalique CM, Zhang T. Magnetohydrodynamic Darcy Forchheimer nanofluid flow over nonlinear stretching sheet. Physica Scripta. 2019;**94**(10):105221

[13] Sheikh M, Abass Z. Effects of thermophoresis and heat generation/absorption on MHD flow due to an oscillatory stretching sheet with chemically reactive species. Journal of Magnetism and Magnetic Materials. 2015;**396**:204-213

[14] Animasaun IL, Ibraheem RO, Mahanthesh B, Babatunde HA. A meta-analysis on the effects of haphazard motion of tiny/nano-sized particles on the dynamics and other physical properties of some fluids. Chinese Journal of Physics. 2019;**60**:676-687

[15] Lund LA, Omar Z, Khan I, Dero S. Multiple solutions of Cu-$C_6$-$H_9NaO_7$ and

$Ag\text{-}C_6\text{-}H_9NaO_7$ nanofluids flow over nonlinear shrinking surface. Journal of Central South University. 2019;**26**(5): 1283-1293

[16] Wang G, Yao Y, Chen Z, Hu P. Thermodynamic and optical analyses of a hybrid solar CPV/T system with high solar concentrating uniformity based on spectral beam splitting technology. Energy. 2019;**166**:256-266

[17] Yu H, Dai W, Qian G, Gong X, Zhou D, Li X, et al. The $NO_x$ degradation performance of Nano-$TiO_2$ coating for asphalt pavement. Nanomaterials (Basel, Switzerland). 2020;**10**(5):897

[18] Guo C, Hu M, Li Z, Duan F, He L, Zhang Z, et al. Structural hybridization of bimetallic zeolitic imidazolate framework (ZIF) nanosheets and carbon nanofibers for efficiently sensing α-synuclein oligomers. Sensors and actuators. B, Chemical. 2020;**309**: 127821

[19] Liu C, Huang X, Wu Y, Deng X, Liu J, Zheng Z, et al. Review on the research progress of cement-based and geopolymer materials modified by graphene and graphene oxide. Nanotechnology reviews (Berlin). 2020; **9**(1):155-169

[20] Yan H, Xue X, Chen W, Wu X, Dong J, Liu Y, et al. Reversible Na+ insertion/extraction in conductive polypyrrole-decorated $NaTi_2(PO_4)_3$ nanocomposite with outstanding electrochemical property. Applied Surface Science. 2020;**530**:147295

[21] Cortell R. Fluid flow and radiative nonlinear heat transfer over stretching sheet. Journal of King Saud University, Science. 2013;**26**:161-167

[22] Shehzad SA, Hayat T, Alsaedi A, Obid MA. Nonlinear thermal radiation in three-dimensional flow of Jeffrey nanofluid: A model for solar energy. Applied Mathematics and Computation. 2014;**248**:273-286

[23] Wakif A. A novel numerical procedure for simulating steady MHD convective flows of radiative Casson fluids over a horizontal stretching sheet with irregular geometry under the combined influence of temperature-dependent viscosity and thermal conductivity. Mathematical Problems in Engineering. 2020;**2020**:1675350

[24] Animasaun I, Raju C, Sandeep N. Unequal diffusivities case of homogeneous–heterogeneous reactions within viscoelastic fluid flow in the presence of induced magnetic-field and nonlinear thermal radiation. Alexandria Engineering Journal. 2016;**55**(2):1595-1606. DOI: 10.1016/j.aej.2016.01.018

[25] Hayat T, Imtiaz M, Alsaedi A, Kutbi MA. MHD three-dimensional flow of nanofluid with velocity slip and nonlinear thermal radiation. Journal of Magnetism and Magnetic Materials. 2015;**396**:31-37

[26] Rasool G, Zhang T. Characteristics of chemical reaction and convective boundary conditions in Powell-Eyring nanofluid flow along a radiative Riga plate. Heliyon. 2019;**5**:e01479

[27] Hayat T, Hussain S, Muhammad T, Alsaedi A, Ayub M. Radiative flow of Powell-Eyring nanofluid with convective boundary conditions. Chinese Journal of Physics- Taipei. 2017; **55**(4):1523-1538

[28] Powell RE, Eyring H. Mechanism for relaxation theory of viscosity. Nature. 1944;**154**:427-428

[29] Jalil M, Asghar S, Imran SM. Self similar solutions for the flow and heat

transfer of Powell-Eyring fluid over a moving surface in a parallel free stream. International Journal of Heat and Mass Transfer. 2013;**65**:73-79

[30] Mushtaq A, Mustafa M, Hayat T, Rahi M, Alsaedi A. Exponentially stretching sheet in a Powell-Eyring fluid: Numerical and series solutions. Z. Naturforsch. A. 2013;**68a**:791-798

[31] Poonia M, Bhargava R. Finite element study of Eyring-Powell fluid flow with convective boundary conditions. Journal of Thermophysics and Heat Transfer. 2014;**28**:499-506

[32] Hayat T, Asad S, Mustafa M, Alsaedi A. Radiation effects on the flow of Powell-Eyring fluid past an unsteady inclined stretching sheet with non-uniform heat source/sink. PLoS One. 2014;**9**:e103214

[33] Hayat T, Imtiaz M, Alsaedi A. Effects of homogeneous-heterogeneous reactions in flow of Powell-Eyring fluid. Journal of Central South University. 2015;**22**:3211-3216

[34] Rasool G, Shafiq A. Numerical exploration of the features of thermally enhanced chemically reactive radiative Powell–Eyring nanofluid flow via Darcy medium over non-linearly stretching surface affected by a transverse magnetic field and convective boundary conditions. Applied Nanoscience. 2023; **13**:229-246

Chapter 3

# Rapoport Revisited: A Review of Scaling Theory of Two Phase Flow in Porous Media and its Application

*Kenneth S. Sorbie*

## Abstract

This paper presents a detailed analytical exposition of the classic 1955 paper by Rapoport on the *rigorous* scaling of two phase flow in porous media. It extends Rapoport's work to anisotropic systems and fills in a number of the missing details in the original paper, which has led to an incomplete appreciation of this seminal work. The various forms of the rigorous force balances (capillary, viscous and gravity) which can exist are shown. Within any given force balance, say viscous/capillary, there are three formulae which emerge but these, are related through the "shape groups" and reduce to one. The same is true for the gravity/capillary and the viscous/gravity force balances, only one of which is then required in addition to the viscous/capillary group. In fact, only two force balances and two shape groups are required (in 3D) and these are intimately related in a way that is explained in detail here for the first time, but which is "hidden" in Rapoport's paper. Most papers which use scaling theory present *approximate* scaling groups for the various force balances, although the correct *forms* of the shape groups are usually broadly correct. However, Rapoport's work on rigorous scaling groups and the full findings from it, also teaches us some important lessons about what can be expected when less rigorous (or approximate) scaling groups are developed. Since there are many different equivalent formula even for rigorous scaling, then a common literature worry may be laid to rest: That is, when some papers give a long list of the various force balance (viscous/capillary, viscous/gravity balance etc.), and then go on to worry about the question, "which is correct"? The exact formulae are not vital, although a correct set of force balances which are consistently linked by the shape groups should be used. However, the precise number of groups is fixed for a given system. There are also some dimensionless quantities, which are sometimes wrongly labelled as "scaling groups" or "force balances". A number of novel applications of Rapoport's scaling theory are also presented, coupled with numerical simulations, to tackle a range of problems in a way which Rapoport may not have anticipated. These examples illustrate the exact scaling that is possible using the Rapoport approach and show some surprising consequences of his results.

**Keywords:** scaling theory, two phase flow in porous media, force balance groups, scaling groups, rigorous scaling, Rapoport

# 1. Introduction

## 1.1 Preliminary discussion: Scaling groups

This paper presents a full analysis along with a number of novel consequences of the classic paper, *Scaling Laws for Use and Design of Water Oil Flow Models*, by Rapoport [1]. This may seem like a strange task given that this paper will celebrate its 70th birthday this year. There are several reasons which motivate me to do this. Principal among them is that Rapoport's contribution in this work is, in my view, vastly under-rated, under applied and under appreciated. Just to take one measure, the Google Scholar entry for Rapoport [1]. claims, at the time of checking (November 2024) 311 citations, about 4.47 per year! Another reason for presenting this work is to give an *exposition*[1] of this work to younger engineers and researchers who will probably not read a paper from this era.

There are further technical reasons for this exposition which will appear throughout the paper. One relates to the balance of the various forces in 2 phase flow: viz. viscous, capillary and gravity. Throughout the literature there are many references that refer to the dimensionless numbers which describe these balances, such as the capillary number, $N_{Ca}$ (viscous/capillary force balance) or the Bond number, $N_{Bo}$ (gravity/capillary force balance); and each of these various numbers appear in myriad forms; e.g. see Shook *et al.* [2], Rahman *et al.* [3], Guo *et al.* [4] etc.). To choose one, let us take the following version of the (dimensionless) capillary number:

$$N_{Ca} = \frac{\mu v}{\sigma}. \tag{1}$$

where $\mu$ = viscosity, $v$ = fluid velocity and $\sigma$ = interfacial tension (IFT) between 2 fluid phases. This version of $N_{Ca}$ is actually a useful brief cypher for referring to the broad direction of changes in force balance in a *qualitative* (and physically correct) manner; e.g. as viscosity or flow velocity increases, then viscous forces get larger, as IFT increases then capillary forces increase. However, the above expression is *not* a "scaling group" in any rigorous sense (defined below); if it were, then in a given system I could double $\mu$ and half $v$ and the system should be identical. In a general non-linear system such as in 2 phase flow in porous media this is simply not the case. The above capillary number simply expresses the *qualitative* direction of the force balance, i.e. higher $\nu$ or $\mu$, or lower $\sigma$ lead to the viscous forces increasing relative to capillary forces. The development of these scaling groups in Rapoport is much more rigorous, within a few clearly stated assumptions.

Another issue with the use of Eq. (1) above for "capillary number" is that it makes no reference to the *size* of the system. This is rather strange since it implies that a system of 1 cm and one of 1 km with a given $\mu$, $v$ and $\sigma$ will have the same capillary number – whereas clearly they do not have the same viscous/capillary force balance ratio. As stated in Eq. (1), this "capillary number" is actually a *microscopic* quantity. This issue is picked up in a recent paper by Guo *et al.* [4] who review over 40 forms of the "capillary number" and they list which versions of this refer to only microscopic quantities and which include a lengthscale parameter. We will return to these issues later in this paper, since an analysis of Rapoport's paper clarifies and resolves it at the Darcy scale.

[1] "Exposition" = setting forth of the meaning or purpose (as of a writing)

### 1.2 Just read Rapoport's paper?

The obvious question that may come to mind is: why do not I just recommend that you read Rapoport's [1] paper? The reason for this is that you may do exactly what I did when this great work was first recommended to me. You read it, you think you understand it and then you move on. But most of us, me included, are not nearly as smart as Rapoport, and in the paper he simply did not "spell out" how some of the results, specifically the exact forms of the scaling groups, turn up as they do. For example, why is a particular choice of scaling group formula "correct", since on closer inspection there appear to be different versions of the various groups? In fact, there is no "correct" choice of scaling group, but the various sets of possible groups are related in a clear consistent manner, as explained later in this paper, but not in Rapoport's original work.

A further reason for this paper is that Rapoport presented the *isotropic* permeability case, as can be seen in **Figure 1** which shows his precise form of the 2 phase incompressible flow equation. Talking with expert colleagues around the world, on many occasions I have heard the complaint that Rapoport's rigorous scaling groups only apply to the isotropic permeability case and, as presented in Rapoport [1], this appears to be true; but it is not. It is very straightforward to extend this to the anisotropic case and it is this more general case that I present here. In fact, this was originally done in 1985 but not published, since we thought it was a fairly obvious and trivial extension of Rapoport's great opus (Evans *et al.* [5])[2].

I believe that Rapoport probably knew either all or much of what I will present in detail below, including the extension to the anisotropic case. However, he assumed that the various "hidden manipulations" which he carried out would be obvious to the rest of us mortals. The main reason I went back to Rapoport's paper in the 1980s-90s was that I decided to use more rigorous scaling theory (explained below) in my own research, and I also wanted to teach it to my students. Someone once said: "*You never really fully understand something until you have to teach it*". Also, on the command of the technical details, Richard Feynman said "*What I cannot create, I do not understand*". I would have to agree with both of these quotes.

As noted above, this is essentially an exposition, but in the course of this I will also point out some novel uses of scaling theory in two phase flow, which Rapoport could not at that time have been aware. For example, Rapoport was not aware that his method could be used very fruitfully to scale *numerical* simulations at different length scales, since this was a topic in its infancy in the 1950s. In fact, scaling theory can be

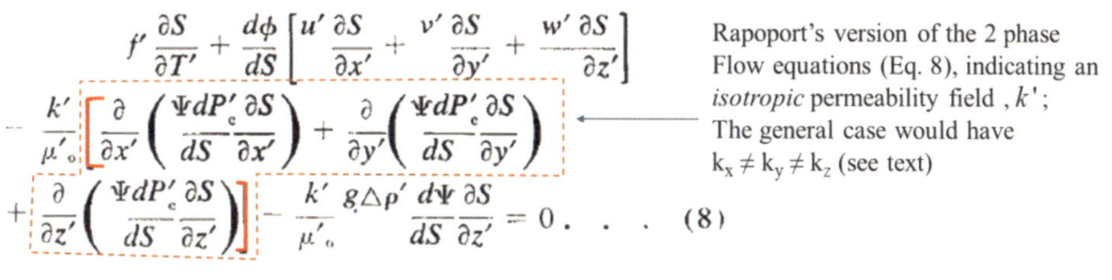


**Figure 1.**
*Rapoport's form of the two phase flow equations, using his original notation (which we change to a more modern form below). Note that he refers to the case where permeability is isotropic, but this is easily generalised (see Eq. (2) below).*

[2] Uploaded as Ancillary Material #1

used to check if your two phase flow simulator (for incompressible flow) is actually working correctly! I will also present some examples of the use of scaling theory to elucidate some aspects of "upscaling" and how the changes in the balance of forces manifest with lengthscale. I would contend that the Rapoport scaling groups are *much* more powerful and useful than *even he* ever thought they were, and they can still be applied in very novel and useful ways. I will try to bring out some of these novel aspects of Rapoport's scaling theory through the examples presented later in this paper.

## 2. A brief review of scaling theory

### 2.1 Rigorous and approximate scaling

One of the broad observations in the literature on Scaling Theory as applied to flow through porous media is that there has been a drift –back and forth over time - from rigorous scaling, to the establishment of "correlating groups", which give some "approximate" scaling. Indeed, there have been changes of what the *intent* of using scaling theory (or dimensionless groups in general) actually is. I first explain the distinction I make between rigorous and approximate scaling as follows.

*Rigorous Scaling:* I would describe rigorous scaling as where we have a governing set of equations (invariably PDEs), and by an algebraically correct set of variable scaling, we arrive at a complete set of (possible) dimensionless scaling groups. If two apparently widely different sized systems have each of these groups with identical values, then the two systems are *exactly* scaled within this mathematical formulation; i.e. they are similar or essentially numerically identical. However, this rigour comes at the cost of making some assumptions such as those made by Rapoport for two phase flow, i.e. that the relative permeabilities at all scales are the same, and also that the scaling of capillary pressure (which can change) is quite restricted in form (see below).

*Approximate Scaling:* For some complex systems, it can be very useful to give some broad indications of the sensitivities of the system to the various groups which describe the flow (and other) physics of that system. When we are simulating a complex process of flow through porous media, then deriving some dimensionless quantities may be useful in grouping (or reducing) variables which will describe a certain type of behaviour. For example, this has been done in some EOR processes to find the sensitivities which govern the underlying mechanisms involved (e.g. Pujol and Boberg [6]; Wood *et al.* [7]). Such quantities are, in my view, really "correlating groups" rather than true scaling groups in the rigorous sense above. However, the two advantages which this approach has are (i) that one may try to "scale everything" (rather than make simplifying assumptions) and (ii) it can be immensely useful to analyse the parameter sensitivities in particular processes (Zhou *et al.* [8]). The attempt to "scale everything" just referred to, tends to be manifest in what I would describe as a "like for like" scaling; that is each feature of the model is separately "scaled". As we will see below, sometimes that quantity is *very* important to maintain in a scaled system, e.g. the viscosity ratio $(\mu_o/\mu_w)$. However, if it is possible to derive the exact quantities required for scaling, then is gives the scaling groups in what is referred to above as their "rigorous form".

Different approaches are required to achieve each of the above routes to scaling quantities or groups. For rigorous scaling, a full mathematical description of the system is required as a transport equation having all the relevant forces present, followed by the application of some relatively tedious but straightforward algebraic manipulation. For approximate scaling, the underlying equations are often available but are not always strictly required, and then one may proceed using the approaches of Dimensional or Inspectional Analysis (Shook [2]; Bridgman [9]; Greenkorn [10]; Loomis and Crowell [11]). There is a huge literature on the previous two approaches and some very good reviews can be found in the porous medium literature as they apply to problems in flow through porous media (Shook et al. [2]). We will not expand further on these procedures here since our focus in on the Rapoport approach.

### 2.2 Scaling theory applied to flow through porous media: Emergence and development

In the earliest work on developing scaling groups for two phase flow in porous media, the general *intent* was to build laboratory scaled models that were exactly scaled to actual or simplified reservoirs. This was done of necessity because, at the time, there were no other ways in which one could make calculations on actual reservoirs. A parallel approach at the time was also to try to build electrical analogues to reservoirs, but that is another (fascinating) story (Bruce [12]; Schaefer [13]; Karplus [14]). Later, the intent of scaling theory varied quite considerably to finding scaling groups that correlated with system performance or could predict recovery, or simply to reduce the number of variables required in order to predict the main features of a given reservoir system or displacement process.

One of the earliest papers on scaling in porous media from idealised oil reservoirs to scaled models was by Leverett *et al.* [15]. Leverett *et al.* gave a clear if slightly wordy early statement of what was meant when two systems are said to be scaled: they said, this is so "*if for each element in one system there is a corresponding element in the second system and the ratios of the numerical magnitudes of corresponding elements are constant for all elements of the same dimensional type*". Leverett *et al.* [15] did not derive any "scaling groups" in their 1942 paper, but instead listed the various features in each of the scaled system which has to be scaled, and in 2 lab scale models (a near well coning model and a layered pack) they matched as many of these as possible for idealised reservoir situations; this is an early example of "like for like" scaling. They also focused on trying to scale properties such as capillary pressure, $P_c(S_w)$, between different unconsolidated sands and 2 phase fluid pairs used in various models, using the famous Leverett J-function.

Engelberts and Klinkenberg [16] discussed scaling of waterflooding using the 6 groups as listed in **Table 1**; their notation is shown as well as that used here, where dip angle α is not used, or required, in the current work.

Since their objective was to correlate oil recovery for various parameter ratios, then these workers were not trying to scale models in the manner which appears later in Rapoport. But the last 2 groups in the above list are clearly related to gravity/viscous and gravity/capillary force balances. Later authors Croes and Schwarz [17] also use

| Engelberts and Klinkenberg | Notation of this work |
|---|---|
| $(\eta_o/\eta_w), 1/h, \alpha, 1/\sqrt{k}, \left(\frac{\Delta\rho g k}{v\eta_w}\right), \left(\frac{\Delta\rho g l\sqrt{k}}{\gamma\cos\theta}\right);$ | $(\mu_o/\mu_w), 1/\Delta z, -, 1/\sqrt{k}, \left(\frac{\Delta\rho g k}{v\mu_w}\right), \left(\frac{\Delta\rho g\Delta x\sqrt{k}}{\sigma_{ow}\cos\theta}\right)$ |

**Table 1.**
*The scaling groups of Engelberts and Klinkenberg [16], where, as well as "like for like" parameter scaling, there are clearly 2 groups representing the gravity/viscous and the gravity/capillary viscous force balances.*

these groups to extend the work of Engelberts and Klinkenberg [16] to examine the correlation of waterflood recovery and these groups. In addition, Croes and Schwarz were also testing out the "competing theories of water displacement" which had been proposed by Buckley and Leverett [18] and Dietz [19].

The broad agenda of the previous decades on scaled models was followed by a number of researchers into the 1960s and 70s. Gaucher and Lindley [20] built scaled stratified (2 layer) 5-spot packs to study waterflooding for both stable $((\mu_o/\mu_w)\approx1)$ and potentially unstable $((\mu_o/\mu_w)\approx4.3)$ two phase displacements. The "direct" scaling groups they used are as shown in **Table 1** along with the Reservoir 1 scaling details, where a and b denote the top and bottom layers of the 2 layer 5 spot model, $i_w$ is the water injection rate, the $k_{o,wr,a}$-like terms are phase permeabilities and other terms $(L, \Delta\rho, h_a, \mu_o, \phi_a, \sigma\ etc..)$ are as expected. They tried to ensure that these 6 "groups" were identical in their reservoir prototype and laboratory model, as shown in **Figure 2** in the extracted case for "Reservoir 1" of their study below (where the original reservoir is a "20 acre 5-spot").

Perkins and Collins [21] addressed the limitation assumed by Rapoport [1] that the relative permeabilities had to be the same in the prototype and model, and they also considered the somewhat restrictive constraint on the scaling of capillary pressure. They claim to propose new scaling criteria which permit having different relative permeabilities and capillary pressures in the prototype and model. We will discuss this development in more detail later in this paper, after the discussion of Rapoport's work.

Reviewing scaling theory in porous media in the early 1960s, Greenkorn [10] gave a very good introduction to the Buckingham π- theorem and the basic principles of scaling theory. A recipe for deriving the exact number of required scaling groups was given, and the more approximate analysis was used to find all of these groups. He then went on to apply this approach to the miscible flow equations in porous media.

Farouq Ali *et al.* [22] in a published proceeding reviewed the problems of applying scaling theory from the lab to the field. Here, they noted that Douglas *et al.* [23] found excellent agreement between the model and the prototype, when all of Rapoport's criteria (scaling groups) were satisfied. If correct, this would be the earliest case of the use of numerical simulation to confirm that the scaling groups are correct – and

| Areal pattern | Symbol | Units Model | Units Prototype | Reservoir 1 (20 acre 5 spot) Model | Reservoir 1 (20 acre 5 spot) Prototype |
|---|---|---|---|---|---|
| Layer b, thickness (botton | $h_b$ | inch | foot | 0.7 | 10 |
| Layer a, abs. perm, | $k_a$ | D | mD | 35 | 15 |
| Layer b, abs. perm, | $k_b$ | D | mD | 279 | 116 |
| Phase permeabilities | $k_{o,wr,a}$ | D | mD | 26 | 11 |
| | $k_{o,wr,b}$ | D | mD | 222 | 92 |
| | $k_{w,or,a}$ | D | mD | 19 | 8 |
| | $k_{w,or,b}$ | D | mD | 174 | 72 |
| Layer a, porosity | $\phi_a$ | - | - | 0.46 | 0.2 |
| Layer , porosity | $\phi_b$ | - | - | 0.47 | 0.2 |
| $1-S_{or,a}-S_{wr,a}$ | | - | - | 0.75 | 0.5 |
| $1-S_{or,a}-S_{wr,b}$ | | - | - | 0.75 | 0.5 |
| Oil viscosity | $\mu_o$ | cP | cP | 165 | 2.17 |
| Water viscosity | $\mu_w$ | cP | cP | 38 | 0.5 |
| O/W density difference | $\Delta\rho$ | g/cm$^3$ | g/cm$^3$ | 1.06 | 0.2 |
| O/W IFT | $\sigma$ | dyne/cm | dyne/cm | 25 | 25 |
| Water injection rates | $i_w$ | cm$^3$/s | bbl/day | 0.053 -0.533 | 40 - 400 |

Gaucher and Lindley (1960) scaling Groups & the "Reservoir 1" scaling

$$\left(\frac{Lk_{w,or,a}\Delta\rho g h_a}{\mu_o i_w}\right); \left(\frac{L\sigma\sqrt{\phi_a k_{w,or,a}}}{\mu_o i_w}\right); \left(\frac{k_{o,wr,a}}{k_{w,or,a}}\right);$$

$$\left(\frac{k_{o,wr,a}}{k_{w,or,b}}\right); \left(\frac{k_{w,or,b}}{k_{w,or,a}}\right); \left(\frac{\mu_o}{\mu_w}\right) \& \left(\frac{\phi_a}{\phi_b}\right)$$

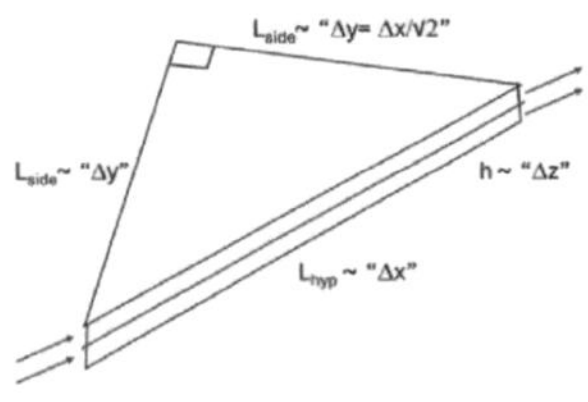


**Figure 2.**
*Scaling of "Reservoir 1" example from Gaucher and Lindley [20]. Again gravity/viscous and capillary/viscous force balance groups can be clearly identified.*

indeed if they are accepted, to verify the numerical codes. However, I cannot find such a confirmation in their (Douglas *et al.*'s) paper. Using the numerical approach of Douglas *et al.*, Nielsen and Tek [24] did examine how 3D and 2D systems at the lab and field scale are able to be scaled numerically. They pointed out that Rapoport's constraint of have the same relative permeabilities and specifically scaled capillary pressure was quite stringent and that "significant deviations" were observed if these conditions were violated. Their work also confirmed that scaled systems (according to Rapoport criteria) gave very similar answers numerically, although these results did not seem to achieve the very high accuracies found by this author (which are commented upon and discussed below).

Work on scaled models continued into the 1970s and van Daalen and van Domselaar [25] derived some slightly different scaling groups to describe flow in tilted horizontal experimental models. As the title suggests, they focused on changing the reservoir geometry of the tilted homogeneous systems, mainly by adjusting the aspect ratio of the models, ($\Delta x/\Delta z$), in the notation of this work) and the tilt angle, in order to derive some similarities. In fact, the previous work by Rapoport already suggested how this system should be scaled much more rigorously, but these authors used their own analysis for this purpose.

Interestingly, in their 1973 survey of progress in the application of reservoir engineering over the previous 25 years, Richardson and Stone [26] devoted only a short 74 word paragraph to scaling theory. They did not mention Rapoport's 1955 paper, although they cited 2 of his other papers. This partly reflects the fact that there were many other excellent and more applied innovations to discuss, but also the sense that the topic had become somewhat "old fashioned" or even less useful by this time. Much was (quite rightly) made in this review of the emerging field of numerical reservoir simulation which in due course was destined to almost *become* "reservoir engineering". No connection was foreseen between scaling theory and numerical simulation, a situation that has persisted. However, interest and activity in scaling theory in porous media did persist beyond that time, much of it driven by the application of enhanced oil recovery (EOR) methods.

An early application of scaling theory to thermal EOR, in fact to steam flood models was reported by Pujol and Boberg [6]. They identified a long list of 20+ similarity (scaling) groups intending to capture both the mass and heat flow quantities relevant for steam flooding. They then used these in order to present an analysis of the various parameters affecting the efficiency of steam flooding. They also give an interesting discussion of the viscous /capillary force balance in oils with very high viscosity to those with medium high viscosity. In the very high viscosity cases, the capillary forces were thought to be negligible.

Work on applying scaling theory to steam flooding continued over the following decades. Doscher [27] reported on a study of this topic noting that, given the complexity of the steam flooding process, there was no universally applicable technique to simultaneously scale all the relationships between the independent and dependent variables. He further observed that the principal deficiency of scaled physical models of a steam drive operation was in modelling the fine details of the lithology of the reservoir. Some progress has been made on rescaling systems with ordered small scale heterogeneity, but this still leaves the problem of knowing exactly what the rock heterogeneity in a reservoir or aquifer system actually is.

There was something of an upturn in the attention paid to scaling theory in the 1990s, some of it focusing on the traditional prototype to model rigorous scaling, but most of it studying the use of scaling groups to determine fluid flow regimes or the correlation of scaling groups to recovery.

Sorbie *et al.* [28] carried out miscible flooding experiments in layered bead packs to study viscous and gravity crossflow and other mechanisms related to polymer flooding and viscous slug breakdown. The Rapoport scaling groups were used directly in this work supplemented by numerical simulation to analyse the experiments. In fact miscible fluids were used since the scaling theory showed that the effects of capillary pressure in the lab should be negligibly small. The main features of the viscous crossflow mechanism could be clearly scaled between these laboratory experiments and the field. However, other features such as the effect of the sub-layer permeability heterogeneity could not be studied experimentally (as foreseen by Doscher [27]), but these were investigated by numerical simulation. The paper also noted that the calculated numerical values of the scaling groups could not in themselves allocate the various contributions, from viscous and gravity forces to crossflow; they had to be *calibrated* by numerical simulation, as explained in the paper. This underlines much of what is discussed throughout this paper about interpreting a particular numerical value of a scaling group, and taking this to be a clear indicator of the dominant force balance.

The work of Shook *et al.* [2] is, I believe, the best application of inspectional analysis in the porous medium literature. In their view, they highlight *two* main problems in the scaling literature (i) that there is often little agreement on the exact *number* of scaling groups required to completely characterise the system; and (ii) they pointed out that there was a lack of consistency in the actual *form* of any specific group. On the latter point, they list a table with 14 forms of the "gravity number"; that is the gravity/viscous force balance. In our view, for the "standard model" of 2 phase incompressible flow, there is no doubt about the number of scaling groups (there are 5 in 3D, 4 in 2D; see below). However, in more complex EOR processes, such as in thermal processes, the situation is much more difficult; see the comments of Doscher [27] referred to above. On Shook *et al.*'s latter point, we have already noted that the capillary number has over 40 forms in the literature [4], and there are many forms of any chosen force balances, and some were reviewed above. Once we have fully analysed Rapoport's paper below, we will see that the exact numerical *form* is not such a great problem. Indeed, even in the rigorous scaling of the exact set of equations used – i.e. the *precise formulae* – are not unique, but one must use a consistent set (see below). However, getting the complete *number* of groups to exactly scale a system is a real issue; but *not*, as noted above, for the 2 phase incompressible flow problem.

Zhou *et al.* [8] presented a scaling analysis of multiphase flow in a simple heterogeneous 2D system. Their analysis in essence correctly showed that 4 groups were sufficient to describe this 2D system; they chose groups to represent the gravity/viscous and capillary/viscous force balances, 1 shape group (there are 2 in 3D) and implicitly an assumed constant $M = \lambda_w/\lambda_o \sim \mu_o/\mu_w$. In fact, the forms they choose were *approximate* in the sense that that they were numbers (not functions – see below). However, the analysis of Rapoport, as explained below, leads to quite similar, but more *rigorous*, groups. However, the *intent* of the work of Zhou *et al.* was to correlate their amalgamated forms of these approximate scaling groups to the observed flow regimes which occur in their fine grid simulations. They summarised the results of this exercise in some phase diagrams of the possible flow regimes correlating the values of their amalgamated groups and the corresponding flow regimes.

Gharbi *et al.* [29] returned to the issue of scaling a 2D miscible flood in a dipping system, including a first order chemical reaction. They applied the inspectional analysis procedure of Shook et al. [2] to arrive at 8 independent groups required to fully scale the system. Again their focus was to determine the effects of these groups on the final recovery. Given the fairly large number of controlling factors (the scaling groups), they use an artificial neural net (ANN) to show that a good prediction of recovery can be made for this system from the values of the various groups.

Work in the same vein continued into the early 2000s applying scaling theory to other more complex reservoir processes, such as gas assisted gravity drainage (GAGD). Grattoni *et al.* [30] proposed a new dimensionless group to describe GAGD which was a linear combination of the three-phase Bond number and capillary number. This was developed heuristically to represent the relative strength of the combined effects of gravity and viscous forces to capillary forces. Again this approximate group was applied to correlate the results of the GAGD bead pack experiments. It is a good example of a useful non rigorous development of a group attempting to capture the central physics of a complex 3 phase process (GAGD), which it did quite well. Later work by Jadhawar and Sarma [31] also applied scaling theory to the study of reservoir scale GAGD processes. They took a rather different approach by listing 10 empirical "scaling groups" to describe GAGD in the reservoir. Although this approach moves even further away from rigorous scaling, it can still be very useful. Jadhawar and Sarma showed that for 3 different reservoir systems that models having the same values of the 10 groups behaved in a very similar (but not *exact*) manner in terms of recovery profiles.

Wood *et al.* [7] used a similar approach to develop 10 heuristic "scaling groups" to predict reservoir behaviour for continuous $CO_2$ flooding or storage. They then used experimental design techniques to construct response surfaces for oil recovery and $CO_2$ storage which gave a good applied screening tool to predict the suitability of reservoirs for these processes. They also checked by numerical simulation that systems having all 10 groups identical performed in a very similar manner. Trivedi and Babadagli [32] also follow a similar line of attack as the previous few papers in this section for scaling miscible flooding in fractured systems. Again, this is a useful approach for correlating scaling groups with recovery in such systems, and a new group describing matrix-fracture flow was introduced. However, this is also moving away from the rigorous scaling we wish to focus on relating to the Rapoport approach.

Guo *et al.* [33] presented a very thorough study on developing the scaling groups to describe polymer flooding of viscous oil. Experimental waterfloods and polymer floods in 2D sandpacks were also carried out in this study. They took a somewhat brute force approach of initially defining every variable in the process as being a potential "scaling group", arriving at 28 "groups" which they subsequently reduced to 8. In my view, 8 groups still seems rather too many as we have managed to model polymer flooding in viscous oils with a rather more limited set of input data than this. In addition, the case of a simplified polymer flood in the examples presented below can be scaled quite well using a subset of the Rapoport waterflood scaling groups alone (see Example 3, Section 8.3; Evans *et al.* [5]).

Rahman *et al.* [3] presented a compendium of the literature on the application of scaling theory to EOR. They did not present a critical analysis of this literature or of the scaling groups presented for the various EOR processes. They highlighted the large number of forms of the scaling groups that have been used. In particular, tables are given of the core scale and reservoir scale capillary number, which implicitly recognises that lengthscale must be of importance, and they list several forms of the gravity number. The very fact that there are so many groups essentially tells us that their exact *form* is not of vital importance, as discussed above (and we will show below in rigorous scaling). A more recent, and much more in-depth, analysis and review of capillary number was presented by Guo *et al.* [4] which lists over 40 forms of the capillary number. Guo *et al.* realised that some definitions of capillary numbers are essentially microscopic and other are macroscopic. At first sight, this does not really make sense if the scaling group is meant to be universal. In this paper, they explicitly acknowledge which versions of the capillary number refer to only microscopic quantities and which include a lengthscale parameter. We will return to this matter below.

## 2.3 Summary and discussion

Virtually all of the early work on the scaling theory of flow in porous media tend to be of the "approximate scaling" approach where a type of "like for like" matching between the model and the prototype were made. In all cases, a clear force balance group (or groups) usually appeared, but again in an approximate form, i.e. as a linear combination of parameters such as k, σ, etc. As discussed above, this has its advantages in its attempt to scale everything at the cost of ending up with "scaling groups" which are essentially heuristic, rather than rigorous. I have focused on these points here since the Rapoport approach leads to rigorous, if more limited, scaling groups. However, having a rigorous (and provably complete) set of scaling groups, then allows us to refer back to these to assess how the less rigorous (linear) groups actually relate back to these. In addition, the distinction in type of "scaling group" is emphasised here for pedagogical reasons, since I observe frequent use of these groups in the literature which refers to certain groups *as it they were rigorous*; e.g. "the capillary number, as given for example by Eq. (1), is $N_{Ca} = 10^{-6}$".

It is noted that many attempts to carry out "scaling" do aim directly to scale some of the nonlinear terms in the flow equations such as relative permeabilities and capillary pressures. However, as we will see below from the scaling of the full equation for two phase flow in porous media, it is not possible to do more than approximate scaling with this particular approach. The fundamental reason for this is that terms such as $\left(\frac{\Delta\rho g\Delta x\sqrt{k}}{\sigma_{ow}\cos\theta}\right)$, the nominal gravity/capillary force balance is intended to be viewed as follows: $\left(\frac{\Delta\rho g\Delta x\sqrt{k}}{\sigma_{ow}\cos\theta}\right) = \left(\frac{\Delta\rho g\Delta x}{\sigma_{ow}\cos\theta/\sqrt{k}}\right)$ where the term in the denominator $\sigma_{ow}\cos\theta/\sqrt{k}$ is a sort of Young-Laplace capillary pressure term; i.e. $P_c = \sigma_{ow}\cos\theta/\sqrt{k}$. However, since it is a *constant*, it cannot scale in the non-linear manner required by the full flow equations. Indeed, the gravity/capillary balance in the above equation is not quite correct since it is the *slope* of $P_c$ (i.e. $(dP_c/dS_w)$) that appears in the rigorous scaling groups of Rapoport; although one might argue that this quantity is intended to be $(dP_c/dS_w) = \left(\sigma_{ow}\cos\theta/\sqrt{k}\right)/\Delta S_w$ with $\Delta S_w \approx 1$. But, as mentioned above such approximate scaling groups do give the *qualitative* direction of the change of forces with system parameters, such as density difference ($\Delta\rho$), flow rate ($\nu$), viscosity ($\mu$) etc.

Finally, I would comment that later work on scaling theory (say from the 1970s onward) still mostly involved approximate scaling, but also its *intent* was usually rather different. The development of scaling groups was proposed in order to correlate them with other system properties, e.g. with flow regimes, final recoveries, the identification of key groups governing the performance of the process (EOR, steam flooding, polymer flooding etc.). In fact, my opinion is that a lot of this work would have been rather better focused if researchers had acquired a more complete appreciation of Rapoport's work. For example, less time might have been spent worrying about the precise formula for the various force balances and trying to search for the "correct" one.

## 3. The two phase flow equations in porous media

There are many forms of the two phase flow equations describing flow in porous media (Peaceman [34]; Aziz and Settari [35]; Pinder and Grey [36]), and here we take the precise version used by Rapoport [1]. In fact, it does not matter what version of these equations one takes as long as all the terms for viscous, capillary and gravity forces are correctly included. The different forms lead to rigorous scaling groups which are slightly different in form, but in fact they are all equivalent. We note here that we are using the "*standard model formulation*" of the incompressible flow equations as implemented in ***all*** numerical multiphase reservoir simulators for modelling either oil reservoirs or two phase flow processes in ground water contaminant flows. The modern literature is awash with many alternative formulations of these equations which may or may not give better descriptions of the flow physics. However, since no widely used two phase numerical simulators use these alternative formulations, we will omit discussion of these here, and use the *standard model*, as assumed by Rapoport.

### 3.1 Rapoport's form of the two phase flow equations: The problem

Rapoport's presents the main saturation equation for two phase incompressible flow in porous media (given in **Figure 1**) is given below in Eq. (2) using slightly more modern notation:

$$\phi\left(\frac{\partial S_w}{\partial t}\right)+\frac{1}{\mu_o}\left[k_x\frac{\partial}{\partial x}\left(\Psi\left(\frac{dP_c}{dS_w}\right)\left(\frac{\partial S_w}{\partial x}\right)\right)+k_y\frac{\partial}{\partial y}\left(\Psi\left(\frac{dP_c}{dS_w}\right)\left(\frac{\partial S_w}{\partial y}\right)\right)+k_z\frac{\partial}{\partial z}\left(\Psi\left(\frac{dP_c}{dS_w}\right)\left(\frac{\partial S_w}{\partial z}\right)\right)\right]$$
$$+\left\{\frac{k_z}{\mu_o}g(\rho_o-\rho_w)\left(\frac{d\Psi}{dS_w}\right)\left(\frac{\partial S_w}{\partial z}\right)\right\}+\left\langle\left(\frac{d\kappa}{dS_w}\right)\left(u\left(\frac{\partial S_w}{\partial x}\right)+v\left(\frac{\partial S_w}{\partial y}\right)+w\left(\frac{\partial S_w}{\partial z}\right)\right)\right\rangle=0 \tag{2}$$

where we have extended Rapoport's form of the 2 phase transport equation to allow for *anisotropy* of the permeability where in general, $k_x \neq k_y \neq k_z$[3] (see **Figure 1**; and Ancillary Material #1; Evans *et al.*, [5]). The functions $\kappa(S_w)$, $\psi(S_w)$ and $P_c(S_w)$ in Eq. (2) are functions of saturation, $S_w$ only, as follows.

$$\kappa(S_w)=\left(1+\left(\frac{k_{ro}\cdot\mu_w}{k_{rw}\cdot\mu_o}\right)\right)^{-1}=f_w(S_w) \tag{3}$$

$$\psi(S_w)=k_{ro}.\kappa=k_{ro}f_w(S_w) \tag{4}$$

where we note that the function, $\kappa(S_w)$, is the same as the normal fractional flow expression which was called $\phi(S_w)$ in Rapoport's paper, and he also used $f$ for porosity; but this notation is too confusing given modern usage. The capillary pressure is given by:

[3] Note that absolute permeability is in fact a *tensor* quantity, denoted $\underline{\underline{k}}$, which in general may have non-zero off diagonal terms. However, it is a symmetric positive definite tensor which is always diagonalizable. Here, we have made the widely applied used assumption that this tensor is diagonal (i.e. the cross terms such as $k_{xy}$ etc. are zero). This will have some bearing on the "shape groups" which we will point out below and in Appendix A.

$$P_c(S_w) = P_o - P_w; \tag{5}$$

In Eq. (2), the capillary terms are given by the expressions in the [...] parentheses, the gravity term is given in the {...} parentheses and the viscous terms are in the $\langle ... \rangle$ parentheses. There is also a corresponding pressure equation but in the incompressible case, absolute pressure is irrelevant as only pressure gradients are required to compute flows. This pressure equation contains no additional information on scaling groups which is not already present in Eq. (2); all the information on viscous pressure gradients is already encoded in Eq. (2) through the two phase Darcy law.

Now consider the scaling task addressed by Rapoport in the title of his paper, as shown in **Figure 3**. At the time of that publication, as noted in the review above, it was a popular activity to try build actual scaled physical models of petroleum reservoirs in order to study their behaviour in the laboratory (e.g. Leverett et al. [15]; Engelberts and Klinkenberg [16]; Croes and Schwarz [17] etc.). Rapoport's objective was to determine the conditions for a "prototype reservoir" and a "model reservoir" to be exactly "scaled"; by "scaled" we mean that in some dimensionless variables the two are actually *identical* in that they have *exactly* the same balance of all forces. This implies *rigorous scaling* in the sense discussed above in Section 2.1. We may think of the prototype reservoir of being at the "field" scale (i.e. it is large) and the model reservoir being at the "laboratory" scale (i.e. it is small), but in fact it does not matter which is which.

As in Rapoport's paper, the process we will consider here is the injection of water into an oil reservoir but the situation is the same as injecting an incompressible (or practically incompressible) gas into a water system, such as supercritical $CO_2$ injection into an aquifer. To be more precise about what we mean by "scaled", we mean that the dimensionless saturation distribution in the prototype reservoir, $S_w(x, y, z, t)$, is identical to the corresponding values in the model reservoir, $S_w(x', y', z', t')$ at a given corresponding dimensionless "time" in pore volumes (PV) throughput. These reservoirs are of course different sizes and may have several different properties but as long as these are correctly scaled the results in each are *identical* at a given number of PV injection throughput. To see what is required for the prototype and model systems to be exactly scaled, we go back to the governing equation, Eq. (2). In this equation the prototype is described in variables $x, y, z, t$ etc. and the model is described in scaled

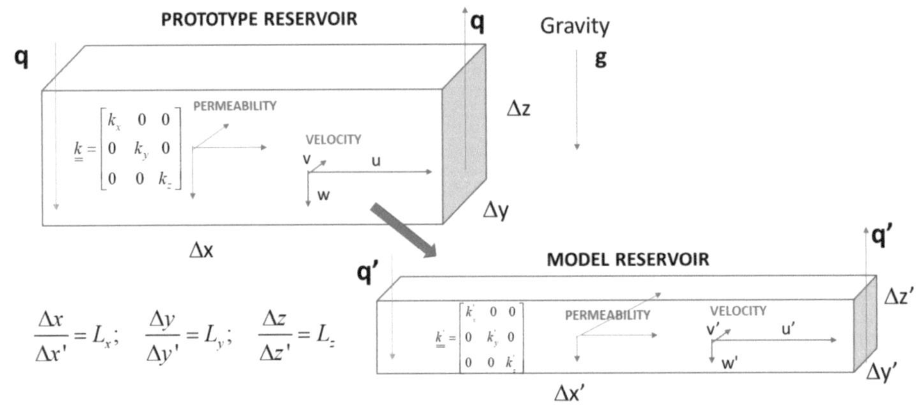


**Figure 3.**
*The general idea of a "Prototype" and a "Model" reservoir, usually to denote the field and laboratory, respectively. The coordinates of the prototype and model are demoted* $x, y$ and $z$ *and* $x', y'$ and $z'$, *respectively.*

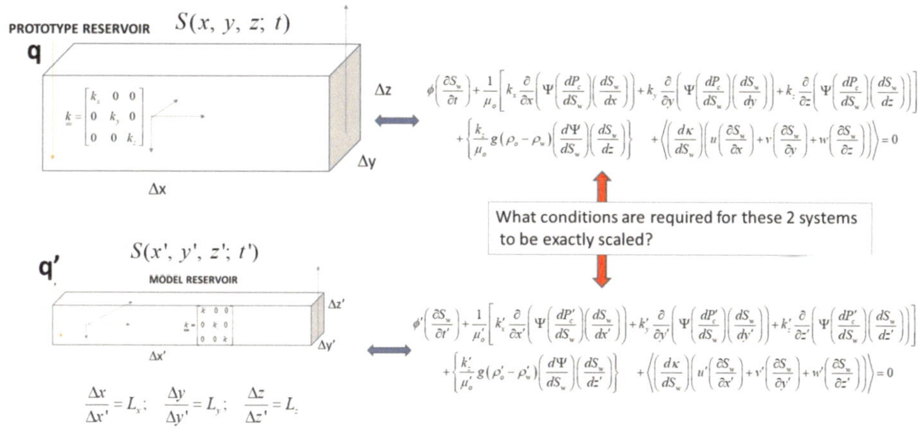


**Figure 4.**
*A schematic illustration of the central question of "rigorous scaling"; what are the conditions required for the two systems, prototype and model, to be exactly scaled?*

variables $x', y', z', t'$ etc.. The model is governed by an *identical* equation to Eq. (2) but in primed variables, $x', y', z', t'$ etc. as shown in Eq. (6) below:

$$\phi'\left(\frac{\partial S_w}{\partial t'}\right)+\frac{1}{\mu_o'}\left[k_x'\frac{\partial}{\partial x'}\left(\Psi\left(\frac{dP_c'}{dS_w}\right)\left(\frac{dS_w}{dx'}\right)\right)+k_y'\frac{\partial}{\partial y'}\left(\Psi\left(\frac{dP_c'}{dS_w}\right)\left(\frac{dS_w}{dy'}\right)\right)+k_z'\frac{\partial}{\partial z'}\left(\Psi\left(\frac{dP_c'}{dS_w}\right)\left(\frac{dS_w}{dz'}\right)\right)\right]$$
$$+\left\{\frac{k_z'}{\mu_o'}g(\rho_o'-\rho_w')\left(\frac{d\Psi}{dS_w}\right)\left(\frac{dS_w}{dz'}\right)\right\}+\left\langle\left(\frac{d\kappa}{dS_w}\right)\left(u'\left(\frac{\partial S_w}{\partial x'}\right)+v'\left(\frac{\partial S_w}{\partial y'}\right)+w'\left(\frac{\partial S_w}{\partial z'}\right)\right)\right\rangle=0 \quad (6)$$

Where the complete set of dashed variables (to be scaled) in Eq. (6) is as follows: $x', y', z', t', \phi', \mu_o', k_x', k_y', k_z', P_c', \rho_o', \rho_w', u', v'$ and $w'$. We note here that the velocities in the prototype ($u, v$ and $w$) and model ($u', v'$ and $w'$) will be different and need not be explicitly "scaled" since they emerge from the chosen volumetric flows and geometry of the system (see below).

Thus, we address the question posed in **Figure 4**: *What are the conditions on the scaled variables that imply that the prototype and model systems are dimensionally identical?* The relative permeabilities, and hence $\kappa(S_w)$ and $\psi(S_w)$, are assumed to be the same in the prototype and the model systems, but $P_c(S_w)$ may vary, as discussed below. The relative permeability assumption, which to some may seem quite drastic, has been revisited in the literature (Perkins and Collins [21]) and we will discuss this issue later in this paper.

### 3.2 The answer: The two phase scaling groups

Before going through the detailed derivation of the scaling groups, we present "the answer" in **Table 2**. This table does not appear in Rapoport's paper, but all the information is there to construct most of it. **Table 2** presents the various dimensionless dependent and independent quantities for both the 3D prototype and model reservoirs, along with the similarity groups that ensure they are identically scaled. The basic assumptions in deriving the quantities in **Table 2** are: (a) that the relative permeabilities, and hence the other saturation dependent functions, $\kappa(S_w)$ and $\psi(S_w)$,

| | | | |
|---|---|---|---|
| **Independent groups** | | | |
| $\frac{x}{\Delta x}$ ; | $\frac{y}{\Delta y}$ ; | $\frac{z}{\Delta z}$ ; | $\frac{t}{\Delta t}$ |
| **Dependent groups** | | | |
| $S_w$ ; | $\frac{P}{\Delta P}$ ; | $\frac{Fluid\ phase\ recovery}{Total\ phase\ in\ place}(fraction)$ | |
| **Similarity groups** | | | |
| *Force balances:* | | | |
| $C_{VC1} = \left(\frac{\mu_o \Delta xq}{k_x \Delta y \Delta z \left(\frac{dP_c}{dS}\right)}\right) = \frac{Viscous\ forces}{Capillary\ forces}$; $C_{GC} = \left(\frac{g\Delta\rho.\Delta z}{\left(\frac{dP_c}{dS_w}\right)}\right) = \frac{Gravity\ forces}{Capillary\ forces}$ | | | |
| *Shape groups:* $C_{S1} = \left(\frac{k_x \Delta y^2}{k_y \Delta x^2}\right)$; $C_{S2} = \left(\frac{k_x \Delta z^2}{k_z \Delta x^2}\right)$ | | | |
| *And* $\left(\frac{\mu_o}{\mu_w}\right) = cons\tan t$ | | | |

*Notes: (i) All Dependent, Independent and Similarity Groups in this table are dimensionless; (ii) the gravity/capillary group requires* **g** *in order to be dimensionless which is "added back" later (see text).*

**Table 2.**
*The Independent, Dependent and Similarity groups for incompressible two phase flow in porous media; after Rapoport [1]. Strictly, this should read "extracted" from Rapoport [1].*

are the same for prototype and model, and (b) reservoir structure remains broadly similar, i.e. aspect ratios may change, a 3 layer system say, remains a 3 layer system but layer thicknesses may change if other adjustments are made, (c) the initial fluid saturation distributions $S_{wi}(x, y, z, t = 0)$ are structurally identical and (d) the boundary conditions are the same (source/sink [well] positions, completions etc.). Under these assumptions then, if all the similarity groups in **Table 2** are held constant, then the prototype and model reservoirs are *exactly* scaled. This means that in suitable dimensionless variables (discussed below) the 2 systems are essentially *identical*.[4] This is what is meant by Rapoport (and in this paper) as rigorous scaling; we do *not* mean finding scaling groups (like $N_{Ca}$ in Eq. (1) above) which give broad indications of the behaviour of the system under certain parameter changes.

As shown in **Table 2**, for incompressible 2 phase flow in 3D, there are exactly 5 similarity groups as follows:

i. *2 force balances* – for example, viscous/capillary ($C_{VC1}$) and gravity/capillary ($C_{GC1}$); the viscous/gravity balance ($C_{VG1}$) is then fixed;

ii. *2 "shape groups"* ($C_{S1}$ and $C_{S2}$) governing reservoir shape and permeability anisotropy; and

iii. *the viscosity ratio* $(\mu_o/\mu_w)$ must be fixed.

The reason for the subscript 1 in the force balance groups arises because there are actually 3 forms of each of the 3 force/balance groups, but they are not independent and any sets can be used. Rapoport does not lay this out explicitly, but this will be explained and demonstrated below. To our knowledge, the formulae for these

---

[4] By "identical" here, we mean in a numerical quantity of order $\sim$1, $S_w$ say, about 6 to 10 decimal point agreement in the numerical simulation model. See later discussion on this point.

interconnected groups are not presented in the literature. However, the author believes that Rapoport must have known about these relations, unlike some later authors who believe in some rather specific forms of the simplified scaling groups.

There is much to say about the precise form of the similarity groups in **Table 2** but we will keep our discussion to two brief comments, at this point. Firstly, the viscous/capillary force balance in Eq. (7) is essentially a more rigorous "capillary number", and it can be seen that[5]:

$$C_{VC1} = \frac{q.\Delta x.\mu_o}{k_x.\Delta y.\Delta z\left(\frac{dP_c}{dS_w}\right)} = \frac{v.\Delta x.\mu_o}{k_x\left(\frac{dP_c}{dS_w}\right)} \tag{7}$$

where the velocity, $v = (q/\Delta y.\Delta z)$, is introduced and this has a form somewhat reminiscent of the simpler dimensionless capillary number, $N_{Ca} = (v\mu/\sigma)$. However, the group, $C_{VC1}$, (which is also dimensionless) is now a little more elaborate and contain the rather awkward term $(dP_c/dS_w)$ which is actually a function of $S_w$; therefore, $C_{VC1}$ is not a simple number. This implies that any "dimensionless number" which is indeed a single number cannot capture a true viscous/capillary force balance in a non-linear system such as this in a rigorous sense, but it may still be qualitatively useful.

Secondly, we note that 2 "shape groups" (in a 3D system), $C_{S1}$ and $C_{S2}$, appear in the dimensionless scaling functions, and these relate to system size and permeability anisotropy. Physically, these have been described as relating to "effective aspect ratio" of the system (Shook *et al.* [2])[6] and we also note in Appendix A how the shape group also keeps the balance of flows in the various directions. These groups allow us to design laboratory scaled models of anisotropic systems using homogenous rocks or bead packs by simply changing the dimensions (shape) of the system. However, they also fulfil another role, not previously discussed in the literature, which is that they provide "links" between the other forms of the force balance similarity groups, as discussed in detail later in this paper.

Note that the 2 shape groups in 3D incorporate anisotropy, which was not addressed in Rapoport's paper. We noted above that in fact the permeability is actually a *tensor* quantity, denoted $\underline{\underline{k}}$, which we assume is diagonal in this work. In fact, the shape groups can be defined for a tensor permeability field and this issue is addressed in Appendix A.

## 4. The approach to solving the scaling problem

### 4.1 Term by term rescaling

We now address the derivation of the scaling groups (in **Table 2**) by manually rescaling *all* the corresponding variables in turn in each system – the prototype

[5] Note that if we write the Darcy law as $v = -(k_x/\mu)(\Delta P_{vis}/\Delta x)$, where $\Delta P_{vis}$ is the viscous pressure drop, then $\Delta P_{vis} = (v\Delta x\mu/k_x)$and Eq. (7) can be viewed in an approximate manner as the ratio of viscous/capillary pressure drops, i.e. $C_{VC1} \approx \Delta P_{vis}/(dP_c/dS_w)$;

[6] Strictly, the "effective aspect ratio", relating to shape groups $C_{S1}$ is often defined as, $R_1 = \left(\frac{k_x}{k_y}\right)^{1/2}\frac{\Delta y}{\Delta x} = \sqrt{C_{S1}}$.

($x, y, z, t$ etc.) and the model (*primed* variables $x', y', z', t'$ etc.). We consider in turn the Lengthscale (L), Flow Rate (Q), etc. and these scaling variables, L, Q etc. (see below) here are important to note, as they appear in the formulae leading to the scaling groups themselves.

We start from the physical size or *Lengthscales (L)* of the two systems. The prototype reservoir which is of size $\Delta x, \Delta y, \Delta z$ , we define the size of the model reservoir as being $\Delta x', \Delta y', \Delta z'$ and the scaling variable of length are given in **Figure 3** as:

$$\frac{\Delta x}{\Delta x'} = L_x; \quad \frac{\Delta y}{\Delta y'} = L_y; \quad \frac{\Delta z}{\Delta z'} = L_z \tag{8}$$

And hence:

$$x' = \frac{x}{L_x}; \quad y' = \frac{y}{L_y}; \quad z' = \frac{z}{L_z}; \tag{9}$$

The *Flow Rates (Q)* are then rescaled as follows:

$$\frac{q}{q'} = Q \Rightarrow q' = \frac{q}{Q} \tag{10}$$

$$\text{Which implies} \rightarrow \frac{q_x}{q_x{}'} = Q_x; \quad \frac{q_y}{q_y{}'} = Q_y; \quad \frac{q_z}{q_z{}'} = Q_z \tag{11}$$

And so it is evident that:

$$\Rightarrow \frac{\mathrm{v}_x}{\mathrm{v}'_x} = \frac{\mathrm{u}}{\mathrm{u}'} = \frac{Q_x}{L_y.L_z}; \quad \frac{\mathrm{v}_y}{\mathrm{v}'_y} = \frac{\mathrm{v}}{\mathrm{v}'} = \frac{Q_y}{L_x.L_z}; \quad \frac{\mathrm{v}_z}{\mathrm{v}'_z} = \frac{\mathrm{w}}{w'} = \frac{Q_z}{L_x.L_y}; \tag{12}$$

$$\text{And so } u' = \left(\frac{L_y.L_z}{Q_x}\right)u; \quad v' = \left(\frac{L_x.L_z}{Q_y}\right)v; \quad w' = \left(\frac{L_x.L_y}{Q_z}\right)w \tag{13}$$

Thus, the rescaled velocities are already determined by fixing the flow rates in the prototype and model, and the boundary conditions which must be the same is both cases.

*Porosities (F) and Permeabilities (K)* may then be scaled as follows:

$$\Rightarrow \frac{\phi}{\phi'} = F; \quad \frac{k_x}{k'_x} = K_x; \frac{k_y}{k'_y} = K_y; \frac{k_z}{k'_z} = K_z; \tag{14}$$

$$\text{And so} \Rightarrow \phi' = \frac{\phi}{F}; \quad k'_x = \frac{k_x}{K_x}; k'_y = \frac{k_y}{K_y}; k'_z = \frac{k_z}{K_z} \tag{15}$$

*Viscosities (M) and Densities (R)* are scaled as follows:

$$\Rightarrow \frac{\mu_o}{\mu'_o} = \frac{\mu_w}{\mu'_w} = M; \Rightarrow \mu'_o = \frac{\mu_o}{M} \text{ and } \mu'_w = \frac{\mu_w}{M} \Rightarrow \quad \frac{\mu_o}{\mu_w} = \frac{\mu'_o}{\mu'_w} \tag{16}$$

$$\Rightarrow \frac{\Delta\rho}{\Delta\rho'} = \frac{\rho_o - \rho_w}{\rho'_o - \rho'_w} = R \Rightarrow \left(\rho'_o - \rho'_w\right) = \frac{(\rho_o - \rho_w)}{R} \tag{17}$$

*Capillarity (B)* is scaled, where we note it is the *slope* of the capillary pressure, $(dP_c/dS)$ which is required (See Eq. (2)), as follows:

$$\Rightarrow \frac{\left(\frac{dP_c}{dS}\right)}{\left(\frac{dP'_c}{dS}\right)} = B; \quad \Rightarrow \left(\frac{dP'_c}{dS}\right) = \frac{1}{B}\left(\frac{dP_c}{dS}\right) \tag{18}$$

From the rescaled parameters so far, it can be shown that the *Time* rescaling to pore volumes (PV) is as follows:

$$\Rightarrow \frac{t}{t'} = \frac{L_x L_y L_z.F}{Q} \quad or \quad t' = \left(\frac{Q}{L_x L_y L_z.F}\right)t \tag{19}$$

The PV equivalence is made clear by taking the $(t/t')$ ratio and inserting the previous scaling expressions which gives:

$$\Rightarrow \frac{t}{t'} == \frac{\left(\frac{x}{x'}\cdot\frac{y}{y'}\cdot\frac{z}{z'}\cdot\frac{\phi}{\phi'}\right)}{\left(\frac{q}{q'}\right)} = \left(\frac{\left(\frac{x.y.z.\phi}{q}\right)}{\left(\frac{x'.y'.z'.\phi'}{q'}\right)}\right) \tag{20}$$

From which it is clear that the ratio of times between the prototype and the model is in PV/PV.

### 4.2 Recasting the governing equations: Model to prototype

The next step in the development is algebraically rather tedious but is actually quite straightforward. Each dashed variable in the governing equation for the model in Eq. (6) (and shown in **Figure 4**), is replaced by the scaling term relating back to the prototype system; for example, from Eq. (14) we find:
$\phi' = \frac{\phi}{F}$ and from Eq. (19), $t' = \left(\frac{Q}{L_x L_y L_z.F}\right)t$, therefore, the first term in Eq. (6) becomes:

$$\phi'\left(\frac{\partial S_w}{\partial t'}\right) = \frac{\phi}{F}\cdot\left(\frac{L_x L_y L_z.F}{Q}\right)\left(\frac{\partial S_w}{\partial t}\right) = \left(\frac{L_x L_y L_z}{Q}\right)\phi\left(\frac{\partial S_w}{\partial t}\right) \tag{21}$$

The expression in Eq. (21) above shows the combination of scaling factors which relates the term in the model flow equations (in dashed variable) to the original prototype; i.e. the term $(L_x L_y L_z/Q)$.

The next terms to be transformed from model (dashed) variables to prototype variables are the *capillary pressure terms* in Eq. (6); first for the x – term as follows:

$$\begin{aligned} \frac{1}{\mu'_o}\left[k'_x\frac{\partial}{\partial x'}\left(\Psi\left(\frac{dP'_c}{dS_w}\right)\left(\frac{\partial S_w}{\partial x'}\right)\right)\right] &= \frac{M}{\mu_o}\left[\frac{k_x L_x}{K_x}\frac{\partial}{\partial x}\left(\Psi\frac{1}{B}\left(\frac{dP_c}{dS_w}\right)\frac{L_x}{1}\left(\frac{\partial S_w}{\partial x}\right)\right)\right] \\ &= \left(\frac{ML_x^2}{BK_x}\right)\frac{1}{\mu_o}\left[k_x\frac{\partial}{\partial x}\left(\Psi\left(\frac{dP_c}{dS_w}\right)\left(\frac{\partial S_w}{\partial x}\right)\right)\right] \end{aligned} \tag{22}$$

And likewise for the capillary terms in y and z in Eq. (6), the transformation is as follows:

$$\frac{1}{\mu_o'}\left[k_y'\frac{\partial}{\partial y'}\left(\Psi\left(\frac{dP_c'}{dS_w}\right)\left(\frac{\partial S_w}{\partial y'}\right)\right)\right]=\left(\frac{ML_y^2}{BK_y}\right)\frac{1}{\mu_o}\left[k_y\frac{\partial}{\partial y}\left(\Psi\left(\frac{dP_c}{dS_w}\right)\left(\frac{\partial S_w}{\partial y}\right)\right)\right] \tag{23}$$

$$\frac{1}{\mu_o'}\left[k_z'\frac{\partial}{\partial z'}\left(\Psi\left(\frac{dP_c}{dS_w}\right)\left(\frac{\partial S_w}{\partial z'}\right)\right)\right]=\left(\frac{ML_z^2}{BK_z}\right)\frac{1}{\mu_o}\left[k_z\frac{\partial}{\partial z}\left(\Psi\left(\frac{dP_c}{dS_w}\right)\left(\frac{\partial S_w}{\partial z}\right)\right)\right] \tag{24}$$

Next we transform the *gravity term*, in Eq. (6), as follows:

$$\begin{aligned}\left\{\frac{k_z'}{\mu_o'}g(\rho_o'-\rho_w')\left(\frac{d\Psi}{dS_w}\right)\left(\frac{\partial S_w}{\partial z'}\right)\right\}&=\left\{\frac{M.k_z}{K_z\mu_o}g\frac{(\rho_o-\rho_w)}{R}\left(\frac{d\Psi}{dS_w}\right)\frac{L_z}{1}\left(\frac{\partial S_w}{\partial z}\right)\right\}\\&=\left(\frac{ML_z}{K_zR}\right)\left\{\frac{k_z}{\mu_o}g(\rho_o-\rho_w)\left(\frac{d\Psi}{dS_w}\right)\left(\frac{\partial S_w}{\partial z}\right)\right\}\end{aligned} \tag{25}$$

And finally we transform the viscous term in Eq. (6), but we already have expressions for $u', v', w'$ since these have been shown (in Eq. (13) above) to be, as follows: $u'=\left(\frac{L_y.L_z}{Q_x}\right)u;\ \ v'=\left(\frac{L_x.L_z}{Q_y}\right)v;\quad w'=\left(\frac{L_x.L_y}{Q_z}\right)w$. Using these substitutions below gives:

$$\begin{aligned}&\left\langle\left(\frac{d\kappa}{dS_w}\right)\left(u'\left(\frac{\partial S_w}{\partial x'}\right)+v'\left(\frac{\partial S_w}{\partial y'}\right)+w'\left(\frac{\partial S_w}{\partial z'}\right)\right)\right\rangle\\&=\left\langle\left(\frac{d\kappa}{dS_w}\right)\left(\left(\frac{L_yL_z}{Q_x}\right)u\left(\frac{\partial S_w}{\partial x'}\right)+\left(\frac{L_xL_z}{Q_y}\right)v\left(\frac{\partial S_w}{\partial y'}\right)+\left(\frac{L_xL_y}{Q_z}\right)w\left(\frac{\partial S_w}{\partial z'}\right)\right)\right\rangle\\&=\left\langle\left(\frac{d\kappa}{dS_w}\right)\left(\left(\frac{L_yL_z}{Q_x}\right)u\frac{L_x}{1}\left(\frac{\partial S_w}{\partial x}\right)+\left(\frac{L_xL_z}{Q_y}\right)v\frac{L_y}{1}\left(\frac{\partial S_w}{\partial y}\right)+\left(\frac{L_xL_y}{Q_z}\right)w\frac{L_z}{1}\left(\frac{\partial S_w}{\partial z}\right)\right)\right\rangle\\&=\left\langle\left(\frac{d\kappa}{dS_w}\right)\left(\left(\frac{L_xL_yL_z}{Q_x}\right)u\left(\frac{\partial S_w}{\partial x}\right)+\left(\frac{L_xL_yL_z}{Q_y}\right)v\left(\frac{\partial S_w}{\partial y}\right)+\left(\frac{L_xL_yL_z}{Q_z}\right)w\left(\frac{\partial S_w}{\partial z}\right)\right)\right\rangle\end{aligned} \tag{26}$$

Thus, combining all of the terms derived above into the model equation (Eq. (6)) becomes the following:

$$\begin{aligned}&\left(\frac{L_xL_yL_z}{Q}\right)\phi\left(\frac{\partial S_w}{\partial t}\right)+\frac{1}{\mu_o}\left[\begin{aligned}&\left(\frac{ML_x^2}{BK_x}\right)k_x\frac{\partial}{\partial x}\left(\Psi\left(\frac{dP_c}{dS_w}\right)\left(\frac{\partial S_w}{\partial x}\right)\right)+\left(\frac{ML_y^2}{BK_y}\right)k_y\frac{\partial}{\partial y}\left(\Psi\left(\frac{dP_c}{dS_w}\right)\left(\frac{\partial S_w}{\partial y}\right)\right)\\&+\left(\frac{ML_z^2}{BK_z}\right)k_z\frac{\partial}{\partial z}\left(\Psi\left(\frac{dP_c}{dS_w}\right)\left(\frac{\partial S_w}{\partial z}\right)\right)\end{aligned}\right]\\&\quad+\left(\frac{ML_z}{K_zR}\right)\left\{\frac{k_z}{\mu_o}g(\rho_o-\rho_w)\left(\frac{d\Psi}{dS_w}\right)\left(\frac{\partial S_w}{\partial z}\right)\right\}\\&\quad+\left\langle\left(\frac{d\kappa}{dS_w}\right)\left(\left(\frac{L_xL_yL_z}{Q_x}\right)u\left(\frac{\partial S_w}{\partial x}\right)+\left(\frac{L_xL_yL_z}{Q_y}\right)v\left(\frac{\partial S_w}{\partial y}\right)+\left(\frac{L_xL_yL_z}{Q_z}\right)w\left(\frac{\partial S_w}{\partial z}\right)\right)\right\rangle=0\end{aligned} \tag{27}$$

Dividing through by the term $(L_xL_yL_z)/Q$ gives the scaled equation:

$$\phi\left(\frac{\partial S_w}{\partial t}\right)+\frac{1}{\mu_o}\left[\begin{array}{l}\left(\frac{ML_xQ}{BK_xL_yL_z}\right)k_x\frac{\partial}{\partial x}\left(\Psi\left(\frac{dP_c}{dS_w}\right)\left(\frac{\partial S_w}{\partial x}\right)\right)+\left(\frac{ML_yQ}{BK_yL_xL_z}\right)k_y\frac{\partial}{\partial y}\left(\Psi\left(\frac{dP_c}{dS_w}\right)\left(\frac{\partial S_w}{\partial y}\right)\right)\\+\left(\frac{ML_zQ}{BK_zL_xL_y}\right)k_z\frac{\partial}{\partial z}\left(\Psi\left(\frac{dP_c}{dS_w}\right)\left(\frac{\partial S_w}{\partial z}\right)\right)\end{array}\right]$$
$$+\left(\frac{MQ}{K_zRL_xL_y}\right)\left\{\frac{k_z}{\mu_o}g(\rho_o-\rho_w)\left(\frac{d\Psi}{dS_w}\right)\left(\frac{\partial S_w}{\partial z}\right)\right\}$$
$$+\left\langle\left(\frac{d\kappa}{dS_w}\right)\left(\left(\frac{Q}{Q_x}\right)u\left(\frac{\partial S_w}{\partial x}\right)+\left(\frac{Q}{Q_y}\right)v\left(\frac{\partial S_w}{\partial y}\right)+\left(\frac{Q}{Q_z}\right)w\left(\frac{\partial S_w}{\partial z}\right)\right)\right\rangle=0 \quad (28)$$

We have reached the point where the messy but straightforward algebra is over, and Rapoport's brilliant stroke is delivered. He answers the question posed in **Figure 4**: what is required such that Eq. (28) (i.e. Eq. (6)) scales exactly to Eq. (2)? The Answer is clearly that the terms of the main scaling ratios [the groups shown in Eq. (28)] must equal unity! That is:

$$\left(\frac{ML_xQ}{BK_xL_yL_z}\right)=\left(\frac{ML_yQ}{BK_yL_xL_z}\right)=\left(\frac{ML_zQ}{BK_zL_xL_y}\right)=\left(\frac{MQ}{K_zRL_xL_y}\right)=1 \quad (29)$$

The velocity terms are already correct by definition, to adjust and acknowledge that the total flows, geometry and boundary conditions correctly and exactly scale the prototype and model velocity flow fields (under the assumptions listed above). In the following section, we will explore exactly what Eq. (29) means.

## 5. The emergence of the two phase scaling groups

To start the process of unfolding what the Rapoport scaling groups mean, we consider, all the various equalities in Eq. (29). Labelling the 4 quantities as A, B, C and D as shown in Eq. (30)

$$\overset{\text{A}}{\left(\frac{ML_xQ}{BK_xL_yL_z}\right)}=\overset{\text{B}}{\left(\frac{ML_yQ}{BK_yL_xL_z}\right)}=\overset{\text{C}}{\left(\frac{ML_zQ}{BK_zL_xL_y}\right)}=\overset{\text{D}}{\left(\frac{MQ}{K_zRL_xL_y}\right)}=1 \quad (30)$$

The 10 possible equalities, divided into 3 Sets, in Eq. (30) above are as follows;
Set 1: (i) A = 1; (ii) B = 1; (iii) C = 1; (iv) D = 1
Set 2: (v) A = B; (vi) A = C; (vii) B=C
Set 3: (viii) A = D; (ix) B = D; (x) C = D
We explore each of these in turn, starting by showing the full details of the simple algebra, but they very quickly start to form a pattern.

### 5.1 The analysis of set 1

*Set 1: (i)* The first equality in Eq. (30) (A = 1) is as follows:

$$\left(\frac{ML_xQ}{BK_xL_yL_z}\right)=1 \quad (31)$$

This is expanded by returning to the basic definitions of the various quantities, where Eq. (31) means the following:

$$\frac{ML_xQ}{K_xL_yL_zB}=\frac{\left(\frac{\mu_o}{\mu_o{}'}\cdot\frac{\Delta x}{\Delta x'}\cdot\frac{q}{q'}\right)}{\left(\frac{k_x}{k_x{}'}\cdot\frac{\Delta y}{\Delta y'}\cdot\frac{\Delta z}{\Delta z'}\cdot\frac{\left(\frac{dP_c}{dS}\right)}{\left(\frac{dP_c{}'}{dS}\right)}\right)}=\left(\frac{\left(\frac{\mu_o\Delta xq}{k_x\Delta y\Delta z\left(\frac{dP_c}{dS}\right)}\right)}{\left(\frac{\mu_o{}'\Delta x'q'}{k_x{}'\Delta y'\Delta z'\left(\frac{dP_c{}'}{dS}\right)}\right)}\right)=1 \tag{32}$$

Which implies that the value of the group: $\left(\frac{\mu_o\Delta xq}{k_x\Delta y\Delta z\left(\frac{dP_c}{dS}\right)}\right)$ must be *the same* in the *prototype* and *model* systems. The physical interpretation of this group is clear; it expresses the viscous/capillary force balance as follows:

$$C_{VC1}=\left(\frac{\mu_o\Delta xq}{k_x\Delta y\Delta z\left(\frac{dP_c}{dS}\right)}\right)=\frac{\textit{Viscous forces}}{\textit{Capillary forces}} \tag{33}$$

*Set 1: (ii)* The second equality in Eq. (30) (B = 1) is as follows, showing the expansion of the terms (in slightly contracted form):

$$\left(\frac{ML_yQ}{BK_yL_xL_z}\right)=\left(\frac{\frac{\mu_o}{\mu_o'}\frac{\Delta y}{\Delta y'}\frac{q}{q'}}{\frac{(dP_c/dS_w)}{(dP_c'/dS_w)}\frac{k_y}{k_y'}\frac{\Delta x}{\Delta x'}\frac{\Delta z}{\Delta z'}}\right)=\left(\frac{\frac{(\mu_o\Delta yq)}{((dP_c/dS_w)k_y\Delta x\Delta z)}}{\frac{(\mu_o\Delta y'q')}{((dP_c{}'/dS_w)k_{y'}\Delta x'\Delta z')}}\right)=1 \tag{34}$$

Thus, to be exactly scaled, the quantity $\frac{(\mu_o\Delta yq)}{((dP_c/dS_w)k_y\Delta x\Delta z)}$ must be kept constant from prototype to the model system and hence:

$$C_{VC2}=\frac{(\mu_o\Delta yq)}{\left(k_y\Delta x\Delta z\left(\frac{dP_c}{dS}\right)\right)}=\frac{\textit{Viscous forces}}{\textit{Capillary forces}} \tag{35}$$

which is simply another form of the viscous/capillary force balance given above as $C_{VC1}$.

*Set 1 (iii):* Corresponding to C = 1, following the previous 2 cases, can now be easily seen to lead to

$$\left(\frac{ML_zQ}{BK_zL_xL_y}\right)=1 \quad\Rightarrow\left(\frac{\mu_o\Delta zq}{((dP_c/dS_w)k_z\Delta x\Delta y)}\right)=\text{constant} \tag{36}$$

The above group is the third form of the viscous/capillary scaling group, $C_{VC3}$, as follows:

$$C_{VC3}=\left(\frac{\mu_o\Delta zq}{\left(k_z\Delta x\Delta y\left(\frac{dP_c}{dS}\right)\right)}\right)=\frac{\textit{Viscous forces}}{\textit{Capillary forces}} \tag{37}$$

There is a clear pattern in the form of the groups, $C_{VC1}$, $C_{VC2}$ and $C_{VC3}$, and these quantities look like they are closely related, which they are. We will return to this once we have worked out all of the possible equalities.

*Set 1 (iv):* The group corresponding to D = 1, is a little different from the above 3 equalities. It can be seen (using the methods above) to lead to:

$$\left(\frac{MQ}{K_z R L_x L_y}\right) = 1 \quad \Rightarrow \left(\frac{\mu_o q}{k_z \Delta\rho \Delta x \Delta y}\right) = \text{constant} \tag{38}$$

And this group is clearly related to the viscous/gravity force balance, $C_{VG1}$, as follows:

$$C_{VG1} = \left(\frac{\mu_o q}{k_z \Delta\rho \Delta x \Delta y}\right) = \frac{\textit{Viscous forces}}{\textit{Gravity forces}} \tag{39}$$

The reader might notice that the viscous/gravity group ($C_{VG1}$) given above is not dimensionless. However, before discussing this matter or the 3 forms of the viscous/capillary scaling groups, we first examine *Set 2* (just below Eq. (30)) of the possible equalities.

## 5.2 The analysis of set 2

*Set 2 (v):* Now consider the following equality (A = B) from Eq. (30) above:

$$\left(\frac{ML_x Q}{BK_x L_y L_z}\right) = \left(\frac{ML_y Q}{BK_y L_x L_z}\right) = 1 \tag{40}$$

After cancellation this leads to:

$$\left(\frac{L_x}{K_x L_y}\right) = \left(\frac{L_y}{K_y L_x}\right) \quad \Rightarrow \left(\frac{K_x L_y^2}{K_y L_x^2}\right) = 1 = \left(\frac{\frac{k_x \Delta y^2}{k_y \Delta x^2}}{\frac{k'_x \Delta y'^2}{k'_y \Delta x'^2}}\right) \tag{41}$$

Thus the following *shape group*, $C_{S1}$, must be a constant from prototype to model:

$$C_{S1} = \left(\frac{k_x \Delta y^2}{k_y \Delta x^2}\right) \tag{42}$$

*Set 2 (vi):* Now consider the following equality (A = C) from Eq. (30) above:

$$\left(\frac{ML_x Q}{BK_x L_y L_z}\right) = \left(\frac{ML_z Q}{BK_z L_x L_y}\right) = 1 \tag{43}$$

Proceeding in exactly the same manner as the previous case, gives:

$$\left(\frac{L_x}{K_x L_z}\right) = \left(\frac{L_z}{K_z L_x}\right) \Rightarrow \left(\frac{K_x L_z^2}{K_z L_x^2}\right) = 1 \Rightarrow \left(\frac{k_x \Delta z^2}{k_{xz} \Delta x^2}\right) = cons\tan t \tag{44}$$

Leads to the second version of the shape group

$$C_{S2} = \left(\frac{k_x \Delta z^2}{k_z \Delta x^2}\right) \tag{45}$$

*Set 2 (vii):* Consider the equality (B = C) from Eq. (30) above:

$$\left(\frac{ML_yQ}{BK_yL_xL_z}\right) = \left(\frac{ML_zQ}{BK_zL_xL_y}\right) = 1 \tag{46}$$

Following the previous 2 cases above, this will give:

$$\left(\frac{L_y}{K_yL_z}\right) = \left(\frac{L_z}{K_zL_y}\right) \Rightarrow \left(\frac{K_yL_z^2}{K_zL_y^2}\right) = 1 \Rightarrow \left(\frac{k_y\Delta z^2}{k_z\Delta y^2}\right) = \text{constant} \tag{47}$$

Which leads to the third shape group, $C_{S3}$, as follows:

$$C_{S3} = \left(\frac{k_y\Delta z^2}{k_z\Delta y^2}\right) \tag{48}$$

Before proceeding to the *Set 3* equalities, we note that, the 3 shape groups above, $C_{S1}$, $C_{S2}$ and $C_{S3}$, are given by:

$$C_{S1} = \left(\frac{k_x\Delta y^2}{k_y\Delta x^2}\right); \quad C_{S2} = \left(\frac{k_x\Delta z^2}{k_z\Delta x^2}\right); \quad C_{S3} = \left(\frac{k_y\Delta z^2}{k_z\Delta y^2}\right) \tag{49}$$

But it is clear that:

$$\frac{C_{S1}}{C_{S2}} = \left(\frac{k_x\Delta y^2}{k_y\Delta x^2}\right)\left(\frac{k_z\Delta x^2}{k_x\Delta z^2}\right) = \left(\frac{k_z\Delta y^2}{k_y\Delta z^2}\right) = \frac{1}{C_{S3}} \Rightarrow C_{S1}.C_{S3} = C_{S2} \tag{50}$$

And hence only 2 of the shape groups are independent in a 3D system, and only 1 applies in a 2D system.

## 5.3 The analysis of set 3

*Set 3 (viii):* Now we see the results of the final 3 equalities starting with A = D in Eq. (30).

$$\left(\frac{ML_xQ}{BK_xL_yL_z}\right) = \left(\frac{MQ}{K_zRL_xL_y}\right) = 1 \tag{51}$$

By cancellation this simplifies to:

$$\left(\frac{L_x}{BK_xL_z}\right) = \left(\frac{1}{K_zRL_x}\right) \Rightarrow \left(\frac{K_zRL_x^2}{BK_xL_z}\right) = 1 \tag{52}$$

Again we recall the fundamental meaning of the various scaling terms and expand these as follows:

$$\Rightarrow \left(\frac{K_zRL_x^2}{BK_xL_z}\right) = \left(\frac{\frac{k_z}{k_z'}\ \frac{\Delta\rho}{\Delta\rho'}\ \frac{\Delta x^2}{\Delta x'^2}}{\frac{k_x}{k_x'}\ \frac{\left(\frac{dP_c}{dS}\right)}{\left(\frac{dP_c'}{dS}\right)}\ \frac{\Delta z}{\Delta z'}}\right) = \left(\frac{\left(\frac{k_z\Delta\rho\Delta x^2}{k_x\left(\frac{dP_c}{dS}\right)\Delta z}\right)}{\left(\frac{k_z'\Delta\rho'\Delta x'^2}{\Delta\rho\left(\frac{dP_c'}{dS}\right)\Delta z'}\right)}\right) = 1 \tag{53}$$

Thus, the quantity that remains the same is, $\left(\frac{k_z \Delta\rho \Delta x^2}{k_x \left(\frac{dP_c}{dS}\right)\Delta z}\right)$, which is clearly a gravity/capillary scaling group:

$$C_{GC1} = \left(\frac{k_z \Delta\rho \Delta x^2}{k_x \left(\frac{dP_c}{dS}\right)\Delta z}\right) = \frac{\textit{Gravity forces}}{\textit{Capillary forces}} \tag{54}$$

We will show that this can be further simplified after working out the last two equalities below.

*Set 3 (ix):* Now setting B = D in Eq. (30):

$$\left(\frac{ML_y Q}{BK_y L_x L_z}\right) = \left(\frac{MQ}{K_z R L_x L_y}\right) = 1 \tag{55}$$

$$\left(\frac{L_y}{BK_y L_z}\right) = \left(\frac{1}{K_z R L_y}\right) = 1; \quad \Rightarrow \left(\frac{K_z R L_y^2}{BK_y L_z}\right) = 1; \quad \Rightarrow \left(\frac{k_z \Delta\rho \Delta y^2}{\left(\frac{dP_c}{dS_w}\right) k_y \Delta z}\right) \tag{56}$$

Which leads to a second form of the gravity/capillary scaling group, $C_{GC2}$, as follows:

$$C_{GC2} = \left(\frac{k_z \Delta\rho \Delta y^2}{k_y \left(\frac{dP_c}{dS_w}\right) \Delta z}\right) \tag{57}$$

*Set 3 (x):* The final equality C = D in Eq. (30) above is given by:

$$\left(\frac{ML_z Q}{BK_z L_x L_y}\right) = \left(\frac{MQ}{K_z R L_x L_y}\right) = 1 \tag{58}$$

$$\left(\frac{L_z}{B}\right) = \left(\frac{1}{R}\right) = 1; \quad \Rightarrow \left(\frac{RL_z}{B}\right) = 1; \quad \Rightarrow \left(\frac{\Delta\rho.\Delta z}{\left(\frac{dP_c}{dS_w}\right)}\right) = \text{constant} \tag{59}$$

And this provides a third, and much simpler form of the gravity/capillary scaling groups, $C_{GC3}$, as follows:

$$C_{GC3} = \left(\frac{\Delta\rho.\Delta z}{\left(\frac{dP_c}{dS_w}\right)}\right) \tag{60}$$

The observant reader will note that the above group $C_{GC3}$ is not dimensionless, and nor are groups $C_{GC1}$ or $C_{GC2}$.above. Actually, as we show in examples below, it does not need to be for scaling purposes. However, the dimensionless form of Eq. (60) is achieved by "adding back" the gravity term, **g**, as shown below, which was effectively cancelled out earlier.

A resolution of this apparent overabundance of scaling groups is presented in the next section.

## 6. Summary and reduction of the scaling groups

In summary, from the total set of possible equalities, we have derived 10 scaling groups and it is probably evident to the reader that some of these will be redundant. First examining the Gravity/Capillarity scaling groups, starting from $C_{GC1}$ we note:

$$C_{GC1} = \left(\frac{k_z \Delta\rho \Delta x^2}{k_x \left(\frac{dP_c}{dS}\right) \Delta z}\right) = \left(\frac{k_z \Delta x^2 \Delta\rho \Delta z}{k_x \Delta z^2 \left(\frac{dP_c}{dS}\right)}\right) = \left(\frac{k_z \Delta x^2}{k_x \Delta z^2}\right)\left(\frac{\Delta\rho \Delta z}{\left(\frac{dP_c}{dS}\right)}\right) = \frac{C_{GC3}}{C_{S2}} \tag{61}$$

And it is also easy to see that:

$$C_{GC2} = \left(\frac{k_z \Delta\rho \Delta y^2}{k_y \left(\frac{dP_c}{dS_w}\right) \Delta z}\right) = \left(\frac{k_z \Delta y^2 \Delta\rho \Delta z}{k_y \Delta z^2 \left(\frac{dP_c}{dS_w}\right)}\right) = \left(\frac{k_z \Delta y^2}{k_y \Delta z^2}\right)\left(\frac{\Delta\rho \Delta z}{\left(\frac{dP_c}{dS_w}\right)}\right) = \frac{C_{GC3}}{C_{S3}} \tag{62}$$

Therefore, from above it is clear that there is just *one* gravity/capillary group that is independent and the simplest form of this group is as follows:

$$C_{GC3} = \left(\frac{\Delta\rho . \Delta z}{\left(\frac{dP_c}{dS_w}\right)}\right) \tag{63}$$

In fact, a dimensional check of this group shows that, although it can be used perfectly well to rescale prototype and model, it is *not* dimensionless (it has dimensions $T^2/L$). This is because we cancelled out $\boldsymbol{g}$ (acceleration due to gravity; dimensions $L/T^2$) in both the prototype and model equations, and so a more physical and dimensionless gravity/capillary scaling group is given below, where its relationship to the other 2 forms of this group are given:

$$C'_{GC3} = \left(\frac{g\Delta\rho . \Delta z}{\left(\frac{dP_c}{dS_w}\right)}\right) = C_{S2}.C'_{GC1} = C_{S3}.C'_{GC2} \tag{64}$$

This form of the group, $C'_{GC3}$, is the only one required to describe the gravity/capillary scaling, and this is given in **Table 2**, simply denoted as $G_{GC} = C'_{GC3}$. Note that we could use *any* other forms of the gravity/capillary group as they all equivalent (i.e. equally "correct"), and the only justification for using this one is simplicity.

Finally we note that, turning to the three forms of the viscous/capillary force balances, $C_{VC1}, C_{VC2}$ and $C_{VC3}$, it is easy to see that:

$$\frac{C_{VC1}}{C_{VC2}} = \left(\frac{\mu_o \Delta x q}{k_x \Delta y \Delta z \left(\frac{dP_c}{dS}\right)}\right)\left(\frac{k_y \Delta x \Delta z \left(\frac{dP_c}{dS}\right)}{\mu_o \Delta y q}\right) = \left(\frac{k_y \Delta x^2}{k_x \Delta y^2}\right) = \frac{1}{C_{S1}} \text{ and} \tag{65}$$

$$\frac{C_{VC1}}{C_{VC3}} = \left(\frac{\mu_o \Delta x q}{k_x \Delta y \Delta z \left(\frac{dP_c}{dS}\right)}\right)\left(\frac{k_z \Delta x \Delta y \left(\frac{dP_c}{dS}\right)}{\mu_o \Delta z q}\right) = \left(\frac{k_z \Delta x^2}{k_x \Delta z^2}\right) = \frac{1}{C_{S2}} \tag{66}$$

$$\Rightarrow C_{VC1} = \frac{C_{VC2}}{C_{S1}} = \frac{C_{VC3}}{C_{S2}} \tag{67}$$

Therefore, there is only *one* independent viscous/capillary scaling group and this can be chosen arbitrarily from the 3 versions, from example we can take:

$$C_{VC1} = \left( \frac{\mu_o \Delta x q}{k_x \Delta y \Delta z \left( \frac{dP_c}{dS} \right)} \right) \tag{68}$$

Thus, the full set of scaling groups is as shown in **Table 2**.

However, we return briefly to the *Set 1 (iv);* D = 1 case (Eq. (39)) leading to the viscous/gravity scaling group:

$$C_{VG1} = \left( \frac{\mu_o q}{k_z \Delta \rho \Delta x \Delta y} \right) = \frac{\textit{Viscous forces}}{\textit{Gravity forces}} \tag{69}$$

This is sometimes written

$$C_{VG1} = \left( \frac{\mu_o q \Delta z}{k_z \Delta \rho \Delta x \Delta y \Delta z} \right) = \left( \frac{\mu_o v_D \Delta z}{k_z \Delta \rho \Delta x} \right) \tag{70}$$

Or including $\boldsymbol{g}$ to make this dimensionless (as described above) then:

$$C'_{VG1} = \left( \frac{\mu_o v_D \Delta z}{k_z g \Delta \rho \Delta x} \right) \tag{71}$$

Where $v_D = \left( \frac{q}{\Delta y \Delta z} \right)$, is a version of the Darcy velocity, and this dimensionless form ($C'_{VG1}$) is used in one of the examples below.

By now, one might suspect that the above group for the viscous/gravity force balance ($C_{VG1}$) is probably redundant, since we already have 2 forces balances, viscous/capillary and gravity/ capillary, and there are only the 3 forces - viscous, capillary and gravity. The likely proof of this is to take a ratio of the other two balances, ($C_{VC1}/C_{GC3}$)to eliminate the capillary forces, for example ($C_{GC3}$ is taken to omit the $\boldsymbol{g}$ for consistency with the earlier development):

$$\frac{C_{VC1}}{C_{GC3}} = \left( \frac{\mu_o \Delta x q}{k_x \Delta y \Delta z \left( \frac{dP_c}{dS} \right)} \right) \left( \frac{\left( \frac{dP_c}{dS_w} \right)}{\Delta \rho . \Delta z} \right) = \left( \frac{\mu_o \Delta x q}{k_x \Delta \rho . \Delta y \Delta z^2} \right) \tag{72}$$

This is not quite there yet, but multiplying numerator and denominator above by $k_z \Delta x$ and rearranging leads to the following:

$$\frac{C_{VC1}}{C_{GC3}} = \left( \frac{\mu_o \Delta x q}{k_x \Delta \rho . \Delta y \Delta z^2} \right) \left( \frac{k_z \Delta x}{k_z \Delta x} \right) = \left( \frac{k_z \Delta x^2 \mu_o q}{k_x \Delta z^2 k_z \Delta \rho . \Delta x \Delta y} \right) = \left( \frac{k_z \Delta x^2}{k_x \Delta z^2} \right) \left( \frac{\mu_o q}{k_z \Delta \rho . \Delta x \Delta y} \right) \tag{73}$$

$$\frac{C_{VC1}}{C_{GC3}} = \left( \frac{k_z \Delta x^2}{k_x \Delta z^2} \right) \left( \frac{\mu_o q}{k_z \Delta \rho . \Delta x \Delta y} \right) = \frac{C_{VG1}}{C_{S2}} \tag{74}$$

Therefore, the viscous/gravity groups (all 3 of them) are already dependent on the existing derived force balances for viscous/capillary and gravity/capillary, and again the shape groups are closely involved in the relationships between them. However, in

| The Scaling Groups | Relationships |
|---|---|
| **Viscous/Capillary force balance:** $C_{VC1} = \left(\frac{\mu_o \Delta x q}{k_x \Delta y \Delta z \left(\frac{dP_c}{dS}\right)}\right)$ $C_{VC2} = \frac{(\mu_o \Delta y q)}{\left(k_y \Delta x \Delta z \left(\frac{dP_c}{dS}\right)\right)}$ $C_{VC3} = \left(\frac{\mu_o \Delta z q}{\left(k_z \Delta x \Delta y \left(\frac{dP_c}{dS}\right)\right)}\right)$ | *1 group* $\Rightarrow C_{VC1} = \frac{C_{VC2}}{C_{S1}} = \frac{C_{VC3}}{C_{S2}}$ |
| **Gravity/Capillary force balance (including g):** $C'_{GC1} = \left(\frac{k_z g \Delta\rho \Delta x^2}{k_x \left(\frac{dP_c}{dS}\right) \Delta z}\right)$ $C'_{GC2} = \left(\frac{k_z g \Delta\rho \Delta y^2}{k_y \left(\frac{dP_c}{dS_w}\right) \Delta z}\right)$ $C'_{GC3} = \left(\frac{g \Delta\rho . \Delta z}{\left(\frac{dP_c}{dS_w}\right)}\right)$ | *1 group* $\Rightarrow C'_{GC3} = C_{S2}.C'_{GC1} = C_{S3}.C'_{GC2}$ |
| **Shape Groups:** $C_{S1} = \left(\frac{k_x \Delta y^2}{k_y \Delta x^2}\right)$ $C_{S2} = \left(\frac{k_x \Delta z^2}{k_z \Delta x^2}\right)$ $C_{S3} = \left(\frac{k_y \Delta z^2}{k_z \Delta y^2}\right)$ | *2 groups (in 3D)* *1 group (in 2D)* $\Rightarrow C_{S1}.C_{S3} = C_{S2}$ |
| The above 4 independent groups, along with $(\mu_o/\mu_w)$, fully define the system and the viscous/gravity (V/G) force balance can be found from these. However, for completeness the V/G group is given below and it is used in one of the examples. | |
| **Viscous/Gravity force balance:** $C_{VG1} = \left(\frac{\mu_o q}{k_z \Delta\rho \Delta x \Delta y}\right)$; $C_{VG2} = \left(\frac{\mu_o q \Delta y}{k_y \Delta\rho \Delta x \Delta z^2}\right)$; $C_{VG3} = \left(\frac{\mu_o q \Delta x}{k_x \Delta\rho \Delta y \Delta z^2}\right)$ | *1 group (not independent of the above)* $\Rightarrow C_{VG1} = C_{S3}.C_{VG2} = C_{S2}.C_{VG3}$ |

**Table 3.**
*The complete set of scaling groups for 2 phase incompressible flow in porous media provided by the Rapoport analysis, showing the relationships between them.*

practical applications, one may wish to work with only viscous and gravity forces in a given scaling application, as in an example below. So these are listed in **Table 3** (without derivation) for completeness and convenience, but it is indicated that only 2 force balances, 2 shape groups and the constant viscosity ratio $(\mu_o/\mu_w)$ are required.

The exposition here brings out the clear derivation of the scaling groups and it also shows the intimate connection between the 2 independent shape groups and the 2 force balances. Another lesson from the Rapoport groups is that there is a broad flexibility around the *exact* formula (or "*forms*") for the scaling groups, even in this more rigorous scaling approach (within its assumptions). It also highlights the rather qualitative nature of any "single number" scaling such as capillary number, Bond number etc., especially in isolation from their related shape groups.

## 7. The Rapoport assumptions for rigorous scaling

The assumptions of the Rapoport approach include two phase fluid incompressibility, the same initial fluid saturation distribution and boundary conditions and the same relative permeabilities (RPs) in the prototype and model systems. This last assumption on the relative permeabilities appear to be rather a strict limitation, and we would perhaps like to be able to "scale out" the RPs between the 2 systems. However, this restriction is rather like the constant $(\mu_o/\mu_w)$ ratio which we list among the required scaling groups. The underlying reason for this is quite easy to see: both the RPs and the $(\mu_o/\mu_w)$ ratio are used to construct the fractional flow function, $f_w(S_w)$, for the system. This in turn determines the Buckley-Leverett shock front height, $S_{wf}$, which is a fundamental feature of the two phase displacement. So it appears that the RP functions would be a very difficult, if not impossible, to rigorously "scale out" from one system to another.

Perkins and Collins [21] proposed a way to address both the relative permeability restriction and also the somewhat limited restriction on the capillary pressure, $P_c(S_w)$; actually, for $P_c$, it is the *slope,* $(dP_c/dS_w)$, that had to accounted for in the scaling groups (**Table 2**). The latter issue was dealt with by Perkins and Collins by assuming a J-function scaling of the capillary pressure. Actually, this is a reasonable approach but does not really relax the constriction on the $P_c$ curve any more than Rapoport does. Also, J-function scaling often does not work very well in systems that are not water wet, which is probably most systems. In Perkins and Collins, the RPs are relaxed by first defining a normalised saturation scale, which is used quite frequently in core analysis where, $\overline{S}_w = (S_w - S_{wc})/(1 - S_{wc} - S_{or})$ and $0 \le \overline{S}_w \le 1$. The newly defined RPs are then end point normalised, as follows;

$$\overline{K}_w\left(\overline{S}_w\right) = K_w/k_{w,or} \text{ and } \overline{K}_o\left(\overline{S}_w\right) = K_o/k_{o,wc} \tag{75}$$

where again $0 \le \overline{K}_w\left(\overline{S}_w\right) \le 1$ and $0 \le \overline{K}_o\left(\overline{S}_w\right) \le 1$, and $k_{w,or}$ and $k_{o,wc}$ are the relative permeabilities of water at residual oil and oil at initial (connate) water, respectively. They then incorporate the various normalisation factors ($k_{w,or}$ , $k_{o,wc}$) into the scaling groups. The scaling requirement is then that the various groups which they list as in Eq. (76):

$$\left(\frac{L_2}{L_1}\right), \left(\frac{L_3}{L_1}\right), \left(\frac{k_{w,or}\cdot\mu_o}{k_{o,wc}\cdot\mu_w}\right), \left(\frac{\sigma_{ow}\cdot k_{w,or}\cdot\mu_o}{\nu.\mu_w L_1}\right)\left(\frac{\phi}{k}\right)^{1/2}, \left(\frac{k_{w,or}.g\Delta\rho L_3}{\nu.\mu_w L_1}\right) \tag{76}$$

are constant between the prototype and model, *and that the transformed quantities* $\overline{K}_w\left(\overline{S}_w\right)$ and $\overline{K}_o\left(\overline{S}_w\right)$ *are also identical*. This does allow a little more flexibility in terms of rescaling the end points of the relative permeabilities, but it is still a fairly restrictive requirement, and it is not too dissimilar to the Rapoport original assumption of holding the simple RP functions constant between prototype and model. For example, if we have a given set of RP functions for the prototype, then we cannot select an arbitrary set for the model and then "scale out" any differences. The reason for this has already been stated above, i.e. that this would change the fractional flow function leading to a different displacement profiles, recoveries and $S_w(x, y, z, t)$ patterns, which are clearly *not* the same between the prototype and the model.

Indeed, Perkins and Collins [21] presented examples of rescaled RPs which showed disappointingly poor agreement between each other for RP functions arising in a prototype reservoir (consolidated sandstones) and unconsolidated sands (as used in lab models). To the author's knowledge, there are no successful rescaling examples in the literature (in the sense of the examples below) using the Perkins and Collins rescaling method.

## 8. Examples of the application of the scaling groups

Here, we present 4 examples of the application of scaling groups to address some problems relating to more modern concerns than their original development intentions. That original concern was to make actual physical models at the laboratory scale of real "reservoir scale situations". Some of these cases have been reported in the literature review, and others will be described below. Example 2 describes viscous/capillary scaling in immiscible viscous fingering. The corollary of this example leads to Example 2.5 which demonstrates one of the novel *consequences* of scaling theory.

There is an additional aspect of the scaling theory developed by Rapoport of which he was probably unaware. This is that the technique can be applied very usefully in carrying of scaled *numerical simulations* of various phenomena. In addition, there are some principles and technical details of such simulations of which many contemporary simulation engineers and researchers may be unaware. Since the early numerical confirmation of the Rapoport scaling groups by Nielsen and Tek [24] referred to above, there has been little work of this type that I have been able to locate. Some researchers did check the validity of their scaling groups using numerical simulations (e.g. Wood *et al.*, [7]) but almost all of the other papers proposing various scaling groups did not.

### 8.1 Example 1: Lab to field scaling of immiscible viscous fingering: Viscous forces only

Consider a system where there are *only* viscous forces acting; i.e. either $P_c = 0$ or, $g$ or $\Delta\rho$ or$\nabla z = 0$. This would occur for example in a 2D horizontal heterogeneous system in which immiscible viscous fingering is taking place. For such a system, a surprising prediction is made by the Rapoport's scaling theory. To show this, we model a lab scale system where the permeability is described by a heterogeneous correlated random field (CRF), as shown in **Figure 5(a)**. We then numerically simulate a 2 phase system in a square laboratory slab $\Delta x = \Delta y = 0.3m$ with water displacing oil (left to right) where the viscosity ratio is $(\mu_o/\mu_w) = 2000$, using a Cartesian grid of size $N_x = N_y = 100$. This displacement process is viscous unstable and results in immiscible viscous fingering, as shown in **Figure 5(b)**. A full description of this simulation, which is based on the modelling of experiments performed by Skauge and coworkers is given in the literature (Beteta *et al.* [37]; Salmo *et al.* [38]); in particular, the specific method to calculate the "fingering relative permeabilities" is described in Sorbie *et al.* [39].

Since there are only viscous forces then the viscous/capillary (and viscous/gravity) force balance group is $C_{VC1} = \infty$, and the system scaling depends only on the shape group, $C_{S1} = \left(k_x\Delta y^2/k_y\Delta x^2\right)$. This implies that if we simply change the size of the system preserving $C_{S1}$, then the system should be identical in similarity terms. For

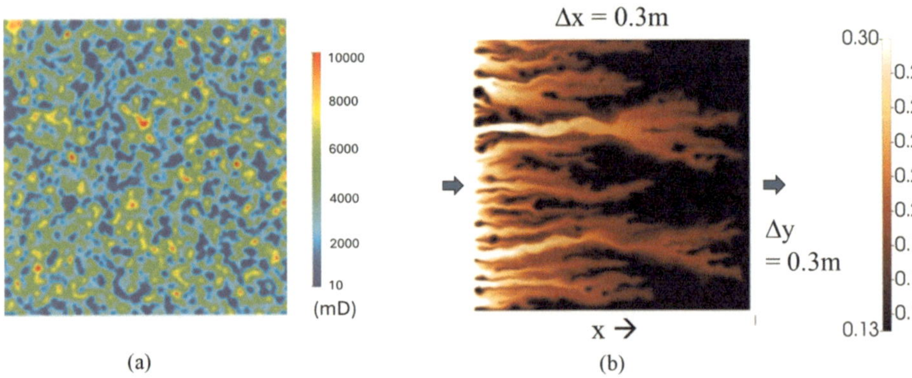


(a) (b)

**Figure 5.**
*Figure shows (a) A random correlated permeability field (dimensionless correlation length, $\lambda_D$ = 0.03), with permeability (k) range of 0.01 to 10 D and an average $k \sim 3$ D; (b) Saturation distribution ($S_w(x,y,t = BT)$) at breakthrough (BT = 0.04 PV) for a viscous dominated 2 phase unstable water → oil displacement in a 2D lab slab experiment with $(\mu_o/\mu_w) = 2000$; adapted from Beteta* et al., *[37].*

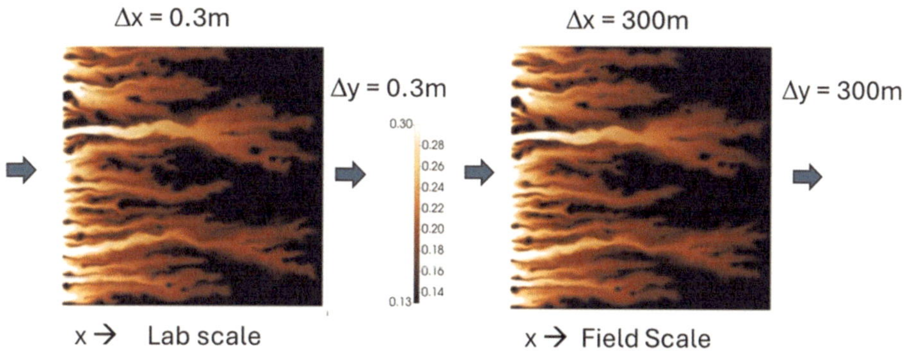


**Figure 6.**
*Scaled lab and field scale model saturation distribution $(S_w(x, y, t = BT))$at breakthrough (BT =0.04 PV) for a viscous dominated 2 phase unstable water → oil displacement in a 2D lab slab and a field scale region experiment with $(\mu_o/\mu_w) = 2000$. Adapted from Beteta* et al. *[37] and Salmo* et al. *[38].*

example, if we keep the expanded shape the same $(\Delta x' = \Delta y' = 300m)$ but keep the permeability field identical, i.e. $k_x$ and $k_y$the same in every pixel, then this would mean the lab and "field" models are identical, as long as we use exactly the same size of grid $(N_x = N_y = 100)$. **Figure 6** shows these 2 calculations using a 100x100 numerical grid at the 2 different lengthscales; a careful inspection by eye shows that these 2 saturation patterns at the same PV are essentially identical. In fact, to ~6 - 7 decimal places the saturation values block by block are indeed identical in the lab and field systems. This result may be a little surprising since numerical simulations would have certain levels of numerical dispersion affecting flow. However, it turns out that the *relative* levels of numerical dispersion are actually identical; i.e. for a simple numerical scheme such as single point upstream weighting, the dispersivity, $\alpha$, is given by $\alpha = \delta x/2$ where $\delta x = \Delta x/N_x$ (where $\delta x$ is the actual grid size, at whatever scale). For dispersity, $\alpha$ (actually $\alpha_x, \alpha_y$ or $\alpha_z$), the numerical dispersion is $D_{num} = \alpha.v$ (and $\alpha = \delta x/2 = \Delta x/(2N_x)$) and the mesh Peclet number, $N_{Pe\_mesh} = v.\Delta x/D_{num}$, thus we can see that $N_{Pe\_mesh} = 2N_x$, independent of the size of the system, and this is the same for each model. Thus, they scale exactly with the system size resulting in virtually identical numerical results for models scaled using the Rapoport similarity groups. Indeed, the author has found that, by taking extra care with the control of wells/ source-sink terms (boundary conditions) and the numerical time truncation error, up to ~10 or more decimal point accuracy can be achieved; this is *exact* scaling.

### 8.2 Example 2: Immiscible viscous fingering: Viscous/capillary scaling

We now discuss a further development of the example in Section 8.1 above, involving viscous/capillary scaling. In this case, we now introduce a capillary pressure into the lab 2D flooding case above which exhibited immiscible viscous fingering. To scale this system, we must now use two groups, one for viscous/capillary scaling $(C_{VC1})$ along with the appropriate x/y shape group ( $C_{S1}$), as follows:

$$C_{VC1} = \left( \frac{\mu_o \Delta x q}{k_x \Delta y \Delta z \left(\frac{dP_c}{dS}\right)} \right) \text{ and } C_{S1} = \left( \frac{k_x \Delta y^2}{k_y \Delta x^2} \right) \tag{77}$$

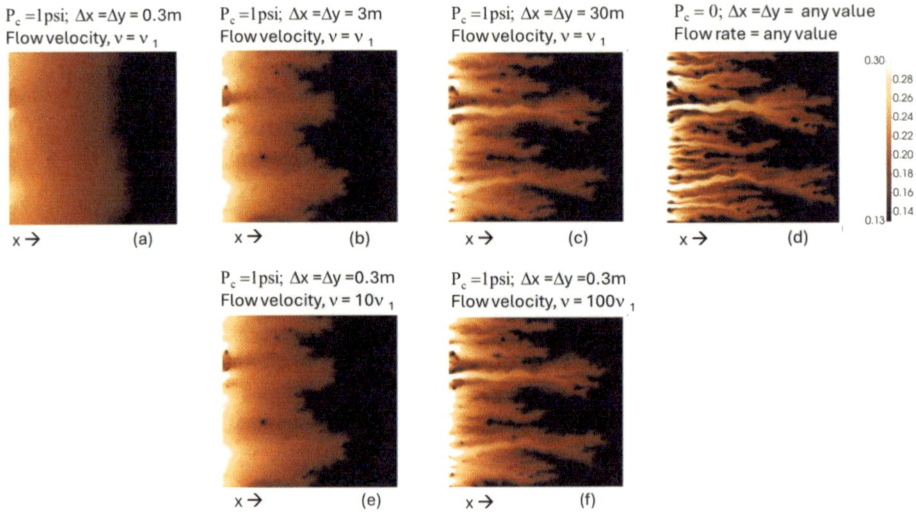


**Figure 7.**
*(a) shows the lab scale example in* ***Figure 6*** *($\Delta x = \Delta y = 0.3$ m; $\nu = \nu_1$) but including a non-zero capillary pressure function with a maximum value of $P_{c,max} = 1$ psi, which has suppressed the viscous fingering at flow velocity, $\nu = \nu_1$; (b)-(d) show the same system but increased in size for the same permeability fields, all with the same flow velocity, $\nu = \nu_1$. (e) and (f) are the experimental slab size cases ($\Delta x = \Delta y = 0.3$ m), but with flow velocities $\nu = 10\nu_1$ and $\nu = 100\nu_1$, respectively. Adapted from Beteta* et al. *[37].*

For the two phase simulations described above, a water-wet $P_c$ curve with a max value of 1 psi at $S_w = S_{wi}$ is introduced into the numerical immiscible viscous fingering calculation for systems of different size, as described in Beteta *et al.* [37]. The results of these calculations are show in **Figure 7(a–d)**. These 4 calculations are *not* similar as the viscous/capillary balance with system size is clearly increasing, as expected from the viscous/capillary scaling group ($C_{VC1}$) since Δx is increasing and $(dP_c/dS_w)$ remains constant. However, if we focus on the 2 cases in **Figure 7(b)** and **(c)**; these two case are the increased system size calculations at the original flow rate of the experiment ($q_1$; giving velocity, $\nu = \nu_1$). The clear effects of capillarity can be seen in both of these calculations. However, it is obvious from the viscous/capillary scale group, $C_{VC1}$, that as long as the product Δx.q remains the same, then the systems are identical (if the shape group is preserved, same relative permeabilities etc.). Thus, if we took the lab scale (capillary dominated) case in **Figure 7(a)** and kept the system size the same ($\Delta x = \Delta y = 0.3$ m) but increased the flow velocity by a factor of x10 ($\nu = 10\nu_1$) as in **Figure 7(e)**, the system should be identical, to **Figure 7(b)**, and we took the same case and increased the flow velocity by a factor of x100 ($\nu = 100\nu_1$) as in **Figure 7(f)**, the system should be identical to the case **Figure 7(c)**. Exactly these calculations have been carried out and the results are shown in **Figure 7(e)** and **(f)**, which are plotted below the figures to which they should be exactly scaled. A close inspection shows that these are indeed identical, as required by Rapoport scaling theory. Again, this agreement is not at an approximate or broadly correlated level, it is virtually exact (accurate to ~6 decimal points).

## 8.3 Example 2.5: A "sense check" on gas/liquid relative permeabilities from scaling theory

This is labelled as *Example 2.5* since it is based directly on the theory from the previous subsection. The scaling theory results presented in Section 8.2 can be used to

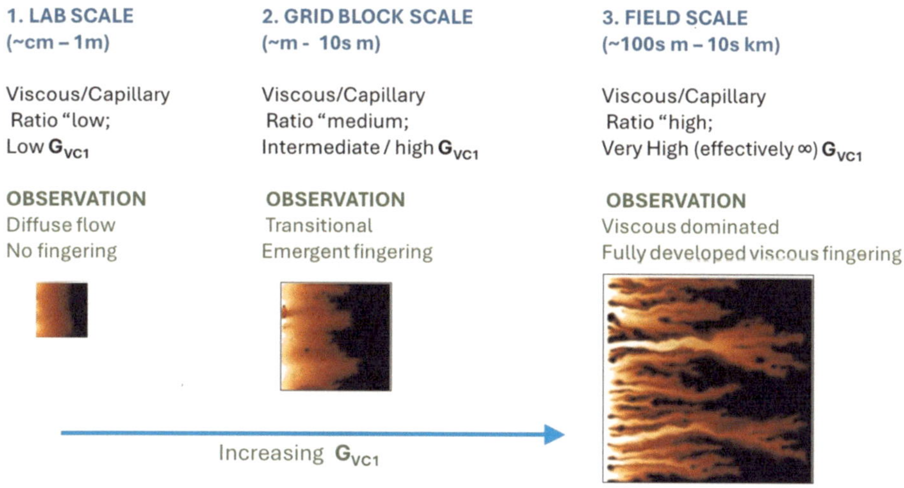


**Figure 8.**
*"Sense check" for RP functions for unstable systems such as for gas displacements e.g. $CO_2 \rightarrow$ water, $CH_4 \rightarrow$ oil, $H_2 \rightarrow$ water etc.*

develop a "*sense check*" for relative permeabilities (RPs) which are derived for systems where we expect to observe viscous fingering *at some lengthscale*; e.g. gas/liquid RPs for $H_2$/water or $CO_2$/water systems. In other words, when the system becomes *viscous dominated*, even in the presence of gravity.

A schematic of this "sense check" for RP functions for unstable systems is shown in **Figure 8**. Currently, RP functions for these systems ($H_2$/water and $CO_2$/water) are being used to assess underground hydrogen storage (UHS) [40] and carbon capture and storage (CCS) projects [41]. The basic prediction of scaling theory is, as shown in **Figure 7** above and implicitly in **Figure 8**, that as a system with a given RP function – which is the only flow function which occurs in the purely viscous dominated case – then, even if Pc is included at the small scale, then as the system size increases, then viscous fingers must "emerge". Thus, if we take any proposed "experimentally derived" RP function for a $H_2$/water, $CO_2$/water or any gas/liquid system then using this in a mildly heterogeneous system (e.g. a CRF with some mild k-variability) in viscous dominated conditions (i.e. with no gravity of capillary forces), the system must show emergent viscous fingering. If it does not, then there is a problem with the RP functions.

In fact, this issue has been studied recently by the author and co-workers, and none of the published $H_2$/water or $CO_2$/water RP functions in the literature show any significant fingering behaviour; see Wang *et al.* [42] and Larki *et al.* [43]. Thus, the currently published experimentally derived relative permeability functions cannot describe the correct gas → water displacement flow physics (i.e. viscous fingering) for the $H_2$/water or the $CO_2$/water systems. This particular "sense check" on the RP functions, which arises indirectly from scaling theory, again could probably not have been anticipated by Rapoport or his contemporaries.

### 8.4 Example 3: Experiment to field polymer flooding case: viscous/gravity scaling

In this example, we describe the design of a two phase 2D vertical cross-section of a scaled laboratory experiment relating to a tertiary polymer flood in a simplified

3 layer reservoir. In the original study from which this example is taken, we used the viscous capillary scaling group to determine that the effect of capillary pressure ($P_c$) was negligible at the field scale (Evans *et al.* [5][7]). Thus, in the calculations at all scales, we set $P_c = 0$, even in the lab experimental simulations; this is done by using miscible fluids as the pseudo "oil" and "water" phases (e.g. Craig, *et al.* [44]; Gaucher and Lindley [20]; Sorbie *et al.* [28]). Thus, in the scaling studies we require only the group describing the viscous/gravity balance ($C_{VG1}$), and the x/z shape group ($C_{S2}$). In this application, we use the following form of $C_{VG1}$ (which is *not* dimensionless, since ***g*** is omitted) and the (dimensionless) shape group $C_{S2}$as follows:

$$C_{VG1} = \left(\frac{\mu_o v_D \Delta z}{k_z \Delta\rho \Delta x}\right) \text{ and } C_{S2} = \left(\frac{k_x \Delta z^2}{k_z \Delta x^2}\right) \tag{78}$$

Firstly note that for viscous/gravity scaling ($P_c$ = 0), then there are just 2 scaling groups in 2D (there would be 3 in 3D, $C_{VG1}$ and 2 shape groups $C_{S1}$ and $C_{S2}$)[8]. Secondly, these scaling groups are just *simple numbers* (not functions as required if $Pc \neq 0$), rather like the usual dimensionless numbers used by many workers, as reviewed above.

To illustrate some issues applying the scaling groups here, we compare results for a lab scale system (Experiment) with 3 slightly different Reservoir systems all of which are identically scaled. The properties of the various systems are given in **Figure 9**, where all cases have the same relative permeabilities as required by the Rapoport procedure. The initial water saturations ($S_{wi}$) and residual oil saturations ($S_{or}$) in each layer are given in the caption to **Figure 9**. Water is injected a given rate such that the rate of advance in the high permeability ($v_D$) layer is specified in **Figure 9**. In the tertiary polymer flood in the experiment and reservoir systems, a non-adsorbing polymer slug of size 0.2 PV is injected with a polymer viscosity at maximum input concentration of 10 cP. Two important points to note in **Figure 9** are that, (i) vertical permeability $k_z$ in the scaling groups is taken as $k_z = k_{x(high)}(k_v/k_h)$, but it could equally be $k_z = k_{x(low)}(k_v/k_h)$; the scaling numbers would be different, but the systems would still be identical, and (ii) in addition, the $C_{VG1}$ scaling groups numbers are in mixed units but this is of no consequence as long as the *same* calculation is made for all models.

Since only viscous and gravity forces are involved then according to Rapoport theory, all these calculations should lead to identical results in dimensionless dependent variables (as in **Table 2**); mainly cumulative oil as a fraction of oil in place vs. PV injected. The calculated 2 scaling values for these are shown in **Figure 9** where the shape group ($C_{S2}$) is dimensionless but the viscous/gravity group ($C_{VG1}$) is *not* dimensionless and is even in mixed units! Thus, the number for $C_{VG1}$ itself means little other than it indicates that the viscous/gravity force balances are the same, as they are almost identical in all systems, as required.

The numerical simulations of the 4 systems in **Figure 9** were carried using a standard numerical simulator treating the system as a 2D vertical x/z cross-section with a grid with $N_x$ = 100, $N_z$ = 20 and $N_y$ = 1. Clearly the Experimental grid size was much smaller than the reservoir grid but the grid blocks in each direction were

7 This is supplied as Ancillary Material #1

8 And the usual Rapoport assumptions of the same ($\mu_o/\mu_w$) and relative permeabilities, etc.

*DOI: http://dx.doi.org/10.5772/intechopen.1010239*

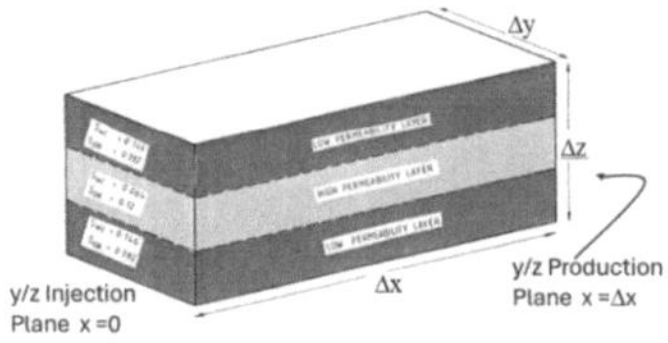


Required scaling groups

$$C_{VG1} = \left(\frac{\mu_o v_D \Lambda^{-}}{k_z \Delta x}\right) \qquad C_{S2} = \left(\frac{k_x \Delta z^2}{k_z \Delta x^2}\right)$$

| Calculation = | 1 | 2 | 3 | 4 | |
|---|---|---|---|---|---|
| Scale = | Experiment | Reservoir | Reservoir | Reservoir | Units |
| Δx | 1.64 | 1000 | 1000 | 1000 | ft |
| Δy | 0.328 | 200 | 200 | 200 | ft |
| Δz | 0.295 | 180 | 216 | 180 | ft |
| $k_x$ (low) | 500 | 500 | 416.67 | 250 | mD |
| kx (high) | 10000 | 10000 | 8333.33 | 5000 | mD |
| $k_v$/kh | 0.1 | 0.1 | 0.144 | 0.1 | - |
| ϕ (low) | 0.2 | 0.2 | 0.2 | 0.2 | - |
| ϕ (high) | 0.38 | 0.38 | 0.38 | 0.38 | - |
| $\mu_o$ | 2 | 2 | 2 | 4 | cP |
| $\mu_w$ | 1 | 1 | 1 | 2 | cP |
| Δρ | 13.21 | 13.21 | 13.21 | 13.21 | lbs/ft3 |
| $V_D$ | 10 | 10 | 10 | 2.5 | ft/day |
| $C_{VG1}$ | 0.000272336 | 0.000272521 | 0.000272521 | 0.000272521 | Viscous/Gravity |
| $C_{S2}$ | 0.323561124 | 0.324 | 0.324 | 0.324 | Shape group |

**Figure 9.**
*The 3 layer experimental model and the 3 reservoir models which should all be identically scaled according to Rapoport. The low permeability layers have identical relative permeabilities with $S_{wi}$ = 0.148 and $S_{or}$ = 0.382, and the high permeability layer has $S_{wi}$ = 0.064 and $S_{or}$ = 0.12; the relevant RP functions are zero at the end points.*

"similar" in the sense they were 1/ $N_x$ and 1/ $N_z$ of the total system length ($\Delta x$) and thickness ($\Delta z$), respectively.

**Figure 10** shows the waterflood results for the 4 cases in **Figure 9**. It is clear that the 4 cases are indeed identically scaled, and in fact the fractional recoveries agree to ~6 to 7 decimal places. The *two* reasons they do not agree to full machine accuracy are due to, (i) the handling of the wells in a lab and field system, which causes a very minor perturbation, and (ii) although the spatial numerical dispersion is identical in each case (see below), there are very small variations in the (much smaller) time truncation error. The saturation distributions as functions of time (PV) also agree and are indistinguishable by eye between the 4 cases.

**Figure 11(a)** show the corresponding oil recovery results for the 3 tertiary polymer injection cases simulated, Experiment and Reservoir calculations 2 and 3, and the waterflood is shown for comparison. Again, the agreement is almost identical numerically between these 3 cases. This precise 6+ decimal place agreement is again due to the fact that since identical grids were used in exactly scaled systems, the

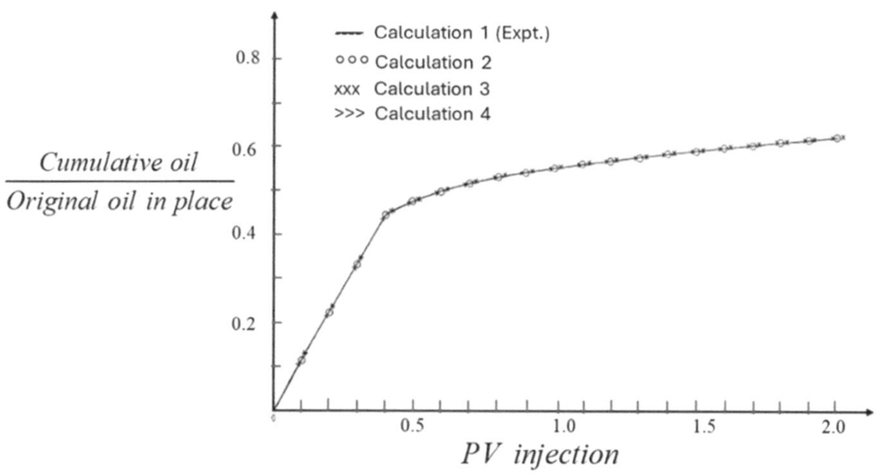


**Figure 10.**
*Comparison the of the oil recoveries in the scaled waterfloods for the Experiment (1) and the 3 Reservoir cases (2-4) in* ***Figure 9***.

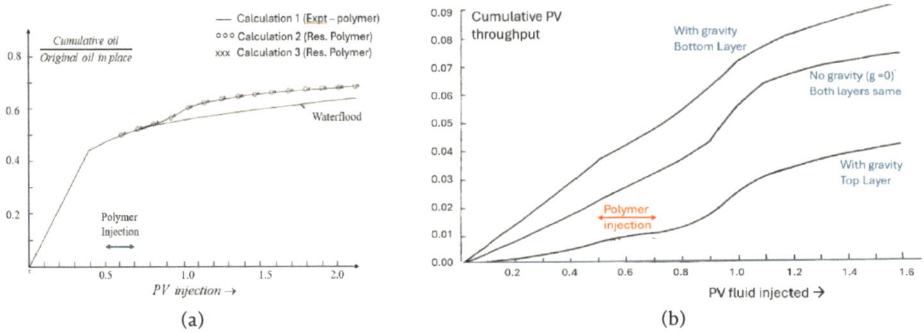


**Figure 11.**
*(a) Comparison the of the oil recoveries in the scaled tertiary polymer floods for the Experiment (1) and the 2 Reservoir cases (2 and 3) in **Figure 9**. (b) The corresponding cumulative PV throughput in the top and bottom low permeability layers for the polymer cases with and without gravity.*

dimensionless numerical dispersion was exactly scaled. This is true for both the water saturation front and the polymer transport in the aqueous phase. Since the polymer transport is governed completely by the combination of viscous (and gravity) forces and the slug spreading in the system is due only to numerical dispersion (which scales also), *no additional scaling groups* are required due to the presence of polymer. However, we note that if other effects of polymer were included in the simulation, e.g. shear thinning or other complex rheology, or polymer adsorption, then these would not be exactly scaled.

**Figure 11(b)** also shows the effect of gravity for all 3 cases in **Figure 11(a)** on the fluid flow and diversion into the lower permeability layers (top and bottom). In the simulation with g = 0, then the throughput of fluid either in the waterflood or diverted in the polymer flood, is identical for both the top and bottom layers. However, with $g \neq 0$, then the bottom layer has more fluid throughput by both gravity and viscous crossflow, but the relatively poor sweep of the top layer is much improved by viscous crossflow.

## 8.5 Example 4: Scaling of subsurface $H_2$ storage with $CO_2$ cushion gas

Finally, coming to more current concerns with issues in the Energy Transition, we describe a study using scaling theory in underground hydrogen storage (Wang *et al.* [45]). This is of course a system which is compressible and therefore the use of the Rapoport groups will be *approximate*, since the incompressibility assumption will be violated. For a purely liquid system, this would probably be a minor matter, but in a system where we are injecting $H_2$ on a "cushion" gas of $CO_2$, it may lead to significant discrepancies in scaling to different size systems. Some other technical point will also come up in the example below.

In this case, we choose a 2D x/z cross-section heterogeneous model example from a recent study (Wang *et al.* [45]) where all 3 forces are present, i.e. viscous, capillary and gravity. This model is shown in **Figure 12** where the smaller model of size Δx by Δz, and the larger model of size $\Delta x' = 3\Delta x$ by $\Delta z' = 3\Delta z$ are shown, and Δy is kept constant. Again the heterogeneity is represented by a correlated random field (CRF) with permeabilities from 10 to 1000 mD and the capillary pressure is related to the local permeability in a grid block by a J-function. This J-function is shown inset in **Figure 13** along with Pc functions for 10, 100 and 1000 mD, and other permeabilities

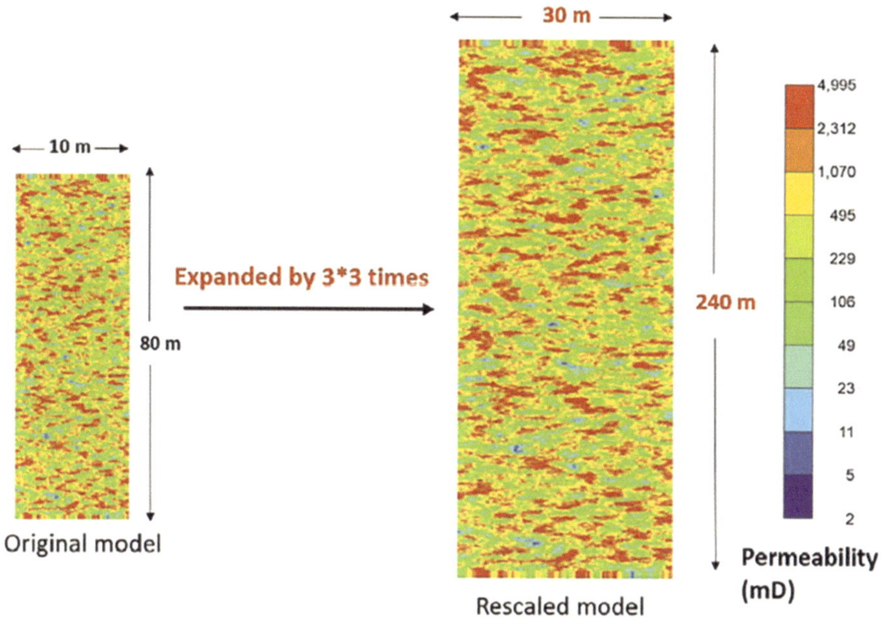


**Figure 12.**
*The x/z vertical cross-sectional model used in the $H_2$ storage over $CO_2$ cushion gas in an aquifer. The smaller model (of size $\Delta x$ and $\Delta z$) is to be rescaled to the larger one where the size is $\Delta x' = 3\Delta x$ and $\Delta z' = 3\Delta z$.*

are scaled accordingly; the (fixed) gas relative permeability with Land function hysteresis (gas trapping) is also shown inset on this figure. As in all cases the same number of grid blocks, $N_x$ and $N_z$, are taken in the x and z directions; that is, the grid block sizes are $\delta x = (\Delta x/N_x), \delta z = (\Delta z/N_z)$ and $\delta x' = (\Delta x'/N_x), \delta z' = (\Delta z'/N_z)$.

Translating the three necessary scaling groups (viscous/gravity, viscous/capillary and 1 x/z shape group in 2D) in [45] into the notation of this paper, we obtain:

$$C_{VG1} = \left(\frac{\mu_o q}{k_z \Delta\rho \Delta x \Delta y}\right); \quad C_{VC1} = \left(\frac{\mu_o \Delta x q}{k_x \Delta y \Delta z \left(\frac{dP_c}{dS}\right)}\right); \quad C_{S2} = \left(\frac{k_x \Delta z^2}{k_z \Delta x^2}\right) \tag{79}$$

Thus, in the larger rescaled model in **Figure 12**, where $\Delta x' = 3\Delta x$ and $\Delta x' = 3\Delta x$, it is clear by simple arithmetic that we can keep the above 3 groups constant and equal to those for the smaller model by simply making $q' = 3q$ and $(dP_c/dS)' = 3(dP_c/dS)$. In the latter case, the rescaling is achieved for every grid block permeability value by simply taking the rescaled J-function where, $J' = 3J$.

The actual injection rates here are not important (details are given in Wang *et al.* [45]), but the $H_2$ storage process is basically to inject a fixed *reservoir*[9] volume of $CO_2$ cushion gas downward displacing water (through a horizontal well at the top of the model). This is followed by a fixed *reservoir* volume of $H_2$, then back produce the $H_2$. Here just 1 cycle is modelled but in $H_2$ storage many cycles are performed (Wang *et al.* [45]; Hematpur *et al.* [40]). In fact, if 1 cycle can be accurately scaled, then the scaling of multiple cycles is achieved.

[9] It is important to note that we discuss the process in *reservoir* volumes. We will return to this shortly.

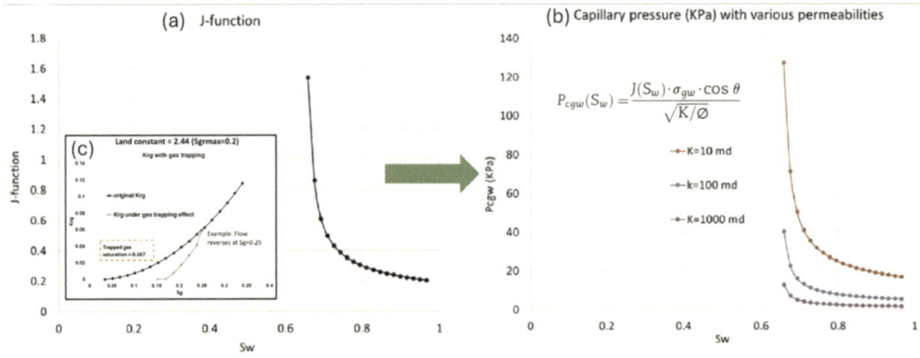


**Figure 13.**
*The input data for the model in **Figure 12** showing (a) the J-function used in the smaller model, (b) some examples of the actual Pc curves for different permeability values and, (inset) (c) the gas relative permeability showing an example of the Land model of gas trapping.*

The base case smaller model (E) results are compared with the rescaled larger model (F) results in **Figure 14**. This figure shows the gas saturation distributions and $H_2$ mole fractions at each stage of the process; Gas saturation is shown at the end of $CO_2$ injection, at the end of $H_2$ injection and at the end of the $H_2$ back production period. The gas $H_2$ mole fraction is shown only for the latter 2 stages. Cases E and F should be identical is the system are exactly scaled in 2 force balances and 1 shape group. The results by eye in **Figure 14** appear to confirm this and it is in fact true to a very good approximation.

However, **Figure 15(a)** shows the $H_2$ purity of the back produced $H_2$ (as it becomes contaminated by produced $CO_2$ cushion gas) and **Figure 15(b)** shows the cumulative $H_2$ recovery for models E and F. The results in **Figure 15** for E and F are

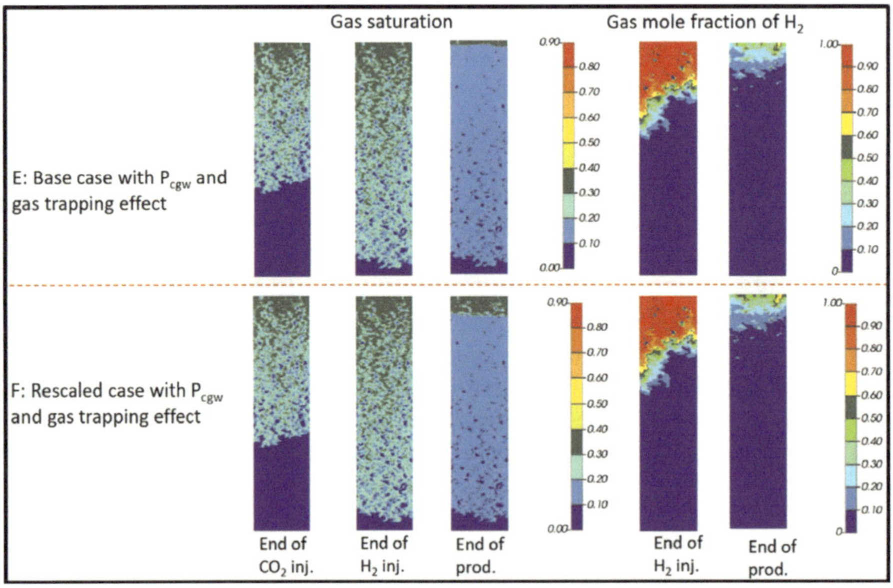


**Figure 14.**
*The gas saturation distributions and $H_2$ mole fractions at each stage of the hydrogen storage process. Gas saturation shown for both the E (smaller) and F (larger) models at the end of $CO_2$ injection, at the end of $H_2$ injection and at the end of the $H_2$ back production period. The gas $H_2$ mole fraction is shown only for the latter 2 stages.*

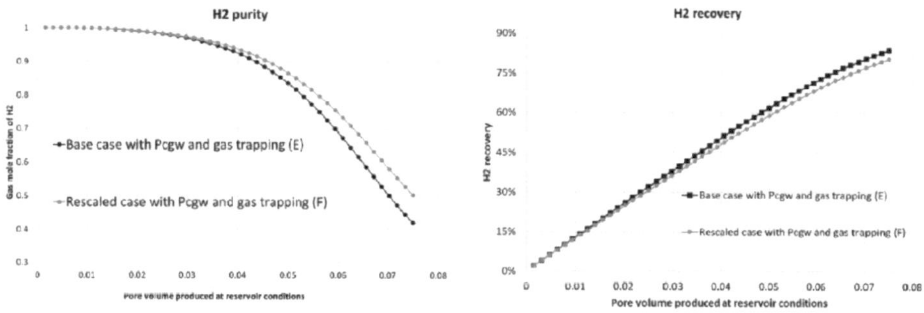


(a) (b)

**Figure 15.**
*For both scaled models (E and F) this figure compares: (a) the $H_2$ purity of the back produced $H_2$ (as it becomes contaminated by produced $CO_2$ cushion gas), and (b) the cumulative $H_2$ recovery.*

reasonably good in the early time in terms of purity but they do start to diverge in the later back production time. The cumulative $H_2$ results for E and F in **Figure 15(b)** are rather less sensitive and show good approximate agreement. The scaling is very good in practical terms but is certainly not exact, as was described for incompressible oil/ water flows.

The reason for this is due to compressibility in the sense that we introduced the same reservoir relative *volumes* in performing the scaling calculation in **Figure 14**. However, the pressures may vary in the 2 systems of different sizes because larger systems may need to inject a larger mass of $H_2$ to achieve the same change in fractional reservoir volume (i.e. the same $H_2$ saturation change $\Delta S_{H_2}$). On back production, this of course leads to larger volumes of surface produced $H_2$ gas and $CO_2$; note that the purity of the larger system (F) is actually above that of the smaller system and this is expected since the mass of gas (for the same volume) is greater.

However, despite the subtlety of the gas compressibility, the system can be scaled (in the reservoir) very well. Thus, approximate scaling can be achieved when looking at fields for $H_2$ storage of different sizes. Also, the differences that one observes during gas production due to compressibility are quite understandable and can be corrected for.

## 9. Summary and conclusions

This paper gives a detailed exposition of the Rapoport scaling theory for incompressible two phase flow in porous media. Rapoport [1] presented an approach which allows us to derive *rigorous* scaling groups, although the original paper did not spell out the full details of the findings. Firstly, it implied that it only applied to isotropic system, and it has been shown in this work that it is easily extendible to anisotropic cases[10]. Secondly, Rapoport's paper did not fully elucidate the fact that each force balance – viscous/capillary, gravity/capillary and viscous/gravity – could appear in *three* forms, but that only one of each was "independent". The forms of the scaling group for any particular force balance (e.g. viscous/capillary) are intimately related

[10] See Evans *et al.* [5] – Ancillary material #1

through the corresponding "shape groups", and this too has not previously been pointed out.

For 2 phase flow in 3D, exactly 5 similarity groups are required to specify the system completely, viz. 2 force balances, 2 shape groups (1 in 2D) and $(\mu_o/\mu_w)$, and this has been shown by other means. If 2 systems had these same 5 groups *completely** identical, then the two systems were exactly scaled (i.e. identical). Although the full relationships between the force balance groups and the shape groups are presented here, the author is fairly sure that Rapoport must have known of these relations. *"Completely" in this context refers to the more complex case where capillary forces are included in the scaling, since the scaling group then contains the quantity, $(dP_c/dS_w)$, and this is a function of $S_w$. Therefore, when a group including capillarity is scaled, the capillary pressure functions in the model $(P_c'(S_w))$ to prototype $(P_c(S_w))$ being scaled must be related as $P_c'(S_w) = B.P_c(S_w)$, where $B$ is a constant, since this ensures a constant value of $(dP_c'/dS_w)/(dP_c/dS_w) = B$ (as in Section 4.1, Eq. (18)). This also means that that the scaling groups including capillarity are strictly *functions*, but in approximate scaling this is usually replaced by some representative *Pc* which then makes the group a constant; e.g. using $p^*$ for the maximum $P_c$ value as in Zhou *et al.* [8]. However, if capillarity is absent ($P_c$ = 0), then only viscous and gravity forces are present. In this case, two phase systems may be rigorously scaled by pure *numbers*, as demonstrated in *Example 3* in Section 8.4. In this (2D) case, it these numbers were the viscous/gravity group $C_{VG1}$ and the shape group $C_{S2}$(along with the other usual assumption of Rapoport).

The Rapoport results for 2 phase flow show that there are 3 rather arbitrary choices of the formula for each of the force balance scaling groups (as noted in **Table 3**), even in *rigorous* scaling. However, although these are exact and rigorous, they must be considered in conjunction with their corresponding shape groups. This tells us something important about the vast numbers of forms of the various force balance scaling groups that are proposed in the literature. Indeed in our review, we noted that several papers commented on the "problem" of having many forms of say the viscous/gravity or viscous/capillary scaling group, and it was frequently hinted that the "correct" one was not clear. Indeed, the recent review by Rahman *et al*. [3], mention this issue, without resolving it, and Guo *et al.* [4] gave a more in depth analysis of the 40+ forms of the capillary number. However, much of this concern is unfounded as Rapoport's results show us that there is no absolutely "correct" forms of these groups, especially when they are derived in a non-rigorous manner by dimensional or inspectional analysis. The implied corollary from Rapoport's approach, and the findings of this paper, is that there are some "wrong" forms of the force balance scaling groups if we are serious about rigour. For example, using the form of the capillary number, $N_{Ca} = (\mu v/\sigma)$ (Eq. (1)), as a simple empiricism can be useful for some experimental situations. However, to identify this form of $N_{Ca}$ as a serious "scaling group" is both inaccurate and misleading.

Another point about even the rigorous scaling groups (in dimensionless form) is that usually the actual *numerical value* they have tells us very little. Obviously, if $G_{VG1} = G_{VC1} = \infty$, the system is totally viscous dominated. However, if the viscous/gravity group $G_{VG1}$ = 1.3 say, it does not tell us very much about how important gravity is in the recovery process. In a different system, which is *not identically scaled* to this, with the same value of $G_{VG1}$, the effect of gravity may be greater or less on recovery. On the other hand, if we change an identically scaled system, by increasing the flow rate, to make $G_{VG1}$ = 2.4, one can say that gravity will be somewhat less important, but still no quantitative statement can be made without performing an actual numerical simulation. Therefore, we conclude that in any application of scaling theory, is still required to perform numerical simulation to *calibrate* the effects of the various scaling groups on the system under study.

The set of examples presented in Section 8 has been chosen to show how Rapoport's rigorous scaling theory can be applied, in conjunction with numerical simulation. This in itself has been rather surprising to many simulation engineers and university researchers to whom I have presented these results. The fact that, when 2 reservoir or lab systems are exactly scaled and then numerical simulations are carried out with the same grid (in numbers of grid blocks) using the same numerical method, then 6+ decimal point accuracy can be seen between the cases. The reason for this is explained in the text in Section 8.1, where it is shown that, since the mesh Peclet numbers are identical, then exactly the same relative numerical dispersion occurs in each case. It is also shown here that quite heterogeneous systems can be exactly scaled as shown in the several correlated random field examples, even including permeability dependent capillary pressure. Exact scaling of course requires that the large system be (geometrically) similar to the smaller one and of course this may not be true in geological formation. If one is using a field scale model in the lab, for example to reproduce some field mechanisms, then it is more practical to reproduce the large scale features (e.g. layers, permeability streaks etc.) and then study the effects of sub-layer heterogeneity by numerical simulation (Sorbie *et al.* [28]).

I hesitate to call the finding on the high accuracy of scaled models "novel", since I have been doing this for the last 30+ years, but at least I am sure that it is not "well known". Indeed, there are very few publication on this topic since Nielsen and Tek [24], which I have been able to find.

## Dedication and acknowledgements

This paper is dedicated to the memory of L.A. Rapoport, one of the great reservoir engineers. The author would also like to thank a number of colleagues for their discussion and insights on scaling theory (and closely related topics) which they have shared with him over some four decades: Larry Lake, Arne Skauge, Mike Shook, Gary Pope, Alan Beteta, Katherine McIver, Marcel Bourgeois, Anne Muggeridge, Martin Blunt, Shehadeh Masalmeh, Randy Seright, Gang Wang, Pål-Eric Øren, Kishore Mohanty, John Fayers, Eric Mackay, Lynn Orr, Karl Stephen, Dag Chun Standnes, Jeroen Groenenboom, Stefen Berg, Diderik Van Batenburg and Peter King. The author would also like to thank the Editor of this book, Professor Huijin Xu, for raising the issue of the scaling groups for tensor permeabilities.

## Conflict of interest

The author declares no conflict of interest, as this work was unfunded.

## A. Physical meaning of the shape group and scaling of tensor permeability

In the main text, permeability is assumed to be anisotropic ($k_x \neq k_y \neq k_z$) which is the general assumption is most numerical simulators. However, permeability is strictly a *tensor* quantity, denoted $\underline{\underline{k}}$, which is represented by a symmetric, positive definite 3x3 or 2x2 matrix in 3D or 2D, respectively. In 2D, the tensor is:

$$\underline{\underline{k}} = \begin{pmatrix} k_{xx} & k_{xy} \\ k_{yx} & k_{yy} \end{pmatrix} \tag{A1}$$

where $k_{xy} = k_{yx}$. This raises the question: *how is the shape group handled if the permeability is a tensor quantity?* To address this we consider the physical meaning of the shape group, $C_{S1} = \left(\frac{k_x \Delta y^2}{k_y \Delta x^2}\right)$. If the system is correctly scaled then we must preserve the ratios of flows in each dimension; in 2D this would be $\left(Q_x/Q_y\right)$ and in 3D it would be $\left(Q_x/Q_y\right)$and $(Q_x/Q_z)$. If the corresponding flows and pressures are correctly scaled then this ratio (in 2D) is easily found using Darcy's Law as follows:

$$\left(\frac{Q_x}{Q_y}\right) = \frac{\frac{1}{\mu}k_x.A_x\frac{\Delta P_x}{\Delta x}}{\frac{1}{\mu}k_y.A_y\frac{\Delta P_y}{\Delta y}} = \frac{\frac{1}{\mu}k_x.\Delta y\Delta z\frac{\Delta P_x}{\Delta x}}{\frac{1}{\mu}k_y.\Delta x\Delta z\frac{\Delta P_y}{\Delta y}} = \left(\frac{k_x.\Delta y^2}{k_y.\Delta x^2}\right)\left(\frac{\Delta P_x}{\Delta P_y}\right) = C_{S1}\left(\frac{\Delta P_x}{\Delta P_y}\right) \tag{A2}$$

from which it is clear that the shape group has a role of preserving the balance of flows. Note that this does not guarantee that the pressure ratio is preserved but that is the job of the *other* scaling groups! Thus, describing $C_{S1}$ (or $\sqrt{C_{S1}}$) as an "effective aspect ratio" (as in Shook *et al.* [2]) is perfectly reasonable, but it is a little more than this. It also gives a clue as to how to approach the issue of tensor permeability rescaling. Darcy's law with tensor permeabilities is as follows:

$$\underline{v} = -\frac{1}{\mu}\underline{\underline{k}}.\underline{\nabla P} = -\frac{1}{\mu}\begin{pmatrix} k_{xx} & k_{xy} \\ k_{yx} & k_{yy} \end{pmatrix}\begin{pmatrix} \partial P/\partial x \\ \partial P/\partial y \end{pmatrix} = \begin{pmatrix} v_x \\ v_y \end{pmatrix} \tag{A3}$$

Following a similar approach to that used in Eq. (A2), it is straightforward to show that the tensor case results in the following ratio of flows (where we have replaced $\partial P/\partial x$and $\partial P/\partial y$by $\Delta P_x/\Delta x$and $\Delta P_y/\Delta y$):

$$\left(\frac{Q_x}{Q_y}\right) = \left(\frac{k_{xx}\Delta y^2\Delta P_x + k_{xy}\Delta x\Delta y\Delta P_y}{k_{xy}\Delta x\Delta y\Delta P_x + k_{yy}\Delta x^2\Delta P_y}\right) \tag{A4}$$

and Eq. (A4) correctly reduces to Eq. (A2) when the off diagonal terms are zero $(k_{xy} = k_{yx} = 0)$. Suppose we rescale the system as discussed in this paper from variables $k_{xx}, k_{xy}, k_{yx}, k_{yy}$, $\Delta x$ and $\Delta y$ to $k'_{xx}, k'_{xy}, k'_{yx}, k'_{yy}$, $\Delta x'$ and $\Delta y'$, where the permeability scaling factors, $f_x$ and $f_y$, are chosen such that $k'_{xx} = f_x.k_{xx}$; & $k'_{yy} = f_y.k_{yy}$. This implies that the size scaling of Δx and Δy must be as follows, $\Delta x' = \Delta x/\sqrt{f_y}$; & $\Delta y' = \Delta y/\sqrt{f_x}$ in order to preserve the diagonal tensor terms (as in Eq. (A2)), and honour the normal form of the shape group, $C_{S1}$. Finally, we must consider how the cross-terms (i.e. $k_{xy}$) are scaled, and a careful inspection of Eq. (A4) shows that if we choose $k'_{xy} = \sqrt{f_x.f_y}.k_{xy}$ (and $k'_{yx} = k'_{xy}$) then in a rescaled system we can easily show that:

$$\left(\frac{Q_x}{Q_y}\right) = \left(\frac{k_{xx}\Delta y^2\Delta P_x + k_{xy}\Delta x\Delta y\Delta P_y}{k_{xy}\Delta x\Delta y\Delta P_x + k_{yy}\Delta x^2\Delta P_y}\right) = \left(\frac{Q'_x}{Q'_y}\right) = \left(\frac{k'_{xx}\Delta y'^2\Delta P_x + k'_{xy}\Delta x'\Delta y'\Delta P_y}{k'_{xy}\Delta x'\Delta y'\Delta P_x + k'_{yy}\Delta x'^2\Delta P_y}\right) \tag{A5}$$

And this is the main requirement of the rescaling.

## Author details

Kenneth S. Sorbie
Institute of GeoEnergy Engineering (IGE), Heriot-Watt University, Edinburgh, UK

*Address all correspondence to: k.sorbie@hw.ac.uk

## References


[1] Rapoport LA. Scaling Laws for use in design and operation of water-oil flow models. Transactions of AIME. 1955;**204**: 143-150

[2] Shook M, Li D, Lake LW. Scaling immiscible flow through permeable media by inspectional analysis. In Situ. 1992;**16**:311-349

[3] Rahman A, Happy FA, Ahmed S, Hossain ME. Development of scaling criteria for enhanced oil recovery: A review. Journal of Petroleum Science and Engineering. 2017;**158**:66-79

[4] Guo H, Song K, Hilfer R. A brief review of capillary number and its use in capillary desaturation curves. Transport in Porous Media. 2022;**144**:3-31. DOI: 10.1007/s11242-021-01743-7

[5] Evans AJ, Sorbie KS, Clifford PJ. Scaling relations for water and polymer floods in stratified systems. In: AEE Winfrith Internal Report. Winfrith, Dorset, UK: UK Atomic Energy Authority; 1985. Available from the author; Ancillary material #1

[6] Pujol L, Boberg TC. Scaling accuracy of laboratory steam flooding models, SPE 4191. In: Presented at the California Regional Meeting of SPE of AIME; 8-10 November 1972; Bakersfield, Calif. Richardson, TX: SPE; 1972

[7] Wood DJ, Lake LW, Johns RT, Nunez V. A screening model for $CO_2$ flooding and storage in Gulf Coast reservoirs based on dimensionless groups, SPE100021. SPE Reservoir Engineering. Jun 2008;**11**(3):513-520

[8] Zhou D, Fayers FJ, Orr FM. Scaling of multiphase flow in simple heterogeneous porous media, SPE/DOE 27833. In: Proceedings of the ASTM, New Orleans, LA. Richardson, TX: SPE; 1993

[9] Bridgman PW. Dimensional Analysis. 2nd ed. New Haven: Yale University Press; 1931

[10] Greenkorn RA. Flow models and scaling Laws for flow through porous media. Industrial & Engineering Chemistry. 1964;**56**:33-37

[11] Loomis AG, Crowell DC. Theory and Application of Dimensional and Inspectional Analysis to Model Study of Fluid Displacements in Petroleum Reservoirs. US Department of the Interior, US Bureau of Mines; 1964

[12] Bruce WA. An electrical device for analyzing oil-reservoir behavior. Petroleum Transactions of the AIME. Dec 1943;**151**(1):112-124

[13] Schaefer H. Extending the application of electric analogy in oil-reservoir studies. Petroleum Transactions of the AIME. Dec 1947;**170** (1):62-68

[14] Karplus WJ. Water-coning before breakthrough, an electronic analog treatment. Petroleum Transactions of the AIME. Dec 1956;**207**(1):240-245

[15] Leverett MC, Lewis WB, True ME. Dimensional model studies of oil field behavior. Transactions of AIME. 1942; **146**:175

[16] Engelberts WF, Klinkenberg LJ. Laboratory experiments on the displacement of oil by water from packs of granular materials. In: Proc. 3rd World Pet. Cong., Section II. Vol. 544. The Hague, Netherlands: World Petroleum Congress; 1951

[17] Croes GA, Schwarz N. Dimensionally scaled experiments and

the theories on the water-drive process. Transactions of AIME. 1955;**205**:35-42

[18] Buckley SE, Leverett MC. Mechanism of fluid displacement in sands. Transactions of AIME. 1942;**146**: 107

[19] Dietz DN. A theoretical approach to the problem of encroaching and bypassing Edgewater. Proceedings of the Koninklijke Nederlandse Akademie van Wetenschappen. 1953;**56-B**:83

[20] Gaucher DH, Lindley DC. Waterflood performance in a stratified, five-spot reservoir -A scaled-model study. Petroleum Transactions of the AIME. Dec 1960;**219**(1):208-215

[21] Perkins FM, Collins RE. Scaling Laws for laboratory flow models of oil reservoirs, paper SPE 1487. Journal of Petroleum Technology. 1960;**12**(8):69-71

[22] Farouq Ali SM, Donohue TAD, Stahl CD. Fluid flow in porous media – Problems in relating experiments to field projects. In: Paper WPC-12216, Presented at the 7th World Petroleum Congress, Mexico City, Mexico. Mexico City, Mexico: World Petroleum Congress; 1967

[23] Douglas J, Peaceman DW, Rachford HH. A method for calculating multi-dimensional immiscible displacement. Petroleum Transactions of the AIME. Dec 1959;**216**(1):297-308. Paper no. SPE-1327-G. DOI: 10.2118/1327-G

[24] Nielsen RL, Tek MR. Evaluation of scale-up Laws for two-phase flow through porous media. SPE Journal. 1963;**3**(2):164-176

[25] van Daalen F, van Domselaar HR. Scaled fluid-flow models with geometry differing from that of prototype. SPE Journal. 1972;**12**(3):220-228

[26] Richardson JG, Stone HL. A quarter century of progress in the application of reservoir engineering. SPE Journal. 1973; **25**(12):1371-1379

[27] Dosher T. Scaled Physical Models of the Steam Drive Process, US DOE Final Report, Project DOE/ET/12075-3. USA: US Depatment of Energy (DOE); 1982

[28] Sorbie KS, Sheb M, Hosseini A, Wat RMS. Scaled miscible floods in layered Beadpacks investigating viscous crossflow, the effects of gravity, and the dynamics of viscous slug breakdown, SPE 20520. In: Proceedings of the 65th Annual Technical Conference and Exhibition of the SPE; 23–26 September 1990; New Orleans, LA. Richardson, TX: SPE; 1990

[29] Gharbi R, Peters E, Elkamel A. Scaling miscible fluid displacements in porous media. Energy and Fuels. 1998; **12**:801-811

[30] Grattoni CA, Jing XD, Dawe RA. Dimensionless groups for three-phase gravity drainage flow in porous media. Journal of Petroleum Science and Engineering. 2001;**29**:53-65

[31] Jadhawar PS, Sarma HK. Scaling and sensitivity analysis of gas-oil gravity drainage EOR, SPE115065. In: Proceedings of the SPE Asia Pacific Oil & Gas Conference and Exhibition; 20–22 October 2008; Perth, Australia. Richardson, TX: SPE; 2008

[32] Trivedi J, Babadagli T. Scaling miscible displacement in fractured porous media using dimensionless groups. Journal of Petroleum Science and Engineering. 2008;**61**:58-66

[33] Guo Z, Dong M, Chen Z, Yao J. Dominant scaling groups of polymer flooding for enhanced heavy oil recovery. Industrial and Engineering

Chemistry Research. 2013;**52**:911-921. DOI: 10.1021/ie300328y

[34] Peaceman DW. Fundamentals of numerical reservoir simulation. In: Developments in Petroleum Science. Vol. 6. New York, USA: Elsevier Scientific Publishing Co.; 1977. eBook, ISBN 9780080868608

[35] Aziz K, Settari A. Petroleum Reservoir Simulation. London: Applied Science Publishers; 1979. DOI: 10.2118/9781613999646

[36] Pinder GF, Gray WG. Essentials of Multiphase Transport in Porous Media. New York: Wiley; 2008. DOI: 10.1002/97804 70380 802

[37] Beteta A, Sorbie KS, Skauge A, Skauge T. Immiscible viscous fingering: The effects of wettability/capillarity and scaling. Transport in Porous Media. 2024;**151**:85-118. DOI: 10.1007/s11242-023-02034-z

[38] Salmo IC, Sorbie KS, Skauge A, Alzaabi MA. Immiscible viscous fingering: Modelling unstable water–oil displacement experiments in porous media. Transport in Porous Media. 2022; **145**:291-322. DOI: 10.1007/s11242-022-01847-8

[39] Sorbie KS, Al Ghafri AY, Skauge A, Mackay EJ. On the modelling of immiscible viscous fingering in two phase flow in porous media. Transport in Porous Media. 2020;**135**:331-359. DOI: 10.1007/s11242-020-01479-w

[40] Hematpur H, Abdollahi R, Rostami S, Haghighi M, Blunt MJ. Review of underground hydrogen storage: Concepts and challenges, (invited review). Advances in Geo-Energy Research. 2023;**7**:111-131

[41] Davoodi S, Al-Shargabi M, Wood DA, Rukavishnikov VS, Minaev KM. Review of technological progress in carbon dioxide capture, storage, and utilization. Gas Science and Engineering. 2023;**117**:1-28

[42] Wang G, Beteta A, Sorbie KS, Mackay EJ. Inherent errors in current relative permeability data for modelling underground hydrogen storage. Interpore Journal. 2025;**2**(1):1-14

[43] Larki SA, Skauge A, Sorbie KS, Mackay EJ. Numerical Simulation of Viscous Fingering in $CO_2$ Storage: Addressing the Limitations of Laboratory-Derived Relative Permeabilities, Submitted for Publication (2024/25)

[44] Craig FF, Sanderlin JL, Moore DW, Geffen TM. A laboratory study of gravity segregation in frontal drives. Transactions of AIME. 1957;**210**:275

[45] Wang G, Pickup GE, Sorbie KS, Mackay EJ. Scaling analysis of hydrogen flow with carbon dioxide cushion gas in subsurface heterogeneous porous media. International Journal of Hydrogen Energy. 2022;**47**:1752-1764

Chapter 4

# A Methodology for Prospecting Fractured Aquifers in Tropical Regions Using Airborne Geophysics

*Ana Luísa Santos Martins, Luiz Henrique Cardoso, Daniela Antunes Lessa, Antonio Santos Sánchez, Ana Letícia Pilz de Castro and Guilherme Augusto Magalhães Gomes*

## Abstract

Investigating the structural architecture of fractured rock masses is crucial, as it can reveal the accumulation of significant volumes of groundwater. However, the development of weathering mantles in tropical regions makes identifying and characterizing these structures difficult. To solve this issue, we present a methodology that explores airborne geophysics, an indirect prospecting method offering excellent cost-effectiveness and fast data acquisition. In this sense, by integrating magnetometry and radiometry with physiographic aspects, remote sensing data, and literature sources, it was possible to map the spatial distribution and main orientations of shallow and deep structures of the fractured aquifer in weathering metasedimentary rock in an important mineral province in southeastern Brazil, the Iron Quadrangle. Furthermore, the locations with the highest potential for groundwater accumulation were identified for future local studies aimed at table water supply for the population. The obtained products and their respective interpretations were validated with field evidence. Therefore, we believe this methodological framework can be applied to similar hydrogeological contexts in other tropical regions.

**Keywords:** iron quadrangle, magnetometry, radiometry, groundwater, fissural porosity

## 1. Introduction

Water has increasingly become the focus of major debates at international environmental conferences (e.g., Eco-92, Rio + 10, Rio + 20). In general, its natural cycle has been impacted by inadequate management. Consequently, table water is becoming scarce worldwide despite being a renewable resource [1]. Therefore, viable and sustainable alternatives for prospecting new water reserves have been necessary.

In this regard, we present a methodology for indirectly prospecting water resources using geophysics, a relatively inexpensive and fast technique for data acquisition. It has shown promising results in detecting rocks, sediments, and structures that accumulate water (e.g., [2–4]).

### 1.1 General context

This study focuses on fractured aquifers, where porosity is associated with the void spaces of open fractures in metasedimentary rocks in tropical regions. In this sense, we investigated an internationally recognized mineral province, the Iron Quadrangle, in southeastern Brazil. This mountainous region covers approximately 7190 $km^2$. It has a rectangular shape, defined by the orientation of saws sustained by quartzites and banded iron formations [5].

Stratigraphically, the crystalline basement comprises Archaean intrusive meta-igneous rocks. Three supracrustal successions have been deposited over it. The first is an Archaean greenstone belt (Rio das Velhas Supergroup). The second (Minas Supergroup) and third (Itacolomi Group) successions are Proterozoic and consist mainly of metasedimentary rocks [5, 6].

The Iron Quadrangle structural architecture records multiple phases of Transamazonian and Brazilian deformation, represented by folds, fractures, and faults, which control major relief features such as escarpments, saws, and valleys in the region [7, 8]. The lithostructural control on the evolution of the Iron Quadrangle landscape is driven by differential erosion due to variations in rock resistance [9].

Overall, the various lithotypes (e.g., phyllites/schists, quartzites, meta-igneous rocks, marbles, banded iron formations) present in the Iron Quadrangle are highly susceptible to chemical weathering [10]. Rock decomposition is further influenced by the mountainous relief, moist tropical climate (high rainfall concentrated in summer combined with temperatures exceeding 30°C), and prolonged exposure to weathering. Consequently, the development of weathering mantles in tropical areas, like that region, makes identifying and characterizing fractured aquifers difficult [11].

The study focus is the fractured quartzite mass of the Itacolomi Group, which stratigraphically underlies the phyllites of the Minas Supergroup. Due to mineralogical variations in the quartzite (quartz and micas), significant weathering is sometimes observed. Field investigations identified two well-defined fracture sets, oriented approximately N60°W and N30°E. Satellite image analysis indicated that these fracture sets extend on a regional scale, revealing zones with higher fracture densities [12].

Fractures and karstic features of the quartzite (Itacolomi Group) were described as contributing to the fractured porosity of the rock mass, characterizing it as an aquifer [13]. The investigation of karstification in this rock mass, along with altimetric differences, slope variations, and even local flow rates, confirms the water potential [14]. In such cases, geophysical surveys can provide subsurface information. It can support mapping the structural architecture of the aquifer and complementing existing studies. Through a regional investigation, this approach can support indicating locations with the highest hydrogeological potential for groundwater extraction.

Thus, the aim is to develop a methodological framework for prospecting the fracture porosity of the quartzite mass using airborne geophysics, integrating magnetometry and radiometry methods. Additionally, we intend to identify potential sites for future detailed studies to assess quantitative aquifer parameters.

### 1.2 Research significance

Groundwater exploitation in fractured bedrock formations can represent a highly viable alternative to abstraction from granular aquifers. When unconfined aquifers associated with the weathered mantles of crystalline and metasedimentary fractured rocks become depleted, groundwater extraction can be extended to deeper zones where interconnected networks of open fractures are present. Alternatively, abstraction strategies can be optimized to ensure consistent resource availability by installing well filters in both the intergranular and fractured zones and adjusting exploitation rates between them based on operational needs.

In this context, identifying promising fractured zones for groundwater development is of critical importance, and the present research is justified on multiple grounds related to sustainable water resource management in tropical regions. The proposed methodology represents an effort to systematically explore fractured aquifers overlain by thick weathered mantles. This was accomplished through a regionally oriented analysis aimed at identifying specific areas with high potential for groundwater accumulation.

Accordingly, the approach is justified by its contributions to the investigation of regional fracture patterns; the understanding of spatial distribution and preferential orientations of structural features at both surface and subsurface levels; the identification of structurally significant orientations not previously reported in the literature for the study area; the qualitative classification of structural features by depth; and the delineation of specific target sites for further groundwater exploration.

## 2. Location of the study area

The selected study area is Ouro Preto, Minas Gerais, Brazil, a historic city designated a United Nations Educational, Scientific and Cultural Organization (UNESCO) World Heritage Site. Unregulated urban expansion, ineffective water resource management, and outdated, inefficient hydraulic structures have negatively impacted the availability and use of surface water resources. Therefore, this study proposes groundwater prospecting as a potential alternative for table water supply.

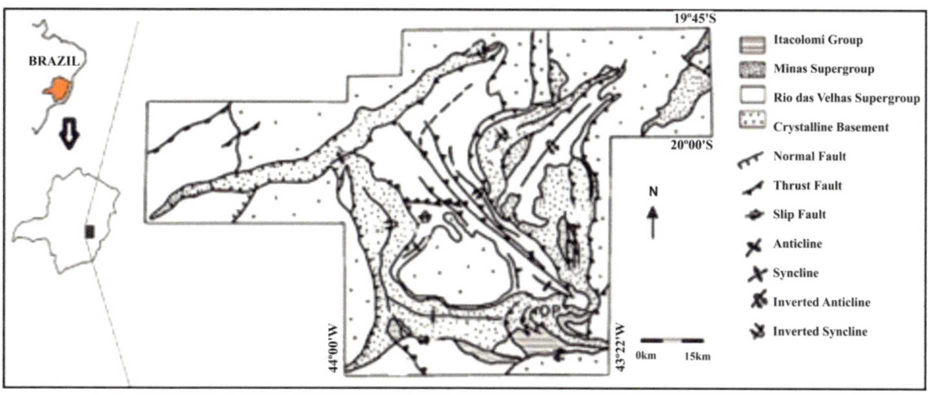


**Figure 1.**
*Location of Ouro Preto (OP) in the Iron Quadrangle. Adapted from the Brazilian Geological Survey [15].*

**Figure 1** presents the location of Ouro Preto in the Iron Quadrangle, along with its main lithostratigraphic units and general structural architecture.

## 3. Magnetometry and radiometry

Magnetometry measures the Earth's magnetic field, which varies due to the physical properties of rocks. Magnetic field anomalies can result from lithological variations in the subsurface. Fluids percolating through planar structures in the subsurface can produce magnetic minerals along fracture walls or as infill material. The structural feature generated by the relative displacement of rock blocks with the same magnitude and orientation of magnetization can be recorded as a magnetic lineament [16].

Radiometry investigates the radiation emitted by geological bodies by measuring their gamma-ray emissions (isotopes of Uranium-$^{238}U$, Thorium-$^{232}Th$, and Potassium-$^{40}K$). This method is sensitive to shallow depths, including exposed rocks and weathering mantles [17]. The clay content of rocks is proportional to these emissions. The mobility of radioisotopes follows the order: thorium < uranium < potassium [18].

Magnetic and radiometric data acquisition can be conducted simultaneously in airborne surveys, offering cost-effective benefits, rapid data collection, and the ability to scan large areas [19]. Lineament analysis through indirect methods is commonly the first step in defining structural trends in fractured aquifers. Lineament orientations can be presented in direction frequency distribution rosettes. This can be integrated with lithological and geomorphological studies, allowing for the association of lineaments with discontinuities such as open fractures [20].

Geophysical data analyses using these methods were conducted to study regional aquifers in the municipality of Itabirito, Minas Gerais, identifying faults at the contact between quartzites of the Moeda Formation and schists of the Nova Lima Group [21]. Regional geophysical data were examined and integrated with geological information to determine the most favorable areas for groundwater exploitation in geothermal regions [22]. Geological data were integrated with lineaments generated through airborne geophysics data processing, and a groundwater favorability map was developed for future studies in the northern portion of the Iron Quadrangle [23].

Magnetometry and radiometry methods were applied to identify lineaments indicative of planar structures associated with gullies in the southern portion of the Iron Quadrangle [24, 25]. The interpreted regional geophysical lineaments were validated through field observations, where metric-scale structures with the same orientation were identified in the investigated areas. The groundwater potential of fractured basement rocks in the Bação Complex of the Iron Quadrangle was assessed by analyzing radiometric and magnetometric lineaments integrated with remote sensing techniques, morphostructural analysis, and geological data [26].

## 4. Methodological framework

The fundamental premise for fractured aquifer prospecting is the identification of planar structures that accumulate and transmit groundwater. In this context, the methodological framework (**Figure 2**) comprised the following six steps.

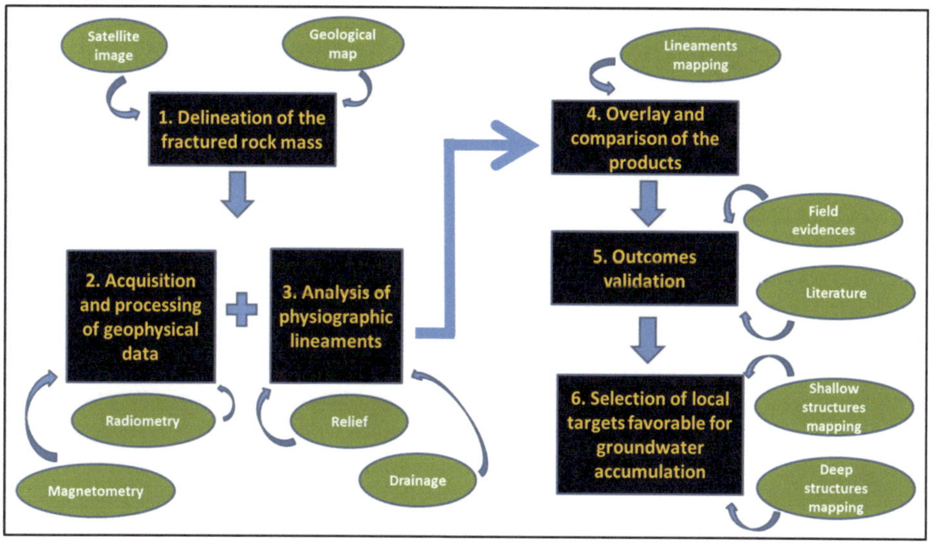


**Figure 2.**
*The flow diagram summarizing the methodological steps of the study.*

### 4.1 Delineation of the fractured rock mass

The first stage of the present methodology involved delineating the quartzite mass in Ouro Preto using satellite images, supported by a georeferenced geological map overlaid in a Geographic Information System. In both satellite images and field observations, the rough texture of the quartzite outcrops contrasts markedly with the smoother texture of the adjacent phyllites.

During field validation, the mineralogical composition of the investigated lithologies was confirmed, along with a preliminary assessment of associated structural features. The quartzites exhibited zones with varying composition—some more quartz-rich, others more micaceous. The micaceous zones were more weathered and exhibited the formation of clay minerals. The alteration of colors ranged from whitish to pinkish hues. The phyllites ranged from sericitic to graphitic in composition, were typically altered, and displayed light to dark gray tones, respectively. In the field, it was preliminarily observed that fractures in the quartzite mass are more numerous, more extensive, and more open than those in the phyllite mass.

### 4.2 Acquisition and processing of geophysical data

The geophysical data were obtained from the Geological Survey of Brazil [15]. They are part of the Eastern Bloc database of the study area. Magnetic and radiometric data were acquired simultaneously. Flight altitude and speed were 100 m and 200 km/h, respectively, with a flight line spacing of 250 m and control line spacing of 2500 m. The flight lines were oriented N30°W, while the control lines were oriented N60°E [27].

The magnetometric survey acquisition rate was 10 measurements per second, using a cesium vapor magnetometer (Geometrics G-822A) with a resolution of 0.001 nT. The radiometric survey acquisition rate was 1 measurement per second, employing a gamma-ray spectrometer (Exploranium GR-820) with 256 spectral channels [27].

The Oasis montaj 9.8.1 software [28] was used for geophysical data processing and the generation of georeferenced maps. Within the defined area, magnetic and radiometric data were processed, including raw data corrections, gridding, frequency-domain analysis using Fast Fourier Transform (FFT), and the creation of various georeferenced maps for subsequent lineament interpretation.

Corrections in magnetic data were applied: International Geomagnetic Reference Field (IGRF) and diurnal variations. The corrected data were interpolated using a minimum curvature method within a 300 m regular grid (below which flight trend influence was observed). A total magnetic field map was generated for subsequent filtering (step-by-step) via the MagMap interface. The second vertical derivative ($d^2z$) was applied to evaluate near-surface anomalies. Horizontal gradients (dx and dy) repalced N-S and E-W trending lineaments. An Analytical Signal Amplitude (ASA) map was created to analyze anomalies surrounding their source bodies.

Radiometric maps have been corrected for cosmic background radiation. They were generated and separated by radioelement channel—Uranium $^{238}U$, Thorium $^{232}Th$, and Potassium $^{40}K$. The corrected data were also gridded using a 300 m mesh. Furthermore, an F-factor map was developed to assess the behavior of the $^{40}K$ isotope relative to the U/Th ratio [29], thereby highlighting the energy response of potassium compared to other elements. High F-factor anomaly values indicate areas with hydrothermal alteration (potassium-enriched), typically associated with discontinuities where heated fluids percolated from high-pressure and high-depth sources [17].

### 4.3 Analysis of physiographic lineaments

The analysis of physiographic lineaments (relief and drainage) was necessary, as their distribution and orientation can reflect the structural architecture of the rock mass. Moreover, these products can be compared with geophysical data, which may aid in distinguishing shallow from deeper structural features. The latter may not manifest as surface relief or drainage features; therefore, integrated geophysical methods provide an indirect way of subsurface evaluation in this context.

Surfer 9.11.947 [30] was employed for the Digital Elevation Model. The cartographic base was developed at a 1:30,000 scale. Georeferencing was based on the Córrego Alegre Datum, Zone 23S [21]. Geophysical maps were integrated into a Geographic Information System as overlapping layers, where structural lineaments and anomalies were interpreted to distinguish the fractured mass from its surroundings.

### 4.4 Overlay and comparison of the products

The interpreted lineament orientations were input into GeoRose 3.0 [31], generating rose diagrams to reflect the statistical distribution of lineament directions. Deep and shallow structural lineaments in the magnetic maps were investigated. The radiometric products, the residual magnetic field map ($d^2z$), and literature-based lineament data indicating surface features were integrated to analyze surface structures.

Lineaments identified in the drainage map (surface watercourses) and the relief contributed to the interpretation of the superficial structural framework of the fractured mass. Dx and dy gradient magnetic maps are derived from horizontal gradient filtering, thus providing information on various depths. The superposition of the geological map and the satellite imagery helped guide the comparison between geophysical and physiographic products.

### 4.5 Outcomes validation

Field investigations were conducted along outcrops of the fractured quartzite and its surroundings. The study focused on fracturing and hydraulic barriers, including lithotype, mineralogy, fracture conditions within the rock mass, lithological contacts, and other relevant geological features. At this stage, the results were cross-validated against existing data from the literature.

### 4.6 Selection of local targets

After overlaying all geophysical products with the physiographic datasets and comparing the distribution and orientation of the interpreted structural lineaments, it was possible to delineate areas with the highest fracture densities. Consequently, zones with the greatest potential for groundwater accumulation were identified, both in the shallower portions and in the deeper sections of the quartzite mass.

### 4.7 Workflow diagram of the adopted methodology

**Figure 2** presents a schematic representation of the methodological workflow for the potential sites prospection for groundwater accumulation in fracture porosity, based on the six steps described.

## 5. Results and discussion

The fractured quartzite mass (Itacolomi Group) was delineated using satellite imagery (**Figure 3**), along with the location of the field investigation points. The quartzite and the underlying rocks' occurrence (the upper phyllites of the Minas Supergroup) were confirmed in the field. A panoramic photograph (**Figure 4**) shows the fractured quartzite mass and lineaments in the relief, with a rough texture, overlying the phyllites (smooth texture). Magnetic maps—ASA, $d^2z$, dx, and dy (**Figure 5**)—are presented with an overlay of the contour (black) delineating the quartzite (Itacolomi Group), the interpreted lineaments (white), and the lineament rosettes.

Analysis ASA map (**Figure 5a**) reveals an inflection at its center, coinciding with the elongated trend of the quartzite mass, which is oriented approximately N60°W. These directions align with structural features identified in previous studies [12]. Low to moderate anomalies (< 0.141 nT/m) within the quartzite area indicate its low magnetic susceptibility, as expected due to the absence of magnetic minerals in its mineralogical composition.

High-value anomalies (> 0.248 nT/m) are concentrated in the phyllite area. However, the cause of this anomaly cannot be conclusively determined, as the mineralogy of the phyllite is not visible to the naked eye, apart from its color, which suggests a carbonaceous and graphitic nature due to its purple and graphite hues, respectively. One hypothesis is that hydrothermal fluids may have magnetized the phyllite, given its predisposition to retain magnetic mineral phases.

The vertical gradient map $d^2z$ (**Figure 5b**) is presented to evaluate shallow structures. These magnetic responses were integrated with radioisotope channel data to enhance the understanding of the near-surface structural framework of the fractured quartzite mass. The $d^2z$ map response exhibits similarities with the ASA map (**Figure 5a**) in terms of the regional N60°W-aligned structures, where additional detailed local features with specific

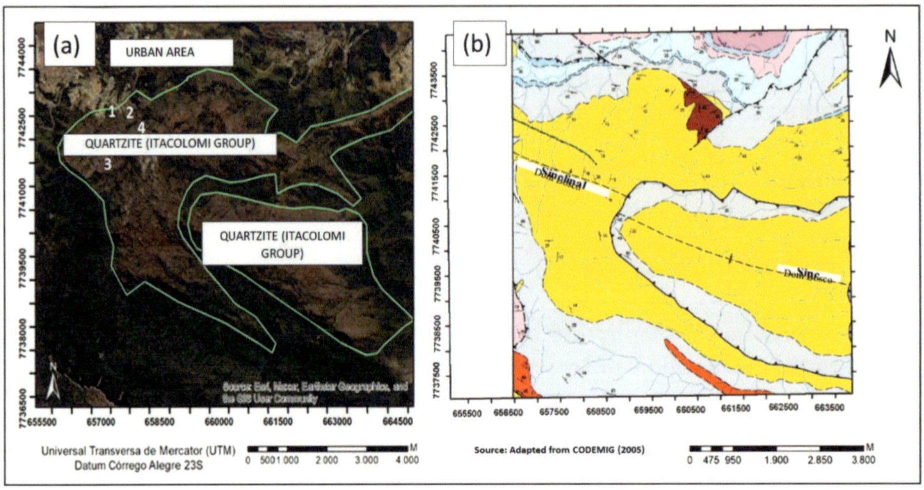


**Figure 3.**
*(a) Study area, delineating the fractured quartzite mass (Itacolomi Group—outlined in green). Adapted from Google Earth [32]; and (b) Geological map of the city of Ouro Preto. The quartzite of the Itacolomi Group is represented in yellow, while the other lithologies generally correspond to top phyllites of the Minas Supergroup. Adapted from the Brazilian Geological Survey [31].*

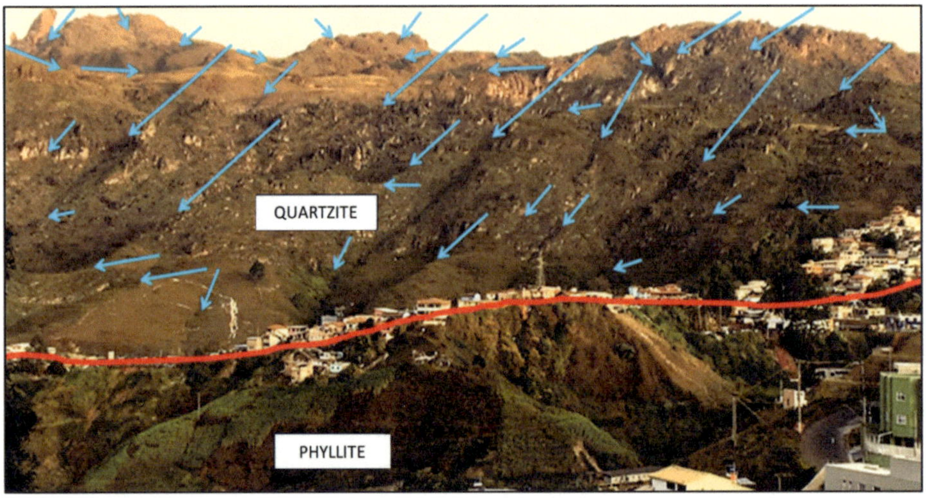


**Figure 4.**
*Panoramic view of the quartzite (Itacolomi Group) overlying the upper phyllites of the Minas Supergroup. Blue arrows indicate major structural features of the quartzite, and the red line marks the contact. Some of the investigated points were observed in the field.*

preferential trends become apparent. The dominant orientations, in order of decreasing frequency as analyzed in the lineament rosette, are N-S, E-W, N45°W to N30°W, and N30°E to N45°E. Thus, this confirms previous data [12].

The dx map (5c) shows high-value anomalies (> 0.04 nT/m) aligned near the N-S direction, emphasizing the regional structural framework of the study area. E-W,

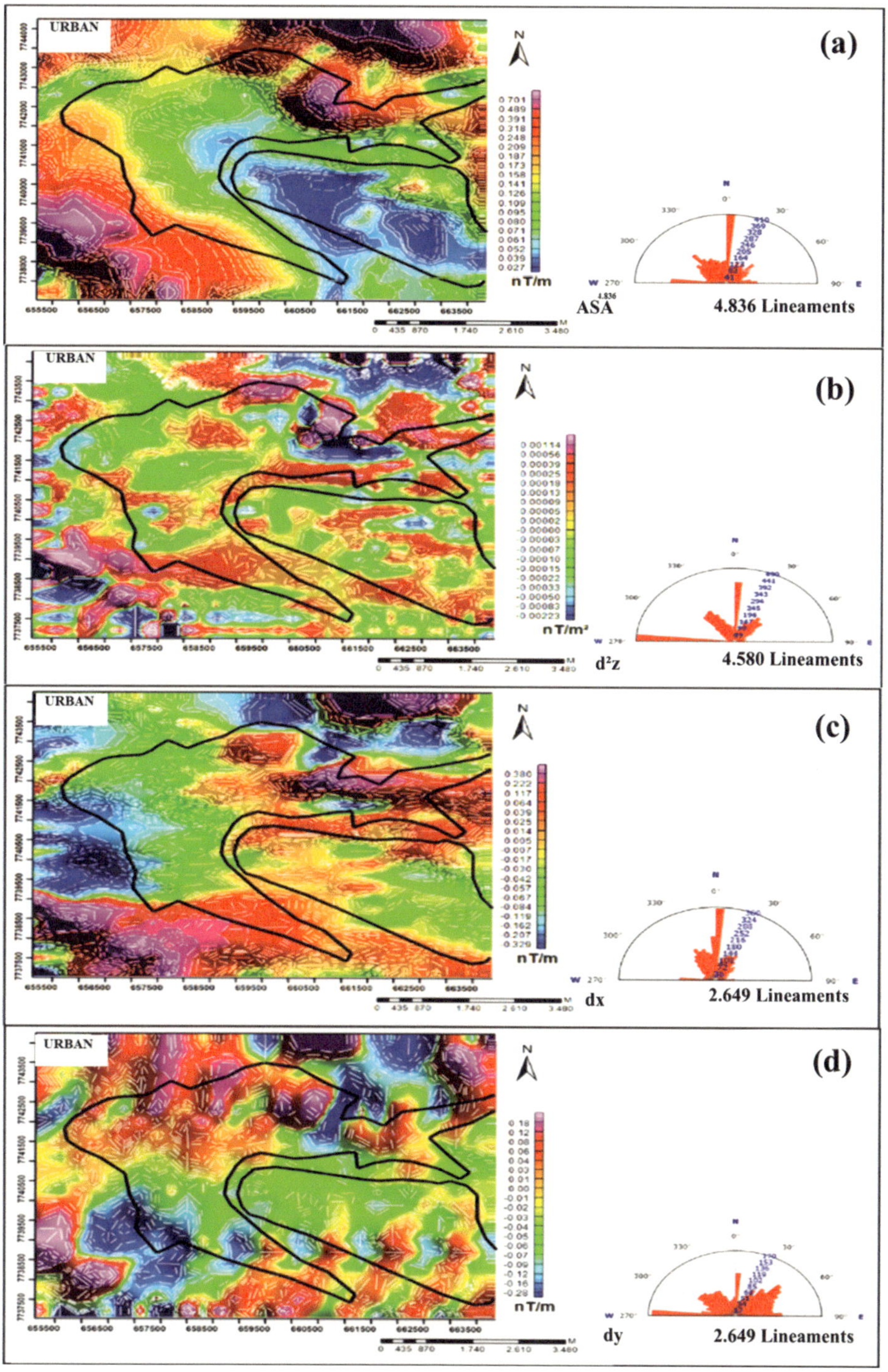


**Figure 5.**
*Magnetic maps: ASA (a), d²z (b), dx (c), and dy (d).*

SW–NE, and SE-NW trends are also observed and, despite being localized, are prominent in the lineament rosette, showing a high frequency of occurrence. In contrast, the dy map (**Figure 5d**) does not depict E-W structuring, which was initially expected, except for a few high-value anomalies (> 0.039 nT/m) aligned in the E-W direction in the eastern part of the study area. Interpretations of the data presented in **Figure 5** are made in light of the geological contexto. In the following stage, magnetic data are integrated with radiometric and physiographic data, and the resulting interpretations are validated. Subsequently, the individual radioisotope channels are presented: uranium (**Figure 6a**), thorium (**Figure 6b**), potassium (**Figure 6c**), and F-factor (**Figure 6d**).

In **Figure 6**, the most significant interpretation was the identification of numerous linear features indicative of discontinuities cutting across the rock mass. The uranium and thorium channel (**Figure 6a, b**) responses exhibit similar patterns, with structural trends comparable to those identified in the $d^2z$ map (**Figure 5b**), as expected. In contrast, the potassium channel (**Figure 6c**) lineaments display a chaotic pattern due to the high mobility and widespread distribution of potassium in the weathering mantle, resulting from mica alteration in the rocks. Consequently, structural trends are obscured. Thus, integrating $d^2z$ responses with uranium and potassium channels confirms the presence of planar geological structures, such as fractures, beneath the weathered mantle.

The Factor F map (**Figure 6d**) further delineates smaller areas within the study region with potential for groundwater accumulation. This map reveals high-value anomalies (> 0.18 cps) predominantly oriented in the SE-NW direction, aligning with the regional structural framework interpreted from other geophysical maps. This suggests that F-factor effectively maps structural features, as the geophysical signatures left by fluids percolated through fractures indicate its relevance for prospecting. Notably, these anomalies are concentrated in the northwestern portion of the study area.

High-value anomalies (represented in magenta) in the thorium (22.59 to 19.10 cps—**Figure 6b**) and potassium (1.79 to 1.59 cps—**Figure 6c**) channels exhibit a curved geometry. This same geometry is also interpreted in the ASA magnetic map, where it is expressed as a medium-amplitude anomaly (0.141 to 0.071 nT/m—**Figure 5a**). When these anomalies are superimposed onto the satellite image (**Figure 3a**) and the geological map (**Figure 3b**), the presence of a geological fold governing the contact zone between the quartzite and phyllite masses is interpreted. This structural configuration also includes the development of a big thrust fault along this contact. These observations highlight the regional-scale structural architecture of the quartzite mass, pointing to the presence of relatively deep-seated features.

Field evidence confirms fluid percolation, as dark minerals (not part of the quartzite's mineralogical composition) mark foliation (**Figure 7a, b**). Vein quartz with fractures filled with an orange-colored oxidized substance and black minerals is observed (**Figure 7c**). Fractures with orange-colored surfaces (**Figure 7d**) suggest fluid interaction with the rock, potentially due to rainwater circulation. Lineaments were traced on the drainage map (**Figure 8a**) and the Digital Elevation Model (**Figure 8b**) to be compared with those interpreted from geophysical data. The age data were sourced from the Bloco 13 database [31]. Interpretations of the data presented in Figure 6 are made in light of the geological contexto. In the following stage, geophysical data are integrated with physiographic information, and the resulting interpretations are validated.

The overlay of maps as layers within the Geographic Information System environment facilitated the integration of generated data. In this context, rose diagrams from all maps were compared, and implications for rock fracturing are discussed based on **Table 1** and **Figure 9**. It presents the preferential orientations of each rose diagram.

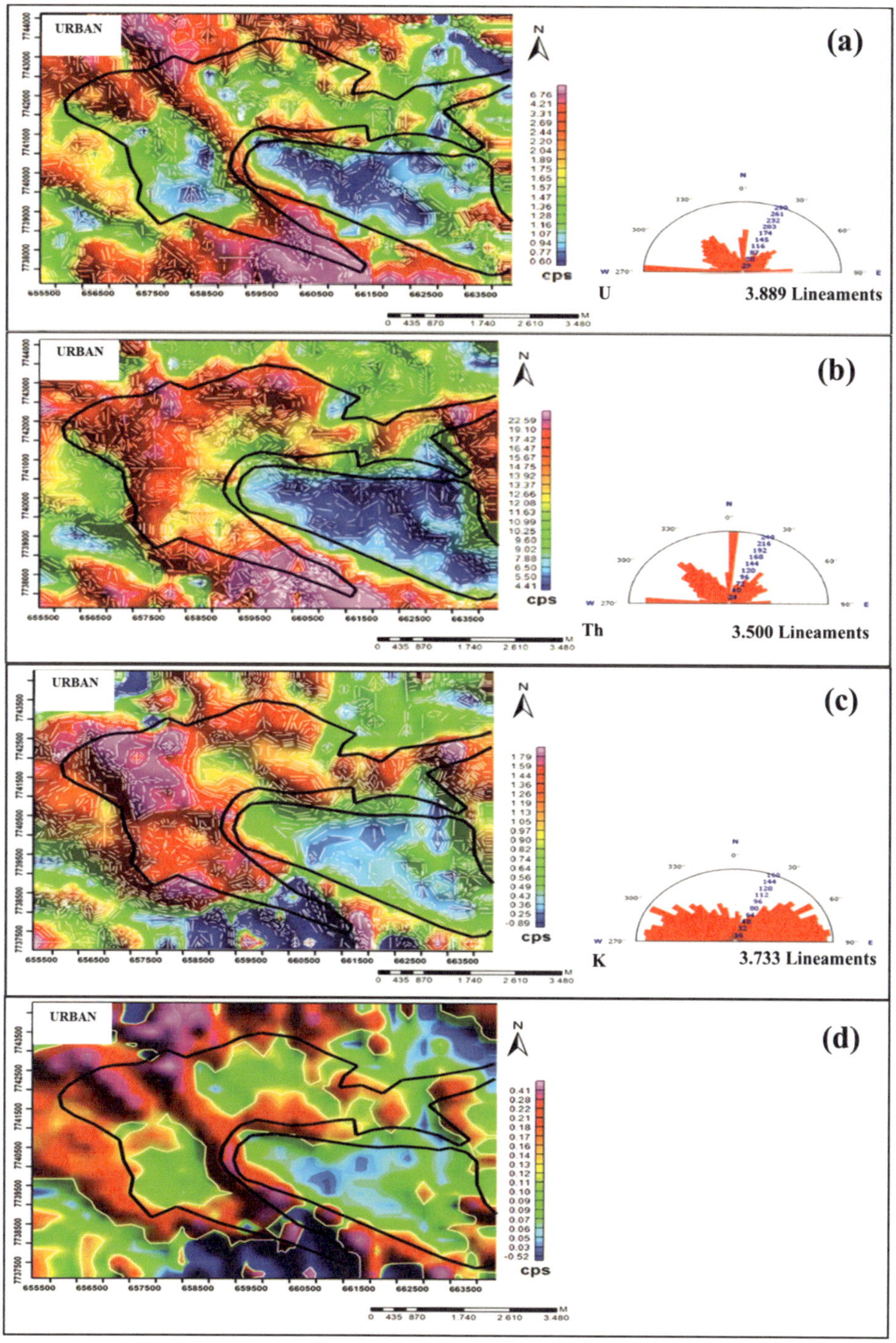


**Figure 6.**
*Maps of uranium (a), thorium (b), and potassium (c) channels, and F-factor (d).*

Field evidence highlights the contrast in fracture characteristics between phyllite and quartzite. In phyllite (Point 1—**Figure 3**), fractures are closed and exhibit signs of folding, indicating a more ductile behavior. In contrast, quartzite (Point 2—**Figure 3**) has

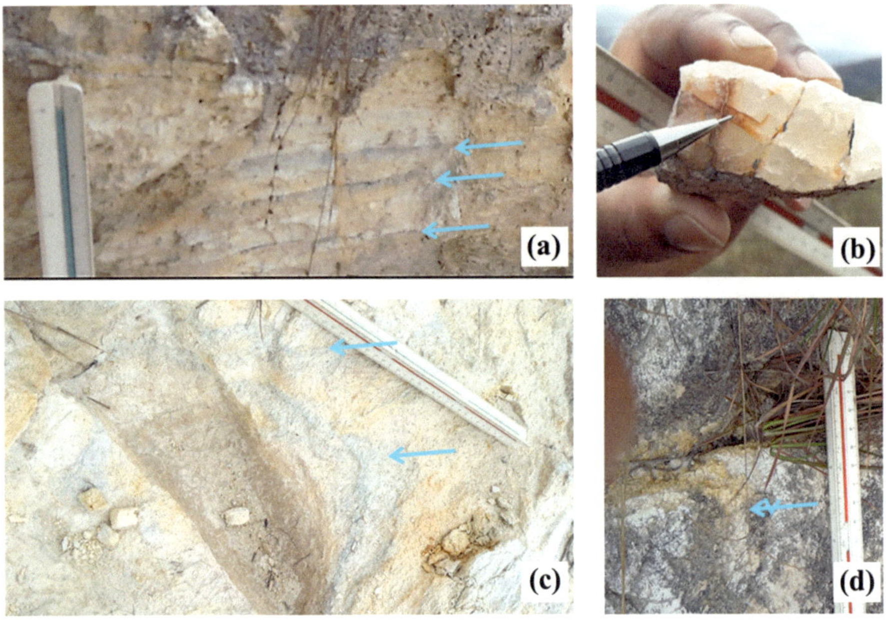


**Figure 7.**
*Field evidence of fluid percolation: Blue arrows indicate dark minerals marking foliation (a, b); quartz with orange and black-colored fractures (c); and fracture surfaces with an orange hue (d).*

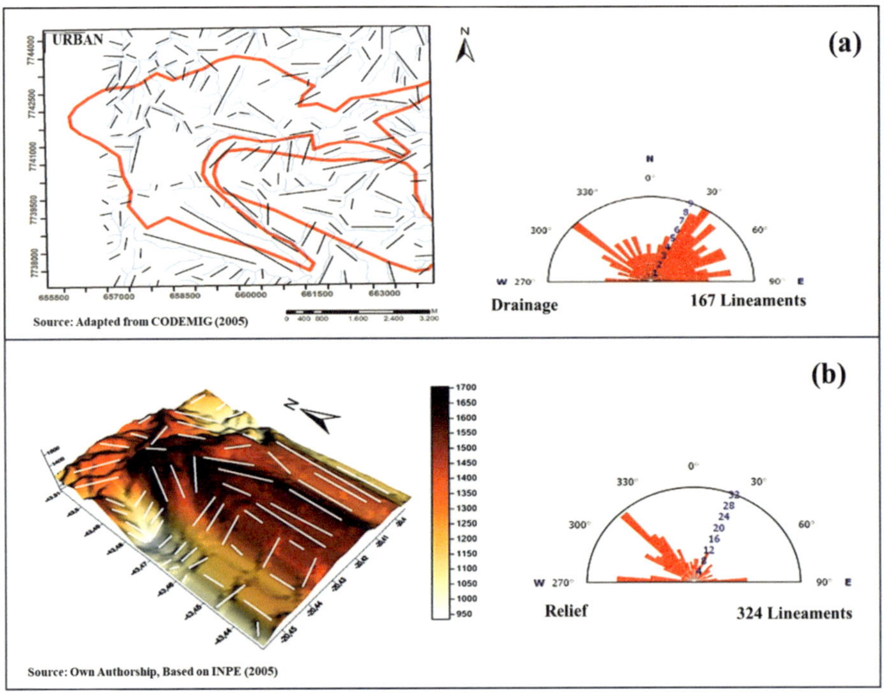


**Figure 8.**
*(a) Drainage map in blue; overlay of the quartzite contour (Itacolomi Group) in red, with interpreted lineaments in black; and (b) 3D Digital Elevation Model generated from the study area's topography, with overlaid interpreted lineaments in white. The lineament rose diagrams show the frequency of lineament orientations.*

| Themes | Main trends |
|---|---|
| ASA | N-S; E-W; N30°–60°W; N30°–60°E |
| dx | N-S; E-W; N30°–60°W; N30°–60°E |
| dy | N-S; E-W; N15°–45°W; N30°–60°E |
| $d^2z$ | N-S; E-W; N45°–30°W; N30°–45°E |
| ASA + dx + dy | N-S; E-W; N30°–60°W; N30°–60°E |
| uranium | N-S; E-W; N30°–60°W; N30°–60°E |
| thorium | N-S; E-W; N30°–60°W; N30°–60°E |
| potassium | without a preferred pattern |
| drainage | N-S; E-W; N45°W; N30°–80°E |
| elevation | E-W; N30°–60°E |

**Table 1.**
*Preferential trends identified in the rose diagrams.*

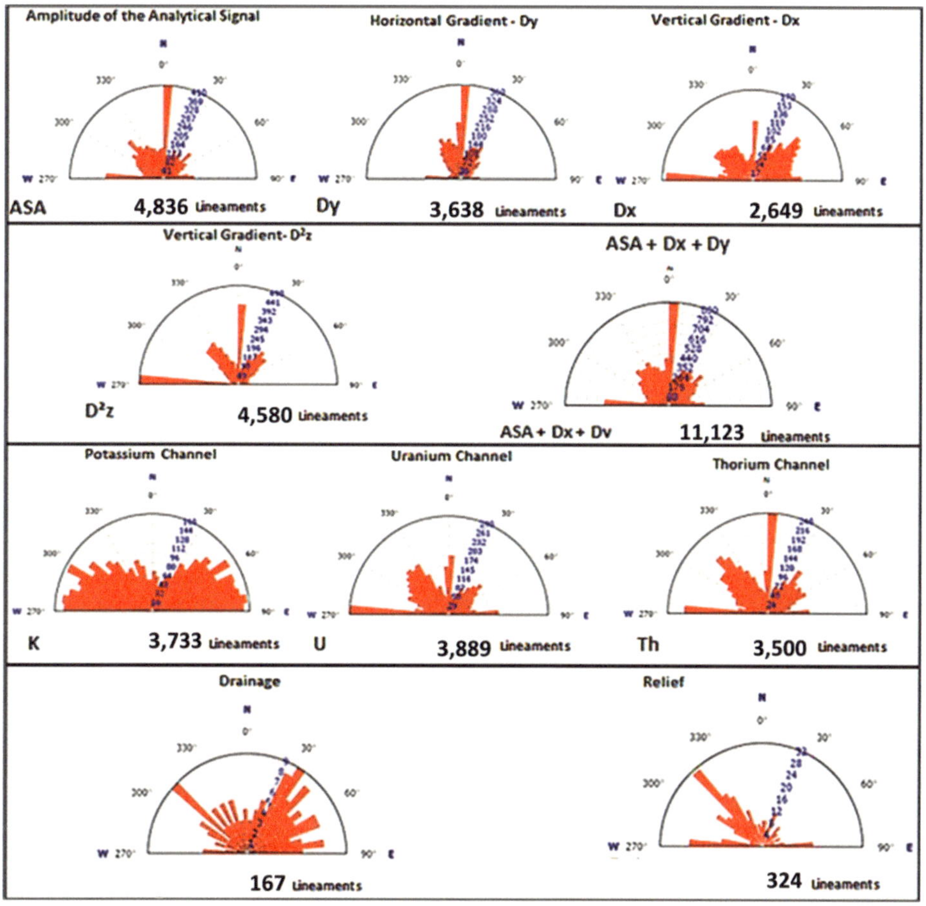


**Figure 9.**
*Rose diagrams of interpreted geophysical and physiographic lineaments.*

higher fracturing, confirming its more brittle nature, with fracture apertures reaching several millimeters. Therefore, considering the intrinsic properties of phyllite, its mineralogical composition, and the observed fracture characteristics, it can be concluded that phyllite acts as a hydraulic barrier at the base of the highly fractured quartzite mass.

Additional field evidence confirms the fracturing of quartzite: rock discontinuities visible along the horizon line (Point 3—**Figures 3** and **10a**) reflect subvertical fractures with decimetric to metric apertures (large fissures in the

**Figure 10.**
*Field evidence in quartzite: subvertical fractures (a) and large-aperture horizontal discontinuities (a, b), indicated by red arrows; erosion following closely spaced, parallel, millimeter-aperture fractures (c, d); and intersections of fractures with diverse orientations, resulting in metric-scale erosion (e).*

rock mass), characterizing very open fractures; horizontal discontinuities (Point 4—**Figures 3** and **10b**) were not detected in the geophysical survey data used in this study. The pronounced erosion evolution trend follows closely spaced parallel fractures (Point 2—**Figures 3**, **10c** and **d**); erosion also follows the topographic slope and the intersection of fractures with diverse orientations (Point 2—**Figures 3** and **10e**).

Comparing the principal orientations of the rose diagrams (**Table 1**), the geophysical maps reveal that the strongest trends align with N-S and E-W azimuths. The geophysical maps corresponding to shallower responses ($d^2z$, thorium, and uranium) exhibit, in addition to these trends, alignments along N30°-60°W and N30°-60°E, which include previously documented orientations [14]. The drainage rose diagram shows similar preferential lineament orientations, primarily N-S and E-W, with E-W in the Digital Elevation Model.

In this sense, it becomes evident that this prospecting campaign discriminated against these additional quartzite fracturing directions. Since geological structuring typically conditions physiography, the similarity between geophysical responses and drainage/elevation is a validation method. Consequently, this integration within the present prospecting campaign corroborates previous groundwater prospecting studies that utilized airborne geophysics [23–26].

Conversely, analyzing the rose diagrams from geophysical maps corresponding to various depths (ASA, dx, and dy), it is observed that their trends are generally consistent with those previously discussed. Thus, it is uncertain whether these trends also extend to greater depths. However, the orientations between N-S and N30°E and between N-S and N15°W are only pronounced in the ASA + dx + dy rose diagram, implying trends occurring at greater depths. The high-value F-factor anomaly (0.18 to 0.41 cps) (**Figure 6d**) indicates that fluids ascended from deeper levels toward shallower crustal levels, confirming the second contribution of this prospecting campaign in delineating both shallow and deep brittle structural frameworks of the quartzite. Discussion of the data from Figure 6 integrated with the other datasets.

Finally, this prospecting campaign's third contribution is identifying potential sites for further groundwater exploration within the quartzite (Itacolomi Group) for the city of

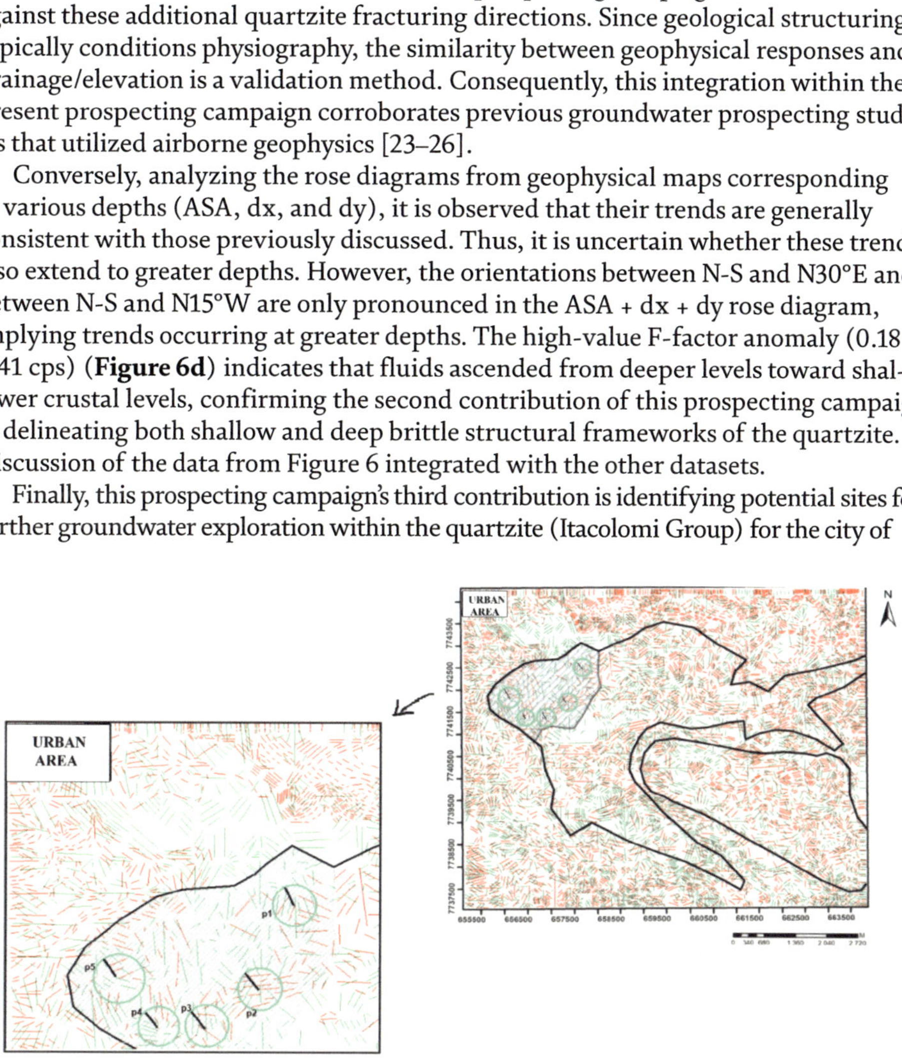


**Figure 11.**
*Locations (blue circles) with potential for continuing groundwater prospecting, considering shallow structures. The reduced area (hatched), red lineaments ($d^2z$), green lineaments (thorium), the black contour of the rock mass (quartzite—Itacolomi Group), and the urban zone are presented.*

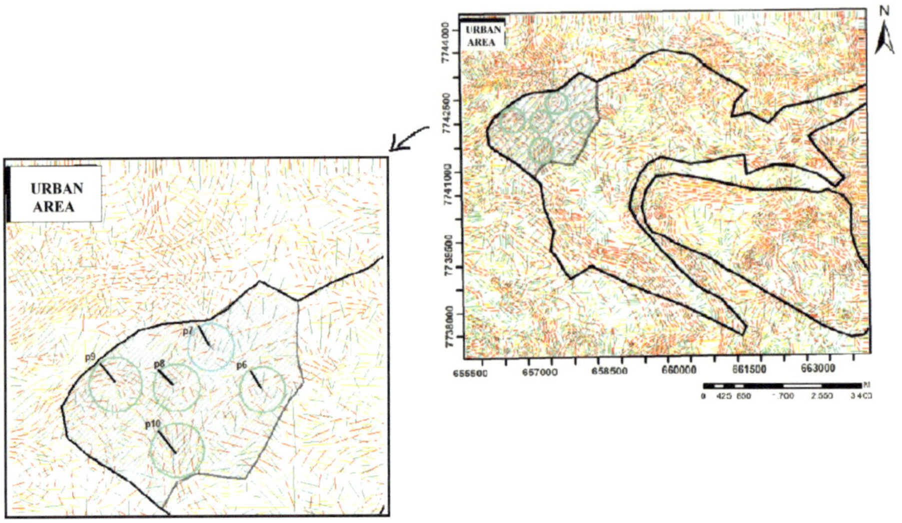


**Figure 12.**
*Locations (blue circles) with potential for continuing groundwater prospecting, considering structures at indistinguishable depths. The reduced area (hatched), red lineaments (ASA), green lineaments (dx), orange lineaments (dy), the black contour of the rock mass (quartzite—Itacolomi Group), and the urban zone are presented.*

| Points | Coordinates (UTM) | |
|---|---|---|
| p1 | 657,699 m E | 7,742,513 m S |
| p2 | 657,375 m E | 7,741,731 m S |
| p3 | 656,901 m E | 7,741,363 m S |
| p4 | 656,467 m E | 7,741,353 m S |
| p5 | 656,102 m E | 7,741,824 m S |
| p6 | 657,689 m E | 7,741,990 m S |
| p7 | 657,172 m E | 7,742,427 m S |
| p8 | 656,828 m E | 7,742,031 m S |
| p9 | 656,188 m E | 7,742,051 m S |
| p10 | 656,811 m E | 7,741,422 m S |

**Table 2.**
*Locations of potential points for local groundwater prospecting within fracture porosity.*

Ouro Preto. These locations are presented in **Table 2**. Analyzing the prospecting locations (**Figures 11** and **12**) and the coordinates of their central points (**Table 2**), it is noteworthy that the intersections between points P5 and P9, as well as between P3 and P10, refine the prospecting scope by enhancing the delineation of shallow structures.

## 6. Conclusion

A fractured quartzite mass was delineated using satellite imagery in conjunction with a georeferenced geological map. Surrounding lithologies are primarily composed of phyllites, which exhibit markedly different hydraulic properties compared to quartzite. This observation was confirmed through field validation. While the fractured quartzite demonstrated significant potential for groundwater storage, the adjacent phyllite units behaved as hydraulic barriers, effectively compartmentalizing the aquifer system.

To overcome the limitations posed by surface cover, airborne geophysical methods were employed, including the acquisition and processing of magnetic and radiometric data. In parallel, a Digital Elevation Model was generated, and the regional drainage network was analyzed to support physiographic lineament interpretation. Structural lineaments were extracted from both geophysical and physiographic datasets, and their spatial distribution and preferred orientations were compared. Field investigations corroborated the presence of intense fracturing in the quartzite mass.

The application of data filtering techniques and the integration of physiographic and geophysical datasets enabled the qualitative discrimination of shallow versus deep structural features. Shallow structures were predominantly oriented along N30°–60°W, N30°–60°E, N–S, and E–W trends. In contrast, deep-seated structures were aligned along directions ranging from N–S to N30°E and N–S to N15°W.

Based on this integrated approach, several promising targets for groundwater exploitation were identified near urbanized areas. Geographic coordinates of these high fracture-density zones were provided as candidates for future, detailed hydrogeological investigations. Overall, the proposed methodology proved effective in resolving the challenge of identifying subsurface structural controls on groundwater accumulation beneath weathered mantles typical of tropical terrains.

To enhance future local hydrogeological studies, it is recommended that fractures be systematically classified into sets to improve the monitoring of preferential subsurface flow paths. The orientations (strike and dip) and specific characteristics of each fracture set (e.g., extent, aperture, spacing, interconnectivity) will facilitate the implementation of an efficient local prospecting methodology.

Therefore, we believe this methodological framework can be applied to similar hydrogeological contexts in other tropical regions. Additionally, it is essential to consult the relevant regulatory agencies before conducting any interventions in aquifers.

## Acknowledgements

The authors thank the Federal University of Ouro Preto, Seequent, and the Geological Survey of Brazil for making this study possible. They also express gratitude for the opportunity to contribute to an important international hydrogeology book.

## Conflict of interest

The authors declare no conflict of interest.

## Other declarations

The authors declare that this work is original. It has never been published and is not currently under evaluation for publication elsewhere.

## Author details

Ana Luísa Santos Martins[1], Luiz Henrique Cardoso[1*], Daniela Antunes Lessa[1], Antonio Santos Sánchez[2], Ana Letícia Pilz de Castro[1] and Guilherme Augusto Magalhães Gomes[1]

1 Federal University of Ouro Preto, Ouro Preto, Brazil

2 Federal University of Bahia, Bahia, Brazil

*Address all correspondence to: luizgeologoop@gmail.com

## References

[1] United Nations Department of Economic and Social Affairs (UNDESA). World Urbanization Prospects. 2018. Available at: https://population.un.org/wup/publications [Accessed: March 24, 2022]

[2] Chandra CA. Groundwater Geophysics in Hard Rock. New Delhi: CRC Press; 2016

[3] Raji WO, Adedoyin ADA. Dam safety assessment using 2D electrical resistivity geophysical survey and geological mapping. Journal of King Saud Uniersity Science. 2020;**32**:1123-1129

[4] Cardoso LH, Bacellar LAP. Characterization with geoelectrical methods of fissural porosity and flow pattern in physical models of fractured rock masses. Journal of Applied Geophysics. 2023;**218**:105201

[5] Alkmim FF, Marshak S. Transamazonian orogeny in the southern São Francisco craton region, Minas Gerais, Brazil: Evidence for paleoproterozoic collision and colapse in the Quadrilátero Ferrífero. Precambrian Research. 1998;**1-2**:29-58

[6] Dorr JVN. Physiographic, Stratigraphic and Structural Development of the Quadrilátero Ferrífero, Brazil. Washington: Professional Paper, USGS/DNPM; 1969

[7] Endo I, Fonseca MA. Sistema de cisalhamento Fundão Cambotas no Quadrilátero Ferrífero, MG: geometria e cinemática. In: SBC/Núcleo MG, editor. Simpósio de geologia de Minas Gerais. Ouro Preto: 6. Semana de estudos da SICEG, 30; 1992. pp. 28-31

[8] Nalini HAJ. Análise estrutural descritiva e cinemática do flanco sul e terminação periclinal do Anticlinal de Mariana e adjacências, região sudeste do Quadrilátero Ferrífero. Master's Dissertation. Brazil: Institute of Geosciences, Federal University of Minas Gerais; 1993

[9] Varajão CAC. A questão da correlação das superfícies de erosão do Quadrilátero Ferrífero, Minas Gerais. Revista Brasileira de Geociencias. 1991;**21**(2):138-145

[10] Salgado AAR, Braucher R, Varajão CAC, Colin F, Varajão AFDC, Nalini HA Jr. Relief evolution of the Quadrilátero Ferrífero (Minas Gerais, Brazil) by means of ($^{10}$Be) cosmogenic nuclei. Zeitschrift für Geomorphologie. 2008;**52**:317-323

[11] Feitosa FAC, Filho JM, Feitosa EC, Demetrio JGA. Hidrogeologia: Conceitos e Aplicações. Vol. 3. Rio de Janeiro: CPRM, UFPE, CTG, Laboratório de Hidrogeologia; 2008

[12] Pereira GG. Caracterização multiescalar de parâmetros de fraturas para avaliação da qualidade geotécnica e hidrogeológica de maciços rochosos. Graduate monograph. Brazil: School of Mines, Federal University of Ouro Preto; 2019

[13] Ferreira A, Lazarin H. Caracterização litoestrutural da região do Pico Itacolomi, Ouro Preto. Graduate monograph. Brazil: School of Mines, Federal University of Ouro Preto; 1993

[14] Andrade LND. Caracterização hidrológica e hidrogeoquímica do Parque Estadual do Itacolomi-Ouro Preto, MG. Master's Dissertation. Brazil: School of Mines, Federal University of Ouro Preto; 2008

[15] CPRM. Geological Survey of Brazil (CPRM). Brazil: Annual Report; 2011

[16] Kearey P, Brooks M, Hill I. An Introduction to Geophysical Exploration. Oxford: Wiley-Blackwell; 2002

[17] Telford WM et al. Applied Geophysics. Cambridge: Cambridge University; 1990

[18] Ellis DV, Singer JM. Well Logging for Earth Scientists. Vol. 2. The Netherlands: Springer; 2008

[19] Luiz JG, Silva LM. Geofísica de prospecção. Belém - Pará: Universitária UFPA e CEJUP; 1995

[20] Carvalho AMD, Fernandes AM, Bacellar LAP, Maurer VC, Melo PR, Borges T. Utilização conjunta de dados de fraturas e lineamentos para indicação de fatores de zoneamento de aquíferos fraturados. Repositório UFOP. Ouro Preto. 2021;**35**(2):47-55

[21] Companhia De Desenvolvimento Do Estado De Minas Gerais (CODEMIG). Projeto geologia do Quadrilátero Ferrífero - integração e correção cartográfica em Sig com nota explicativa. CODEMIG: Belo Horizonte; 2005

[22] Coelho DM. Análise geofísica na prospecção de água no vale do Ribeirão Aredes, no município de Itabirito, Minas Gerais. Graduate Monograph. Brazil: School of Mines, Federal University of Ouro Preto; 2018

[23] Oliveira NCDM et al. Aplicação do método AHP para confecção de modelo de favorabilidade hídrica subterrânea de aquíferos fraturados. Águas Subterrâneas. 2020;**34**(1):90-102

[24] Ribeiro ECS. Análise geofísica aplicada a estudo da voçoroca na região da Chapada subdistrito de Ouro Preto, MG. Graduate Monograph. Brasil: Escola de Minas, Universidade Federal de Ouro Preto; 2017

[25] Cardoso LH, Bacellar LAP, Martins ALS, Bragioni DO. Análise geofísica regional do controle litoestrutural em voçorocas da porção centro-sul do Quadrilátero Ferrífero, Minas Gerais, Brasil. Revista Brasileira De Geomorfologia. 2022;**23**(2):1357-1378

[26] Brito TPD, Bacellar LAP, Barbosa MSC, Barella CF. Favorabilidade hidrogeológica do Complexo Bação, Quadrilátero Ferrífero-MG. Águas Subterrâneas; 2021

[27] Lasa, Engenharia E Prospecções S.A. Relatório final do levantamento e processamento dos dados magnetométricos e gamaespectrométricos. Levantamento Aerogeofísico de Minas Gerais, Area 2 – Projeto Pitangui-São João del Rei-Ipatinga. Secretaria de Estado de Minas e Energia (SEME), Governo de Minas Gerais. 2001

[28] Geosoft/Seequent. Oasis Montaj (versão 9.8.1) [Accessed: January 15, 2021]

[29] Hoff R, Rolim SBA, Neto ACB. Mapeamento aerogamaespectrométrico da alteração hidrotermal associada à mineralização no distrito fluorítico de Santa Catarina, Brasil. Revista Brasileira de Geofisica. 2004;**22**(1):45-55

[30] Golden. Surfer (versão 9.11.947) [Accessed: June 04, 2022]

[31] Yong Technology Inc. Geo Rose (versão 0.3.0). 2022

[32] Google Earth. 2022 [Accessed: June 04, 2022]

Chapter 5

# Study of Fluid Flow Nature of Micropolar Fluid in a Vertical Channel

*Shreedevi Kalyan and Ashwini Sharan*

## Abstract

This study presents a comprehensive analysis of micropolar fluid flow within a vertical channel, focusing on the unique behavior of such fluids due to their microstructural effects and microrotational motion. Micropolar fluids, characterized by the presence of micro-elements capable of rotation, exhibit non-Newtonian behavior significantly different from classical Newtonian fluids. The governing equations for conservation of mass, linear momentum, and angular momentum are formulated based on Eringen's micropolar fluid theory and solved under appropriate boundary conditions. The influence of key parameters such as viscosity coupling number, micropolar material constants, and channel geometry on the velocity and microrotation profiles is investigated. Numerical simulations reveal that microrotation significantly alters the flow structure, especially near the channel walls, and the presence of micropolar effects leads to a reduction in the velocity boundary layer thickness. These findings are essential for applications involving complex fluids in engineering and biomedical systems, where enhanced control overflow characteristics is required.

**Keywords:** micropolar fluid, vertical channel, non-Newtonian flow, microrotation, Eringen's theory, boundary layer, velocity profile, fluid mechanics

## 1. Introduction

### 1.1 Objective and scope

*Fluid mechanics* deals with different types of fluids and their properties, which are either in motion or in static conditions. Fluid mechanics is concerned with many problems: mathematical modeling, formulation of problems, solutions, dimensionless numbers, governing equations, and graphs with analytical results.

First, we must gain knowledge about fluids and their physical properties through the subject of fluid mechanics. Second, we concentrate on the mathematical model, which contains the three basic equations of conservation of physical systems: conservation of mass, momentum, and energy.

The third part is the solution. For this, we have to solve the fluid problem using analytical methods with the help of software such as Mathematica, MathType, Origin, etc. The results are then plotted through graphs using Origin software.

The *Multifluid model* may consist of two or more chemical components, such as liquid–liquid or liquid–gas, simultaneously. This phenomenon is very complex and has demanded exhaustive efforts in both computational and theoretical challenges in the research area. The presence of multiple fluids, including their various properties, is found internally or externally in our daily life. Multifluid flows often arise in our environment, fluid-fuel interactions, engine fuel injection systems, and in the field of engineering.

"The first scientist to utilize differential equations to analyze the phenomenon of a linear shear stress-to-shear rate relationship in fluids was Isaac Newton."

Newton's laws of motion are used to predict velocity, acceleration, mass, position of an object, and the forces acting on it using Newton's equations. The phenomenon of the linear relationship between shear stress and shear rate in fluids has extensive applications in rheology, hydrology, biology, medicine, and various engineering fields.

A type of non-Newtonian fluid called the *Jeffrey fluid* can explain the properties and traits of fluid flow during stress relaxation. Many researchers have done work on Newtonian fluids, while we observe very few studies related to biological, physiological, and industrial aspects. According to Newton's rule of viscosity, determining the characteristics of non-Newtonian fluids is very difficult in rheological fields.

*Porous material* has been significantly used in the past decade in several fields such as chemical separations, filtration, purification of water, industrial catalysts, adsorbents, etc. Porous materials have outstanding demand due to their chemical and physical properties, despite their limited natural accessibility. Porous materials are categorized by the size of their voids:

- *Microporous* – voids size less than 2 nm
- *Mesoporous* – void sizes between 2 and 50 nm
- *Macroporous* – voids size more than 50 nm.

## 1.2 Literature survey

The intuitive topic of *vortex* in fluid mechanics is a very important subject. A vortex is a region in a fluid which revolves around its own axis or an axis line, albeit the axis line may be curved or straight. The nature of vortices is circular and rapidly spinning. We observe these vortices in the wake of a boat, eddies, tornadoes, cyclones, smoke rings, and even in coffee. Vortices take the form of stirred fluids and represent a type of turbulent flow in fluids.

In *fluid dynamics*, vortices are distinguished into two types:

1. *Forced vortices* (caused by some external force)
2. *Free vortices* (occur naturally in fluid flow).

These vortex forces are determined in a cylindrical shape. The *viscosity* of the fluid relates to the concept of "thickness." Vorticity is denoted by $\omega$.

Agafontesva and Kuibin [1] describe the evolution and formation of *helical vortex filaments* caused by the influence of viscosity. They investigated vortex formation and its movement rate in viscous fluids, concluding that during diffusion, helical vortices become very thin under excessively high Reynolds numbers.

The *Grashof number* provides a correlation between viscous, inertial, and buoyancy forces in fluid flow and helps determine the nature of flow—whether turbulent, laminar, or transitional. If the boundary layer is turbulent, the value of the Grashof number is high; for laminar flow, it lies within a certain lower range.

Anuj Gupta and Thakur [2] studied natural convective heat transfer in a steady state, enclosed in various shapes with laminar flow and constant temperature. Various Grashof numbers were considered for evaluating heat transfer rate, aspect ratio, and Nusselt number for effective study and design simulation.

*Fully developed flow*: The flow of any fluid is said to be fully developed in a pipe or channel when the velocity profile along the axial direction does not vary. The fluid flow from the boundary to the inlet is influenced by the fluid's viscosity. In laminar flow, the development is smooth, while in turbulent flow, the fluid may assume a helical shape. This can be mathematically represented as:

$$\partial u/\partial x = 0 \tag{1}$$

According to *Yasushi Takeda* [3], the velocity profile can be developed using ultrasonic measurements at PSI. This technique is advantageous for flow efficiency mapping, applicability to opaque fluids, and recording the spatiotemporal velocity field.

*Material parameters* are specific nodes that allow modification of a material's appearance without recompiling the material to view changes. In fluid mechanics, commonly used non-designable parameters include Reynolds, Prandtl, Brinkmann, Froude, Euler, and Mach numbers.

*Cheng et al.* [4] studied the effects of material parameters on *fluidlastic isolator* performance and concluded that design parameters, such as density, viscosity, loss factor, and rubber stiffness, influence the fluidlastic properties. Fluid density is affected by frequency, which increases due to rubber stiffness.

## 2. Basic equations, boundaryconditions, and dimensionless parameters

### 2.1 Basic equations

In Fluid Mechanics, the three fundamental physical principle for natural system of laws of conservation are considered to be:

1. Conservation of mass or Continuity equation

2. Conservation of momentum or Equation of motion

3. Conservation of energy or Energy equation

#### *2.1.1 Conservation act for the mass equation*

According to this rule, mass cannot be produced or destroyed within an isolated system but can only be transformed between systems. This law can be conveyed in the form of differentiation by using equation of continuity in continuum and fluid mechanics.

#### 2.1.1.1 For compressible flow

$$\frac{\partial \rho}{\partial t} + \nabla\left(\rho \vec{u}\right) = 0 \qquad (2)$$

Where.

$\rho, \vec{u} \rightarrow$ Fluid density and velocity.
$t \rightarrow$ Time.
$\nabla \rightarrow$ Divergence.

#### 2.1.1.2 For steady incompressible flow

$$\nabla.\vec{u} = 0 \qquad (3)$$

Where $\frac{\partial}{\partial t}$ and $\nabla.\vec{u}$ represent the local derivative and convective derivative, respectively.

Local derivative is a quantity changes in time at a fixed point and convective derivative is quantity changes because of movement of the fluid element from one point to another in the fluid flow field, where the characteristics of fluid movement vary.

### 2.1.2 Conservation act for the momentum equation

The rate of change of linear momentum is equal to the body forces and surface forces acting on a fluid, according to Newtons second law of motion, which is the source of the rule of conservation of momentum equation.

$$\rho\left[\frac{\partial \vec{u}}{\partial t} + \left(\nabla.\vec{u}\right)\vec{u}\right] = -\nabla p + \rho g + \mu \nabla^2 \vec{u} \qquad (4)$$

$$\rho \frac{D\vec{u}}{Dt} = \rho \vec{F} - \nabla p + \mu \nabla^2 \vec{u} \qquad (5)$$

Where.

$\frac{D}{Dt} = \frac{\partial}{\partial t} + \left(\nabla.\vec{u}\right)$ is a material derivative describe the temporal rate of a physical quantities change, such as heat or momentum, for a material element subjected to a space and time dependent velocity field.

### 2.1.3 Conservation act for the energy equation

The first rule of thermodynamics is referred to as the amount of energy is neither created nor destroyed. This law describes the different form of energy.

$$\rho C_p\left[\frac{\partial T}{\partial t} + \left(\nabla.\vec{u}\right)T\right] = k\nabla^2 T + \phi \qquad (6)$$

$\phi$ is known as dissipation function, it is measured of heat generated due to frictional forces

$$\phi = \mu\left(\frac{\partial V_i}{\partial x_j} + \frac{\partial V_j}{\partial x_i}\right)\frac{\partial V_i}{\partial x_j} \tag{7}$$

## 2.2 Boundary conditions

### *2.2.1 No-slip conditions*

This condition asserts that at any solid surface the relative velocity of the (Newtonian) fluid with respect to the surface is zero.

### *2.2.2 Slip conditions*

A slip condition is as a symmetry condition, the wall has no effect on the velocity in planes parallel and close to the wall.

### *2.2.3 The fundamental physical mechanism of heat transfer*

#### *2.2.3.1 Conduction heat transfer*

A method for transferring heat from a high-temperature area to a low-temperature area between two different or within a medium indirect physical contact is called conduction heat transfer.

#### *2.2.3.2 Convection heat transfer*

Convection heat transfer is the process of heat transfer by the bulk movement of molecules within the fluids such as gases and liquids. The initial heat transfer between the object and the fluid takes place through conduction, but the bulk heat transfer happens due to the of the fluid.

#### *2.2.3.3 Radiation heat transfer*

The process of thermal transfer between two objects in the absence of any transmitting medium. This is an electromagnetic wave phenomenon.

#### *2.2.3.4 Fourier law of heat conduction*

This law states that the negative gradient of temperature and the time rate of heat transfer is proportional to the area at right angles of that gradient through which the heat flows. This rule has a differential form that can be expressed as:

$$Q = -KA\frac{dt}{dx} \tag{8}$$

Where.

$Q \rightarrow$ Heat conduction.
$K \rightarrow$ Thermal conductivity.
$A \rightarrow$ Area.
$\frac{dt}{dx} \rightarrow$ Temperature gradient.

Basically, thermodynamic deals with the conservation of applicable energy into heat energy, and it is based on some important concepts like the Zeroth law, principle of conservation of energy (first law) and thermodynamic second law. Fundamentally, the phenomenon of heat transfer is non-equilibrium.

Fundamental principle of diffusion is that the net flow of molecules occurs from the higher to the low region concentration under the prevalence of concentration gradient.

*2.2.4 Diffusion or mass transfer depends on the conditions*

i. Driving force

ii. The distance in the direction of mass transfer

iii. Diffusivity coefficient

Mass transfer causes a species in a mixture due to concentration gradient, it is a numerous important application in food processing.

*2.2.5 Fundamental types of mass transfer are distinguished below*

i. Diffusion in a quiescent medium

ii. Mass transfer in laminar and turbulent flow

iii. Exchange of mass transfer between two regions.

*2.2.5.1 Fick's law of diffusion*

The diffusion rule is governed by Fick's law. According to Fick's law, the molar flux causes by diffusion is proportional to concentration gradient.

$$J = -D\frac{d\phi}{dx} \tag{9}$$

where, $J$ is diffusion flux, $D$ is the diffusivity, $\phi$ is concentration and $x$ is position.

## 2.3 Dimensionless parameters in heat transfer

*2.3.1 Grash of number*

The Grashof number defines the proportions of buoyant to viscous forces acting on a fluid, buoyant force is the force that raises the fluid's temperature and density falls as a result of this fluid rising and viscous force that can oppose the motion of fluid albeit it helps to find prevalence of fluid flow in the system of laminar flow. It is a unit less quantity that describes the velocity distributions in the system of natural convection in heat transfer. Grashof Number can be represented by symbol *Gr*.

$$GR_T = \frac{Buoyancy\ Force}{Viscous\ Force} = \frac{gh^3\beta_T(T_{wall} - T_\infty)}{\nu^2}, GR_C = \frac{gh^3\beta_C(\Delta C)}{\nu^2} \tag{10}$$

where.

$GR_T \& GR_C$ - Thermal & mass Grashof number.

$g \rightarrow$ Gravitational acceleration.

$\beta_T \& \beta_C \rightarrow$ Thermal & concentraion expansion coefficients.

$\Delta C \rightarrow$ Difference in concentrations.

$T_{wall} \rightarrow$ Temperature at the wall.

$T_\infty \rightarrow$ Bulk temperature.

$\nu \rightarrow$ Kinematic viscosity.

This above expression is for heat transfer in natural convection (it results from a shift in temperature when the fluid's density changes) in vertical plates. Similarly, for concentration, due to mass transfer or concentration gradients in fluids.

### *2.3.2 Reynolds number(Re)*

The definition of Reynolds number is the proportion of inertia force to viscous force in a flowing fluid. It is a unit less quantity and Reynolds number predicts the fluid flow patterns in a pipe, whether it is laminar or turbulent.

$$\mathrm{Re} = \frac{Inertia\ Force}{Viscous\ Force} = \frac{\rho u L}{\mu} \tag{11}$$

where.

$\mathrm{Re} \rightarrow$ Reynolds number.

$\rho \rightarrow$ Mass density.

$u \rightarrow$ Speed of fluid flow (mean velocity).

$L \rightarrow$ Characteristic length.

$\mu \rightarrow$ Dynamic viscosity of the fluid.

The fluid movement is regarded as laminar the Reynolds number must be less than 2000 and for turbulent the flow should be greater than 2000.

### *2.3.3 Hartmann number*

Hartmann number initially introduced by Julius Hartmann (1881–1951) of Denmark. Mathematically the Hartmann number can be expressed as the relation of the electromagnetic force to viscous force.

$$Ha = \frac{Electromagnetic\ Force}{Viscous\ Force} = BL\sqrt{\frac{\sigma}{\mu}} \tag{12}$$

where.

$Ha \rightarrow$ The Hartmann number.

$B \rightarrow$ The magnetic field intensity.

$L \rightarrow$ The characteristic length scale.

$\mu \rightarrow$ The dynamic viscosity.

### *2.3.4 Prandtl number*

It is described as the relationship between the temperature diffusivity and the momentum diffusivity (kinematic viscosity). It is indicated by the symbol Pr and it is given as:

$$\Pr = \frac{\textit{Momentum Diffusivity}}{\textit{Thermal Diffusivity}} = \frac{\nu}{\alpha} = \frac{\mu/\rho}{(k/C_p\rho)} = \frac{C_p\mu}{k} \quad (13)$$

where.
$\nu, \alpha \rightarrow$ Momentum and thermal diffusivity.
$\mu \rightarrow$ Dynamic viscosity.
$k \rightarrow$ Thermal conductivity.
$C_p \rightarrow$ Temperature at steady pressure.
$\rho \rightarrow$ Density.

### 2.3.5 Eckert number

The advective mass transfer to heat dissipation potential relationship is known as the Eckert number. It is represented by the symbol *Ec* albeit it is given as:

$$Ec = \frac{\textit{Advective mass transport}}{\textit{heat disspiation}} = \frac{V^2}{C_p \Delta T} \quad (14)$$

where.
$V \rightarrow$ localized fluid movement rate.
$C_p \rightarrow$ Temperature at steady pressure.
$\Delta T \rightarrow$ Temperature difference between the wall and the surrounding area.

### 2.3.6 Brinkmann number

It is the ratio between heat produced by viscous dissipiation and heat transported by molecular conduction. Mathematically, it is denoted by *Br*

$$Br = \frac{\mu u^2}{k(T_w - T_0)} = \Pr.Ec \quad (15)$$

where.
$\mu \rightarrow$ Dynamic viscosity.
$u \rightarrow$ Fluid flow velocity.
$k \rightarrow$ Thermal conductivity.
$\Pr \rightarrow$ Prandtl number.
$T_w and T_0 \rightarrow$ Wall and bulk fluid temperature.
$Ec \rightarrow$ Eckert number.

### 2.3.7 Nusselt number

The Nusselt number is used to describe the ratio of thermal energy convected to the fluid to the thermal energy conducted within the fluid. It is designated by the symbol *Nu*

$$Nu = \frac{\textit{heat transfer}_{\textit{Convective}}}{\textit{heat transfer}_{\textit{Conductive}}} = \frac{hL}{k_f} \quad (16)$$

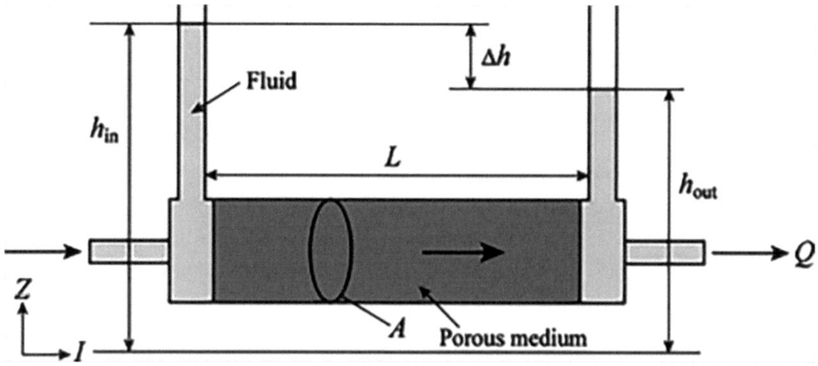


**Figure 1.**
*Darcy Model.*

where.
$h \rightarrow$ Convective heat transmission coefficient.
$k_f \rightarrow$ Fluid's thermal transmission.
$L \rightarrow$ Characteristic length.

*2.3.8 Darcy's law*

Darcy's law is an equation that describes the linear relationship between the flow rate to the pressure differences $\nabla p$ when the fluid flows in porous media, which can also be named as linear flow law.

$$Q = -\frac{k}{\mu}(\nabla p) \tag{17}$$

where.
$Q \rightarrow$ Instantaneous flow rate (total discharge).
$k \rightarrow$ Permeability.
$\mu \rightarrow$ Dynamic viscosity of the fluid.
$\nabla p \rightarrow$ Pressure gradient.
Darcy's law is one of the basic equations used in hydrology to describe the movement of water through an aquifer and shows that the discharge is proportional to the hydraulic gradient and the aquifer material (**Figure 1**).

One darcy is equivalent to $9.869233 \times 10^{-13}\ m^2$, this conversion is usually approximated as $1\mu m^2$.

*2.3.9 Magnetohydrodynamic (MHD)*

Magnetohydrodynamic (magnetofluid dynamics or hydromagnetics) is the study of the dynamics of electrically conducting fluids. Examples of such fluids include plasmas, liquid metals, and salt water or electrolytes. The word magnetohydrodynamic is derived from *magneto*—meaning magnetic field; *hydro*—meaning liquid; and *dynamic*—meaning movement. The field of MHD was initiated by Hannes Alfven, for which he received the Nobel Prize in physics in 1970. The fundamental concept behind MHD is that magnetic fields can induce currents in a moving conductive fluid, which in turn creates forces on the fluid and changes the magnetic field itself (as

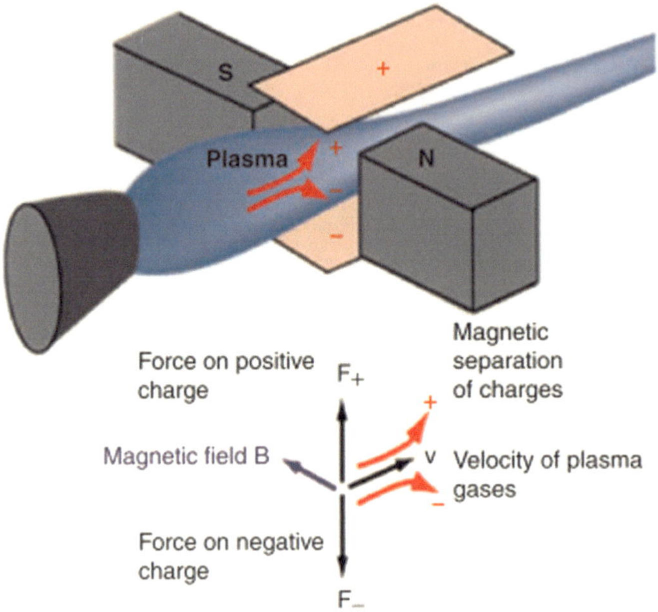


**Figure 2.**
*MHD Model.*

shown figure). The set of equations that describe MHD are a combination of the Navier–Stokes equations of fluid dynamics and Maxwell's equations of electromagnetism. These differential equations have to be solved simultaneously, either analytically or numerically (**Figure 2**).

The principal of MHD power generation is based on Faraday's law of electromagnetic induction, which states that the when a conductor and a magnetic field moves relative to each other, then the voltage is induced in the conductor, which results in the flow of current across the terminals.

## 3. Analysis of micropolar fluid flow nature in a vertical channel

### 3.1 Introduction

The Navier–Stokes theory, while highly successful in modeling conventional Newtonian fluids, assumes that the fluid elements are structureless points. This means it cannot account for fluids that contain microstructures or internal degrees of freedom such as rotation, which are characteristic of complex fluids like liquid crystals, polymeric suspensions, blood, and colloidal fluids.

These microstructured fluids exhibit behaviors such as micro-rotational effects, size-dependent interactions, and asymmetric stress distributions, which the classical theory fails to describe. For example, the Navier–Stokes model cannot accurately represent the behavior of granular materials, foams, or biological fluids, where individual particles or cells may rotate independently of the bulk fluid flow.

Eringen's micropolar fluid theory introduces the concept of microrotation vectors and couple stresses, thus allowing each fluid element to undergo both translational and rotational motion. This enhancement enables the theory to capture the effects of microstructure on macroscopic fluid behavior, providing a more accurate framework for analyzing flows with internal angular momentum or size-dependent phenomena. Including such clarification will help readers appreciate the necessity of micropolar fluid research and the advancements it offers over classical fluid mechanics.

Using examples such as animal blood, muddy water, colloidal fluids, lubricants and chemical suspensions. Eringen [5] suggested that the theory of micropolar fluids to address the shortcomings of Navier–stokes theory in describing the correct behavior of fluids of some kinds of fluids with microstructures. Six degrees of freedom, three for translation and three for microrotation of microelements, are present in the mathematical theory of micropolar fluids. In the recent books [5–7], Lukaszewicz [8], as well as the review pieces Ariman et al. [6, 7], there are in-depth analyses of the theory and applications.

The swirling of a fluid around a centre area is known as a vortex. Mixed convection is defined as circumstances in which the heat transfer mechanisms of forced convection and natural convection coexist. Due to its many real-world uses over the past 20 years, such as cooling nuclear reactors and electronic components as well as turbine rotor blade internal cooling systems, mixed convection in a vertical heated channel has attracted a lot of attention.

Studies of the heat convection movement in micropolar fluids have primarily concentrated on flat Ahmadi [9], Yucel [10] or regular surfaces Balram and sastry [11] and Lien and Chen [12] and Rahaman et al. [13–17] Jena and Mathur [18].

Chamkha et al. [19] investigated numerical and analytic solutions of the developing laminar free convection of a micropolar fluid in a vertical parallel plate channel with asymmetric heating. There are numerous uses for fluid flow and heat transfer in the chemical and nuclear sectors. Designing a two-fluid heat-transfer system for use in space necessitates an understanding of the processes involved in heat and mass transmission as well as fluid mechanics in a low-gravity environment. For the design of two fluid systems, it is crucial to identify the fluid flow area and calculate the pressure drop void fraction, quality reaction, and transfer coefficients for each of the two fluids.

Malashetty and Leela et al. [20] examined the Hartmann flow feature of two fluids in horizontal fluids. Malashetty et al. [21] and Malashetty et al. [22, 23] have investigated two phase flow and heat transfer in an inclined channel. The classical no-slip criterion for velocity may not always hold and fluid slippage may occur at the solid boundary, according to several studies conducted in the past century by Happal and Brenner [24], Hutchins et al. [25], Barrat and Bocquet [26], Pit et al. [27], Shreedevi et al. [28–35] have analyzed the fluid flow behavier in the presence of chemical reaction parameter and got the clear imporetence of Hartman number. Numerous scholars have demonstrated in the literature that the no-slip boundary condition may result in singular or unrealistic behavior (e.g. Thompson and Robbins [36], Koplik and Banavar [37] and Richardson [38]. Navier et al. [39, 40] has suggested a general slip boundary condition that permits the possibility of fluid slip at a solid boundary. We have discovered that the slip coefficient, which is a constant of proportionality, is inversely proportional to the tangential relative velocity of the fluid at a place on the solid boundary and the tangential stress acting at that point. Ashmawy [41], Yang [42], Sherief et al. [43], Ashmawy [44], Hayat et al. [45], Khan et al. [46], micropolar fluids Lok et al. [47] and Ashmawy [48], Maxwell fluids Vieru and Zafar [49] and ferro fluids Khan et al. [50], have all made extensive use of the slip condition recently,Gyro-viscosity parameter was limited by Ahmadi [9] and [51].

One might wonder how this assumption will play out, particularly for steady fluid flows. When studying time-dependent fluid flows, this restriction may be used to enable the field equations. Eringen [5] predicted the correct behavior of viscous fluid flows in the limiting situation when micro-structure effects become insignificant and micro rotation reduces to the angular velocity.

## 3.2 Problem description

### *3.2.1 Physical setup*

- A vertical channel bounded by two parallel plates separated by a distance.
- Region I contains a micropolar fluid characterized by microrotation effects, non-Newtonian behavior, and additional dynamic properties.
- Region II contains a viscous Newtonian fluid with standard properties such as velocity and pressure gradients.
- Assume steady-state, incompressible flow with possible thermal and mass diffusion (**Figure 3**).

### *3.2.2 Micropolar fluid section*

- *Vortex-like rotations:* You can observe circular or rotational movements along the pathways. These rotations represent the micro-rotational behavior unique to micropolar fluids, such as certain liquid crystals or polymeric suspensions.
- *Particle rotation visualization:* The particles in micropolar fluids are not just flowing linearly; they also rotate, adding another level of complexity to their motion.

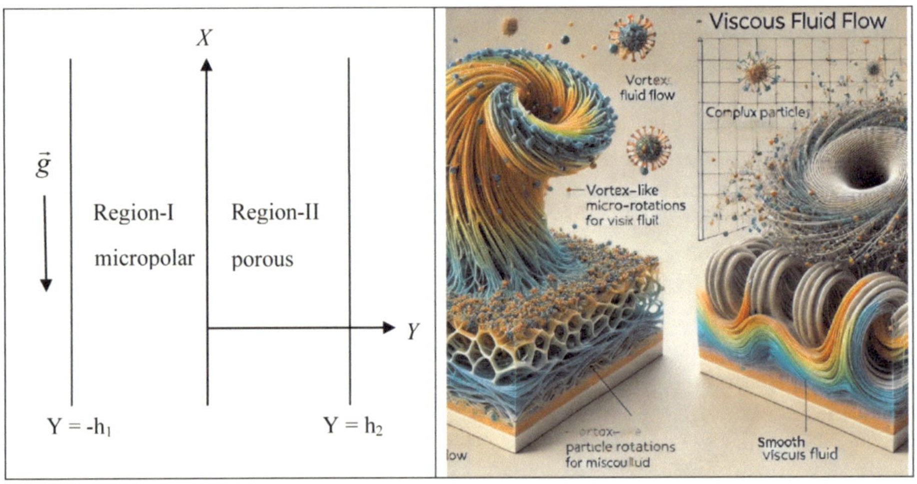


**Figure 3.**
*Physical Configuration. Physical model of micropolar fluid flow alongside viscous fluid flow.*

### 3.2.3 Viscous fluid section

- *Smooth streamlines:* The fluid pathways are sleek and continuous without rotational elements. This represents typical viscous flows, where particles move smoothly without any individual particle spin.

- *Laminar flow:* The parallel lines without disruptions emphasize the well-known laminar nature of viscous fluid dynamics.

### 3.2.4 Color coding and labels

- Color differentiation helps distinguish between micropolar and viscous fluid behaviors, making it easier to visualize complex *versus* smooth flows.

- Labels (or imagined markers) may describe the flow behaviors, Reynolds number implications, or torque-related aspects of micropolar fluids.

The governing equations for momentum, angular momentum, energy and concentration under the mentioned conditions result in,

#### 3.2.4.1 Region-I

$$\rho_1 g\beta_{T1}(T_1 - T_{w2}) - \frac{dp}{dX} + (\mu_1 + \kappa)\frac{d^2U_1}{dY^2} + \rho_1 g\beta_{C1}(C_1 - \overline{C}_2) + \kappa\frac{dn}{dY} = 0 \tag{18}$$

$$\nu\frac{d^2n}{dY^2} - \kappa\left(2n + \frac{dU_1}{dY}\right) = 0 \tag{19}$$

$$\frac{d^2T_1}{dY^2} = 0 \tag{20}$$

$$D_1\frac{d^2C_1}{dY^2} = 0 \tag{21}$$

#### 3.2.4.2 Region-II

$$\rho_2 g\beta_{T2}(T_2 - T_{w2}) - \frac{dp}{dX} + \mu_2\frac{d^2U_2}{dY^2} + \rho_2 g\beta_{C2}(C_2 - \overline{C}_2) - \frac{\mu_2 U_2}{\kappa_2} = 0 \tag{22}$$

$$\frac{d^2T_2}{dY^2} = 0 \tag{23}$$

$$D_2\frac{d^2C_2}{dY^2} = 0 \tag{24}$$

The velocity, temperature, microrotationand concentration boundary and interface parameters are,

$$U_1(-h_1) = 0, U_2(h_2) = 0, U_1(0) = U_2(0), (\mu_1 + \kappa)\frac{dU_1}{dY}(0) + \kappa n(0) = \mu_2\frac{dU_2}{dY}(0),$$

$$T_1(-h_1) = Tw_1, T_2(h_2) = Tw_2\ , T_1(0) = T_2(0), \frac{dT_1}{dY}(0) = \frac{dT_2}{dY}(0),$$

$$C_1(-h_1) = \overline{C_1}, C_2(h_2) = \overline{C_2}, C_1(0) = C_2(0), D_1\frac{dC_1}{dY}(0) = D_2\frac{dC_2}{dY}(0),$$

$$n(-h_1) = 0, \frac{dn}{dY}(0) = 0 \tag{25}$$

We now present the non-dimensional variables listed below,

$$u_i = \frac{U_i}{\overline{u_1}}, y_i = \frac{Y_i}{h_i}, \theta_1 = \frac{T_1 - T_{w2}}{T_{w1} - T_{w2}}, \theta_2 = \frac{T_2 - T_{w2}}{T_{w1} - T_{w2}}, \phi_1 = \frac{C_1 - \overline{C_2}}{\overline{C_1} - \overline{C_2}}, \phi_2 = \frac{C_2 - \overline{C_2}}{\overline{C_1} - \overline{C_2}},$$

$$\Delta T = T_{w1} - T_{w2}, \Delta C = \overline{C}_1 - \overline{C}_2, Gr = \frac{g\beta_{T1}{h_1}^3(T_{w1} - T_{w2})}{{\upsilon_1}^2},$$

$$Gc = \frac{g\beta_{C1}{h_1}^3(\overline{C_1} - \overline{C_2})}{{\upsilon_1}^2},\ \mathrm{Re} = \frac{\overline{U_1}h_1}{\upsilon_1}, H = \frac{nh_1}{\overline{U_1}}, p = \frac{{h_1}^2}{\mu_1\overline{U_1}}\frac{dp}{dX},$$

$$K = \frac{\kappa}{\mu_1},\ \sigma^2 = \frac{{h_2}^2}{\kappa_2} \tag{26}$$

and $K = \frac{\kappa}{\mu_1}$ is the Region-I material attribute. We see that applies to a stream that is viscous or Newtonian.

Eqs. (17)–(23) are amended to include the non-dimensional factors (25) we get

### 3.2.4.3 Region-I

$$(1+K)\frac{d^2u_1}{dy^2} + K\frac{dH}{dy} + GR_T\theta_1 + GR_c\phi_1 - p = 0 \tag{27}$$

$$\frac{d^2H}{dy^2} - \frac{2K}{2+K}\left(2H + \frac{du_1}{dy}\right) = 0 \tag{28}$$

$$\frac{d^2\theta_1}{dy^2} = 0 \tag{29}$$

$$\frac{d^2\phi_1}{dy^2} = 0 \tag{30}$$

### 3.2.4.4 Region-II

$$\frac{d^2u_2}{dy^2} + GR_T mnh^2 b_t\theta_2 + GR_C mnh^2 b_c\phi_2 - mh^2p - \sigma^2u_2 = 0 \tag{31}$$

$$\frac{d^2\theta_2}{dy^2} = 0 \tag{32}$$

$$\frac{d^2\phi_2}{dy^2} = 0 \tag{33}$$

The non-dimensional form of boundary and interface conditions of Eq. (24) becomes,

$$u_1(-1) = 0, u_2(1) = 0, u_1(0) = u_2(0), (1+K)\frac{du_1}{dy}(0) + KH(0) = \frac{1}{mh}\frac{du_2}{dy}(0)$$

$$\theta_1(-1) = 1, \theta_2(1) = 0, \theta_1(0) = \theta_2(0), \frac{d\theta_1}{dy}(0) = \frac{1}{kh}\frac{d\theta_2}{dy}(0),$$

$$\phi_1(-1) = 1, \phi_2(1) = 0, \phi_1(0) = \phi_2(0), \frac{d\phi_1}{dy}(0) = \frac{d}{h}\frac{d\phi_2}{dy}(0),$$

$$H(-1) = 0, \frac{dH}{dy}(0) = 0 \tag{34}$$

Before delving into the solution method, it is important to justify the choice of employing second-order differential equations and specific integration techniques in the analysis. The governing equations for micropolar fluid flow, derived from Eringen's theory, naturally lead to coupled nonlinear second-order differential equations due to the inclusion of microrotation and asymmetric stress terms. These equations are particularly well-suited for capturing the complex interactions between rotational and translational motion in microstructured fluids—interactions that cannot be modeled using first-order approximations or purely empirical models. Compared to other numerical or analytical approaches, the second-order differential equation method provides a robust framework for modeling the underlying physics with higher fidelity. Additionally, integration techniques like the finite difference method or Runge–Kutta methods are preferred due to their proven stability and accuracy in solving boundary value problems common in fluid dynamics. Thus, the chosen method is not only mathematically appropriate but also physically consistent with the complex behavior of micropolar and Newtonian fluid flows within the defined geometry and boundary conditions.

### 3.3 Method of solutions

We are solving Eqs. (26) and (32) by using second order differential equation solving method and integration technique then we obtain solutions for both areas by utilizing the boundary and interface conditions (33). Detail solution and constant values are given in appendix.

#### *3.3.1 Region-I*

$$\theta_1 = C_1 y + C_2 \tag{35}$$

$$\phi_1 = B_1 y + B_2 \tag{36}$$

$$H = P_2 A_1 + P_3 + P_4 y + P_5 y^2 + C_5 Cosh[P_1 y] + C_6 Sinh[P_1 y] \tag{37}$$

$$u_1 = A_2 + q_1 A_1 y + q_2 y + q_3 y^2 + q_4 y^3 + q_5 C_5 Sinh[P_1 y] + q_6 C_6 Cosh[P_1 y] \tag{38}$$

#### *3.3.2 Region-II*

$$\theta_2 = C_3 y + C_4 \tag{39}$$

$$\phi_2 = B_3 y + B_4 \tag{40}$$

$$u_2 = q_7 + q_8 y + A_3 Cosh[\sigma y] + A_4 Sinh[\sigma y] \tag{41}$$

## 3.4 Results and discussion

Analytical solutions have been derived for the system of differential Eqs. (26)–(32), which govern the flow of an incompressible micropolar fluid confined between two vertical plates under the specified boundary and interface conditions (33). Closed-form solutions, as shown in Eqs. (34)–(40), have been visualized graphically to study the impact of various physical parameters. While the constants used in these solutions are not presented explicitly, they can be readily determined. The graphical results provide insights into the variations in linear velocity, microrotational velocity, and temperature in response to changes in key parameters.

### *3.4.1 Effects on velocity and microrotational velocity*

#### *3.4.1.1 Mixed convection parameter (λ)*

**Figure 4a** and **b** show that an increase in λ enhances the buoyancy force, promoting fluid motion more significantly in the micropolar fluid region than in the porous region. This results in higher linear velocity and increased microrotational velocity. The effect is more pronounced when the micropolar fluid is replaced by a Newtonian (viscous) fluid.

#### *3.4.1.2 Mass Grashof number (Grm)*

As depicted in **Figure 5a** and **b**, an increase in the mass Grashof number amplifies the buoyancy forces, causing a notable rise in the velocity profiles in both regions. The microrotational velocity also increases in response to the enhanced mass-driven convection.

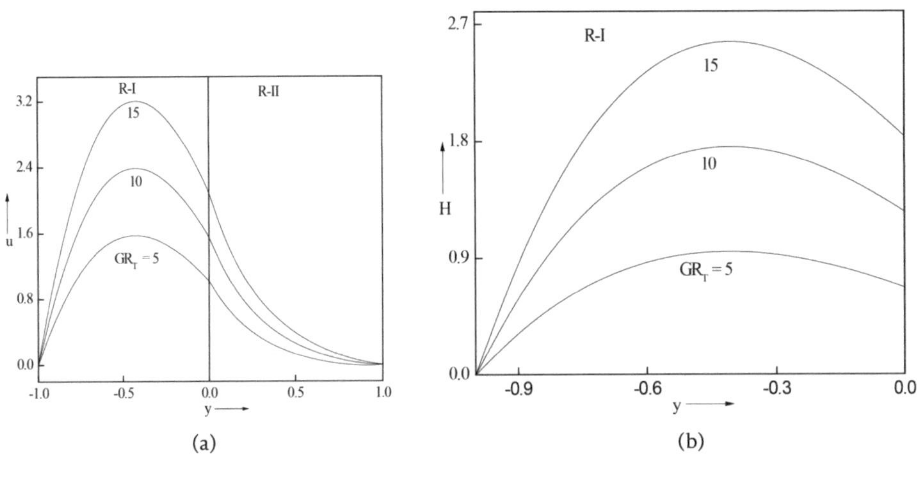


**Figure 4.**
*(a) $GR_T$ Effects on Velocity Patterns, and (b) $GR_T$ Effects on Microrotational Velocity Patterns.*

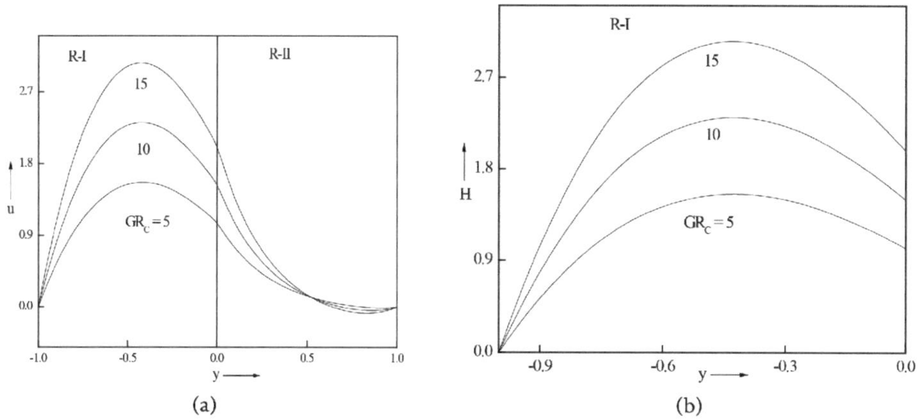


**Figure 5.**
*(a) $GR_C$ Effects on Velocity Patterns, and (b) $GR_C$ Effects on Microrotational Velocity Patterns.*

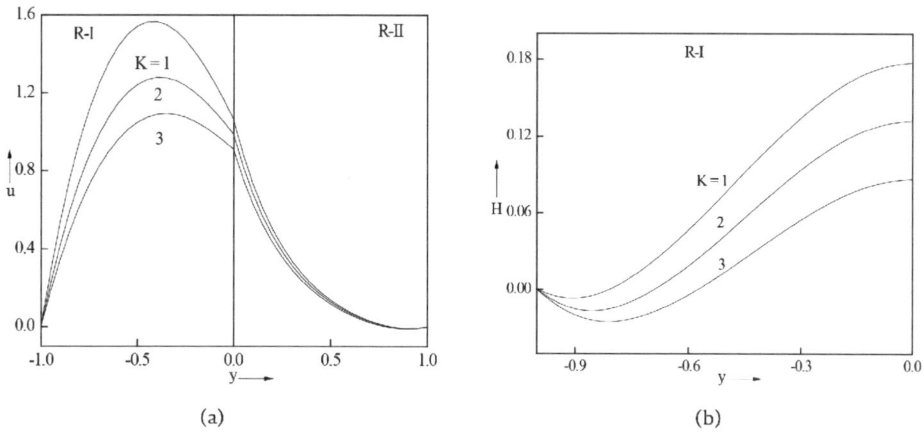


**Figure 6.**
*(a) K Effects on Velocity Patterns, and (b) K Effects on Microrotational Velocity Patterns.*

### *3.4.1.3 Material parameter (K)*

**Figure 6a** and **b** reveal that increasing K results in a reduction of both linear and microrotational velocities in both regions. In Region I, the velocity profiles gradually evolve into a parabolic shape as K increases, indicating a stabilizing effect on the flow field.

### *3.4.1.4 Viscosity ratio (m)*

As shown in **Figure 7a** and **b**, higher values of the viscosity ratio lead to decreased linear and microrotational velocities. This can be attributed to the greater resistance imposed by the increased micropolar viscosity in the fluid.

### *3.4.1.5 Width ratio (h)*

**Figure 8a** and **b** demonstrate that increasing the width ratio leads to a reduction in both velocity and microrotation. The wider channel spacing imposes a drag-like effect, thereby reducing the overall flow intensity in the system.

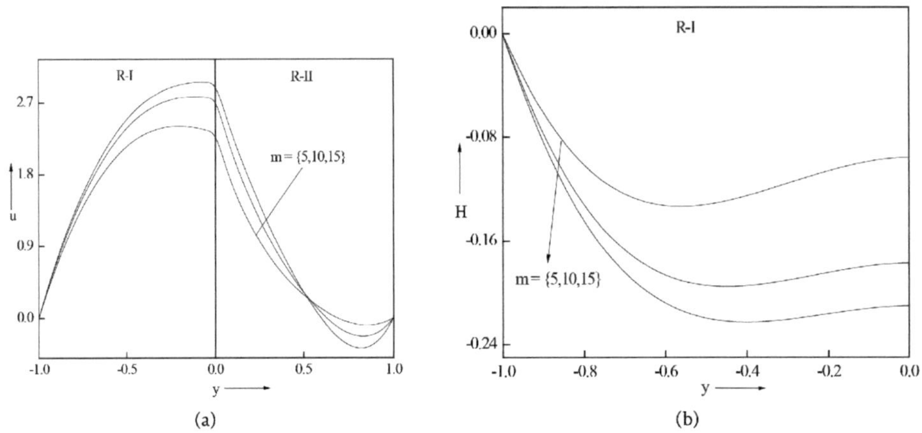


**Figure 7.**
*(a) m Effects on Velocity Patterns, and (b) m Effects on Microrotational Velocity Patterns.*

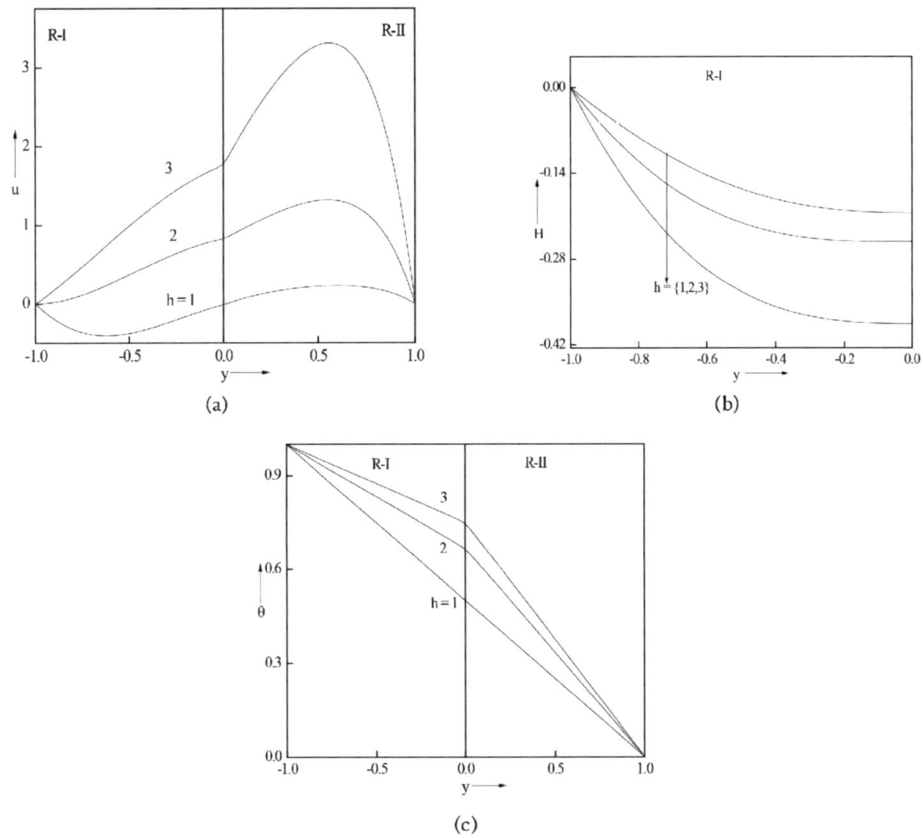


**Figure 8.**
*(a) h Effects on Velocity Patterns, (b) h Effects on Microrotational Velocity Patterns, and (c) h Effects on Temperature Patterns.*

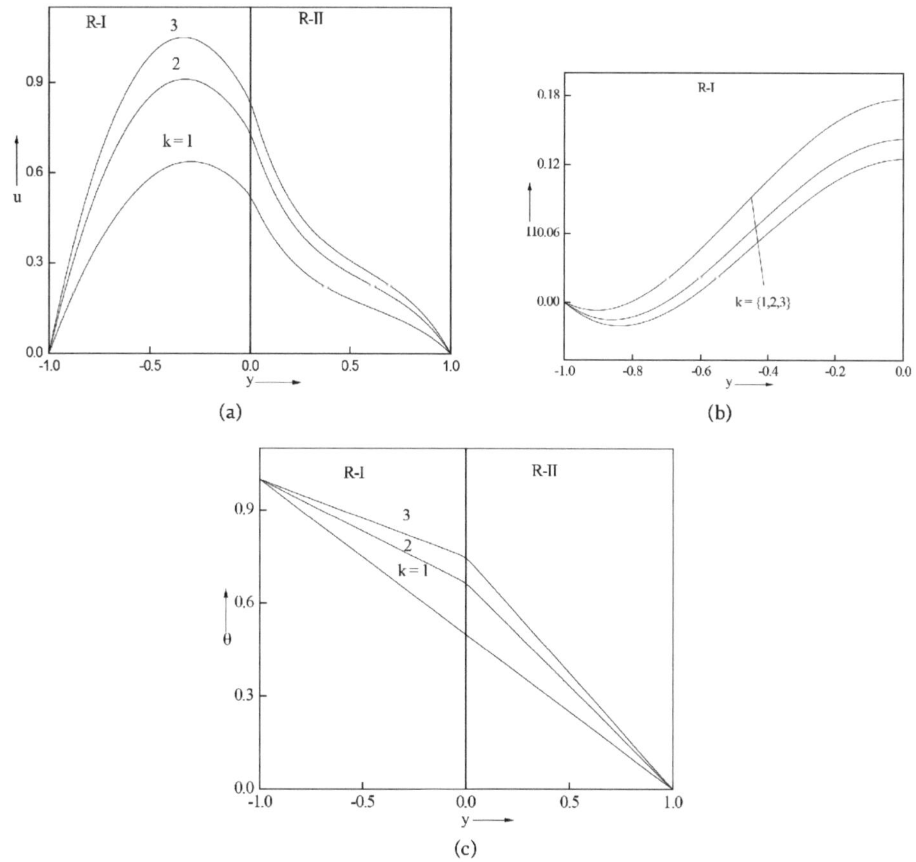


**Figure 9.**
*(a) k Effects on Velocity Patterns, (b) k Effects on Microrotational Velocity Patterns, and (c) k Effects on Temperature Patterns.*

### *3.4.2 Effects on temperature*

#### *3.4.2.1 Conductivity ratio (k)*

**Figure 9a–c** depict the influence of the conductivity ratio on linear velocity, microrotational velocity, and temperature. An increase in the conductivity ratio enhances heat conduction in the fluid, which can influence the thermal boundary layer and the temperature distribution. As thermal energy diffuses more effectively, temperature gradients smooth out, influencing both the microrotational and linear flow behavior.

## 4. Conclusions

This study reveals that increases in the thermal and mass Grashof numbers markedly enhance the velocity, temperature, and concentration profiles, thereby intensifying buoyancy-driven fluid motion. The results also underscore the significant impact of vortex viscosity, illustrating how micropolar characteristics influence and regulate the fluid flow dynamics.

Additionally, the analysis shows that variations in porous medium properties and chemical reaction parameters contribute to a reduction in both temperature and velocity fields. Notably, Region I exhibits higher velocity amplitudes compared to Region II, highlighting the distinct behavior of micropolar fluids relative to their Newtonian counterparts.

These findings contribute meaningful insights into the behavior of complex fluids and have practical implications for fluid mechanics, thermal management systems, and chemically reactive flows. Moreover, they lay a strong foundation for future investigations aimed at advancing technologies in engineering, biomedical applications, and energy systems.

## Appendix

$$
\begin{aligned}
G_1 or GR_T &= \frac{Gr}{\mathrm{Re}}, G_2 or GR_C = \frac{Gc}{\mathrm{Re}}, a_1 = -GR_T m\, h^2 n b_t, a_2 = -GR_C m\, h^2 n b_c, a_3 = m h^2 P,\\
P_1 &= \sqrt{\frac{4K}{(2+K)} - \frac{2K^2}{(2+K)(1+K)}}, P_2 = \frac{2K}{(2+K)(1+K)}(-G_1 C_2 - B_2 G_2 - P),\\
B_1 &= -\left(\frac{d}{d+h}\right), B_2 = \left(\frac{h}{d+h}\right), B_3 = \left(-\frac{h}{d+h}\right), B_4 = \frac{h}{d+h},\\
C_1 &= -\frac{1}{1+kh}, C_2 = \frac{kh}{1+kh}, C_3 - -\frac{kh}{1+kh}, C_4 - \frac{kh}{1+kh}, P_3 = -\frac{-2K G_1 C_1 - 2K G_2 B_1}{(2+K)(1+K)},\\
P_4 &= -\frac{2K}{P_1 P_1 (2+K)}, P_5 = -\frac{P_2}{P_1 P_1}, P_6 = -\frac{P_3}{P_1 P_1}, P_7 = -\frac{2P_3}{P_1 P_1 P_1 P_1},\\
q_1 &= 1 - \frac{K P_4}{(1+K)}, q_2 = -\frac{K P_7}{(1+K)}, q_{14} = C_6 q_3 = \frac{-P - K P_5 - G_1 C_2 - G_2 B_2}{2(1+K)},\\
q_4 &= -\frac{(G_1 C_1 + G_2 B_1)}{2(1+K)} - \frac{P_6 K}{(1+K)}, q_5 = -\frac{K}{P_1(1+K)}, q_6 = -\frac{K}{P_1(1+K)}\\
q_7 &= \frac{-a_3 - a_1 C4}{\sigma^2}, q_8 = \frac{-a_2 B_3 - a_1 C_3}{\sigma^2}, q_9 = q_2 - q_3 + q_4, q_{13} = P_5 - P_6 - P_7 q_{10}\\
&= mh(1+K)q_1 + K P_4, q_{11} = mh(1+K)P_1 q_5 + K, q_{12} = mh(1+K)q_2 + K P_7 - q_8,
\end{aligned}
$$

$$
A_1 = \left(\begin{array}{l}
-\frac{-q_{13} - q_{14} Sinh[P_1]}{P_4} + (Cosh[P_1](-(-q_{13} - q_{14} Sinh[P_1])(q_1 \sigma Cosh[\sigma] + \\
q_{10} Sinh[\sigma] + P_4(-\sigma(-q_7 - q_8 + (-q_{14} q_6 + q_7 - q_9 + q_{14} q_6 Cosh[P_1]) Cosh[\sigma]) \\
+q_{12} Sinh[\sigma]))))/(P_4(-Cosh[P_1](q1 \sigma Cosh[\sigma] + q_{10} Sinh[\sigma] + P_4(q_5 \sigma \cosh[\sigma] \\
Sinh[P_1] + q_{11} Sinh[\sigma])))
\end{array}\right),
$$

$$
A_2 = \left(\begin{array}{l}
-(q_1 q_7 \sigma Cosh[P_1] + q_1 q_8 \sigma Cosh[P_1] + q_1 q_{14} q_6 \sigma Cosh[P_1] Cosh[\sigma] \\
-q_1 q_7 \sigma Cosh[P_1] Cosh[\sigma] - P_4 q_5 q_7 \sigma Sinh[P_1] - P_4 q_5 q_8 \sigma Sinh[P_1] - \\
P_4 q_{14} q_5 q_6 \sigma Cosh[\sigma] Sinh[P_1] + P_4 q_5 q_7 \sigma Cosh[\sigma] Sinh[P_1] + q_1 q_{11} q_{13} Sinh[\sigma] \\
+P_4 q_{11} q_9 Sinh[\sigma] + q_1 q_{12} Cosh[P_1] Sinh[\sigma] - P_4 q_{11} q_{14} q_6 Cosh[P_1] Sinh[\sigma] - \\
q_{10} q_9 Cosh[P_1] Sinh[\sigma] + q_{10} q_{14} q_6 Cosh[P_1]^2 Sinh[\sigma] + q_1 q_{11} q_{14} Sinh[P_1] Sinh[\sigma] - \\
4 q_{12} q_5 Sinh[P_1] Sinh[\sigma] - q_{10} q_{13} q_5 Sinh[P_1] Sinh[\sigma] - q_{10} q_{14} q_5 Sinh[P_1]^2 Sinh[\sigma])/ \\
(q_1 \sigma Cosh[P_1] Cosh[\sigma] - P_4 q_5 \sigma Cosh[\sigma] Sinh[P_1] - P_4 q_{11} Sinh[\sigma] + q_{10} Cosh[P_1] Sinh[\sigma])
\end{array}\right)
$$

$$A_3 = \left( \begin{array}{l} -(-q_1 q_7 \sigma Cosh[P_1] - q_1 q_8 \sigma Cosh[P_1] + P_4 q_5 q_7 \sigma Sinh[P_1] + P_4 q_5 q_8 \sigma Sinh[P_1] \\ -q_1 q_{11} q_{13} Sinh[\sigma] - P_4 q_{11} q_{14} q_6 Sinh[\sigma] + P_4 q_{11} q_7 Sinh[\sigma] - P_4 q_{11} q_9 Sinh[\sigma] - \\ q_1 q_{12} Cosh[P_1] Sinh[\sigma] + q_{10} q_{14} q_6 Cosh[P_1] Sinh[\sigma] + P_4 q_{11} q_{14} q_6 Cosh[P_1] Sinh[\sigma] \\ -q_{10} q_7 Cosh[P_1] Sinh[\sigma] + q_{10} q_9 Cosh[P_1] Sinh[\sigma] - q_{10} q_{14} q_6 Cosh[P_1]^2 Sinh[\sigma] - \\ q_1 q_{11} q_{14} Sinh[P_1] Sinh[\sigma] + P_4 q_{12} q_5 Sinh[P_1] Sinh[\sigma] + q_{10} q_{13} q_5 Sinh[P_1] Sinh[\sigma] + \\ q_{10} q_{14} q_5 Sinh[P_1]^2 Sinh[\sigma] / (-q_1 \sigma Cosh[P_1] Cosh[\sigma] + P_4 q_5 \sigma Cosh[\sigma] Sinh[P_1] + \\ P_4 q_{11} Sinh[\sigma] - q_{10} Cosh[P_1] Sinh[\sigma]) \end{array} \right)$$

$$A_4 = \left( \begin{array}{l} -(P_4 q_{11} q_7 + P_4 q_{11} q_8 - q_{10} q_7 Cosh[P_1] - q_{10} q_8 Cosh[P_1] + q_1 q_{11} q_{13} Cosh[\sigma] + \\ P_4 q_{11} q_{14} q_6 Cosh[\sigma] - P_4 q_{11} q_7 Cosh[\sigma] + P_4 q_{11} q_9 Cosh[\sigma] + q_1 q_{12} Cosh[P_1] Cosh[\sigma] - \\ q_{10} q_{14} q_6 Cosh[P_1] Cosh[\sigma] - P_4 q_{11} q_{14} q_6 Cosh[P_1] Cosh[\sigma] + q_{10} q_7 Cosh[P_1] Cosh[\sigma] - \\ q_{10} q_9 Cosh[P_1] Cosh[\sigma] + q_{10} q_{14} q_6 Cosh[P_1]^2 Cosh[\sigma] + q_1 q_{11} q_{14} Cosh[\sigma] Sinh[P_1] - \\ P_4 q_{12} q_5 Cosh[\sigma] Sinh[P_1] - q_{10} q_{13} q_5 Cosh[\sigma] Sinh[P_1] - q_{10} q_{14} q_5 Cosh[\sigma] Sinh[P_1]^2) / \\ (-q_1 \sigma Cosh[P_1] Cosh[\sigma] + P_4 q_5 \sigma Cosh[\sigma] Sinh[P_1] + P_4 q_{11} Sinh[\sigma] - q_{10} Cosh[P_1] Sinh[\sigma]) \end{array} \right)$$

$$C_5 = \left( \begin{array}{l} -(-(-q_{13} - q_{14} Sinh[P_1])(q_1 \sigma Cosh[\sigma] + q_{10} Sinh[\sigma] + P_4(-\sigma(-q_7 - q_8 + (-q_{14} q_6 + q_7 \\ -q_9 + q_{14} q_6 Cosh[P_1]) Cosh[\sigma] + q_{12} Sinh[\sigma]))/(-Cosh[P_1](q_1 \sigma Cosh[\sigma] + q_{10} Sinh[\sigma]) \\ +P_4(q_5 \sigma Cosh[\sigma] Sinh[P_1] + q_{11} Sinh[\sigma])) \end{array} \right)$$

$$C_6 = -\frac{P_5}{P_1}$$

## Author details

Shreedevi Kalyan* and Ashwini Sharan
Department of Mathematics, Sharnbasva University, Kalaburagi, Karnataka, India

*Address all correspondence to: kalyanshreedevi@gmail.com

## References

[1] Agafontseva MV, Kuibin PA. On the viscosity influence on a helical vortex filament evolution. EPJ Web of Conferences. 2015;**82**:1-3

[2] Gupta A, Thakur H. Attic space convection analysis with full-blown heat condition with different possible geometries. Advances in Fluid and Thermal Engineering. 2019;**8**:627-636. DOI: 10.1115/IMECE2016-65065

[3] Takeda Y. Velocity profile measurement by ultrasonic doppler method. Experimental Thermal and Fluid Science. 1995;**10**:444-453

[4] Cheng QY, Deng JH, Feng ZZ, Qian F. Analysis of material parameter effects on fluidlastic isolators performance. Materials Science and Engineering. 2017;**292**(012039):1-7

[5] Eringen AC. Microcontinuum Field Theories: II. Fluent Media. New York, NY: Springer-Verlag; 2001

[6] Ariman T, Turk MA, Sylvester ND. Microcontinuum field mechanics—A review. International Journal of Engineering Science. 1973;**11**: 905-929

[7] Ariman T, Turk MA, Sylvester ND. Applications of microcontinuum field mechanics. International Journal of Engineering Science. 1974;**12**:273-291

[8] Lukaszewicz G. Micropolar fluids: Theory and applications. Birkhäuser. 1999:1-253. DOI 10.1007/978-1-4612-0641-5

[9] Ahmadi G. Self-similar solution of incompressible micropolar boundary layer flow over a semi-infinite plate. International Journal of Engineering Science. 1976;**14**:639-646

[10] Yucel A. Mixed convection in micropolar fluid over a horizontal plate with surface mass transfer. International Journal of Engineering Science. 1989;**27**: 1593-1608

[11] Balaram M, Sastry VUK. Micropolar free convection flow. International Journal of Heat and Mass Transfer. 1973; **16**:437-441

[12] Lein FS, Chen TM. Analysis of a free-convection micropolar boundary layer about a horizontal permeable cylinder at a non-uniform thermal condition. ASME Journal of Heat Transfer. 1990;**112**: 504-506

[13] Rahaman MM, Eltayeb IA, Mujibur Rahman SM. Thermo-micropolar fluid flow along a vertical permeable plate with uniform surface heat flux in the presence of heat generation. Thermal Science. 2009;**13**(1):23-36

[14] Rahaman MM. Convective flows of micropolar fluids from radiate isothermal porous surfaces with viscous dissipation and joule heating. Communications in Nonlinear Science and Numerical Simulation. 2009;**14**(7): 3018-3030

[15] Rahaman MM, Sultana T. Radiative heat transfer flow of micropolar fluid with variable heat flux in a porous medium. Nonlinear Analysis: Modelling and Control. 2008;**13**(1):71-87

[16] Rahaman MM, Sultana T. Transient convective flow of micropolar fluid past a continuously moving vertical porous plate in the presence of radiation. International Journal of Applied Mechanics and Engineering. 2007;**12**(2): 497-513

[17] Rahaman MM, Sattar MA. MHD convective flow of a micropolar fluid

past a continuously moving vertical porous plate in the presence of heat generation/absorption. ASME Journal of Heat Transfer. 2006;**128**(2): 142-152

[18] Jena SK, Mathur MN. Free-convection in the laminar boundary layer flow of thermomicropolar fluid past a non-isothermal vertical flat plate with suction/injection. Acta Mechanica. 1982;**42**:227-238

[19] Chamkha AJ, Grosan T, Pop I. Fully developed free convection of a micropolar fluid in a vertical channel. International Communications in Heat and Mass Transfer. 2002;**29**: 1021-1030

[20] Malashetty MS, Leela V. Magnetohydrodynamic heat transfer in two-phase flow. International Journal of Engineering Science. 1992;**30**:371-377

[21] Malashetty MS, Umavathi JC. Two-phase magnetohydrodynamic flow and heat transfer in an inclined channel. International Journal of Multiphase Flow. 1997;**22**:545-560

[22] Malashetty MS, Umavathi JC, Prathap Kumar J. Two-fluid magnetoconvection flow in an inclined channel. International Journal of Transport Phenomena. 2000;**3**:73-84

[23] Malashetty MS, Umavathi JC, Prathap Kumar J. Convective magnetohydrodynamic two-fluid flow and heat transfer in an inclined channel. Heat and Mass Transfer. 2001;**37**: 259-264

[24] Happel J, Brenner H. Low Reynolds Number Hydrodynamics. Hague, Boston, Lancaster: Noordhoff International Publishing, Martinus Nijhoff Publishers; 1973. Academic Publishers Group; 1983

[25] Hutchins DK, Harper MH, Felder RL. Slip correction measurements for solid spherical particles by modulated dynamic light scattering. Aerosol Science and Technology. 1995;**22**:202-218

[26] Barrat J, Bocquet L. Large slip effect at a nonwetting fluid–solid interface. Physical Review Letters. 1999;**82**: 4671-4674

[27] Pit R, Hervet H, Léger L. Direct experimental evidence of slip in hexadecane–solid interfaces. Physical Review Letters. 2000;**85**: 980-983

[28] Kalyan S, Ramarao Y, Mallikarjun P. Effect of first-order chemical reaction on fully developed natural convection of micropolar fluid in a vertical channel. Journal of Applied Science and Computations. 2018;**5** (12):1-12

[29] Kalyan S, Sharan A, Metri R. Note on energy transfer of Jeffrey fluid flow on mixed convection in presence of chemical reaction parameter. International Journal of Advanced Science and Technology. 2019;**28**: 328-349

[30] Kalyan S, Saraswathi H. Impact of magnetic field on Jeffrey fluid flow with baffle. European Chemical Bulletin. 2023;**12**:5408-5418. DOI: 10.31838/ecb/ 2023.12.s3.608

[31] Kalyan S, Sharan A, Chamkha AJ. Heat and mass transfer of two immiscible flows of Jeffrey fluid in a vertical channel. Heat transfer—Asian. Research. 2022;**52**:267-288

[32] Saraswathi H, Kalyan S, Chamkha AJ. Steady Jeffrey fluid through porous media in presence of a baffle in a vertical channel. Journal of Nanofluids. 2023;**12**:1644-1651

[33] Kalyan S, Mng'ang'a J. Numerical study of electrically conducting MHD fluids in a vertical channel with Jeffrey fluid flow and first-order chemical reaction. Thermal Science and Engineering. 2023;**6**:1-13

[34] Madhu J, Kalyan S, Gudagi Y, Varun Kumar RS, Kumar R, Sureshkumar S. Fluid sustainability by the effect of microrotational flow and chemical reactions in a vertical channel. International Journal of Modelling and Simulation. 2024:1-12. DOI: 10.1080/02286203.2024.2319008

[35] Kalyan S, Soliman HAA. Analysis of micropolar immiscible fluid flow on electrically conducting free convective flow in a vertical channel in the presence of chemical reaction. International Journal of Advanced Engineering and Business Sciences. 2021;**2**:25-35

[36] Thompson PA, Robbins MO. Simulations of contact line motion: Slip and the dynamic contact angle. Physical Review Letters. 1989;**63**:766-769

[37] Koplik J, Banavar JR. Corner flow in the sliding plate problem. Physics of Fluids. 1995;**7**:3118-3125

[38] Richardson S. On the no-slip boundary condition. Journal of Fluid Mechanics. 1973;**59**:707-719

[39] Navier CLMH. Mémoire sur les lois du mouvement des fluides. Mémoires de l'Académie Royale des Sciences de l'Institut de France. 1823;**1**:414-416

[40] Willmott G. Dynamics of a sphere with inhomogeneous slip boundary conditions in stokes flow. Physical Review E. 2008;**77**:055302-055305

[41] Ashmawy EA. Slip at the surface of a general axisymmetric body rotating in a viscous fluid. Archives of Mechanics. 2011;**63**:341-361

[42] Yang F. Slip boundary condition for viscous flow over solid surfaces. Chemical Engineering Communications. 2010;**197**:544-550

[43] Sherief HH, Faltas MS, Ashmawy EA. Stokes flow between two confocal rotating spheroids with slip. Archive of Applied Mechanics. 2012;**82**:937-948

[44] Ashmawy EA. Unsteady Couette flow of a micropolar fluid with slip. Meccanica. 2012;**47**:85-94

[45] Hayat T, Shehzad SA, Qasim M. Slip effects on the unsteady stagnation point flow with variable free stream. Walailak Journal of Science and Technology. 2013;**10**:385-394

[46] Khan WA, Khan ZH, Rahi M. Fluid flow and heat transfer of carbon nanotubes along a flat plate with Navier slip boundary. Applied Nanoscience. 2013;**4**:1-9

[47] Lok YY, Pop I, Ingham DB. Oblique stagnation slip flow of a micropolar fluid. Meccanica. 2010;**45**:187-198

[48] Ashmawy EA. Unsteady rotational motion of a slip spherical particle in a viscous fluid. ISRN Mathematical Physics. 2012;**2012**:1-8

[49] Vieru D, Zafar A. Some Couette flows of a Maxwell fluid with wall slip condition. Applied Mathematics & Information Sciences. 2013;**7**:209-219

[50] Khan ZH, Khan WA, Shah IA. MHD boundary layer slip flow and heat transfer of ferrofluid along a stretching cylinder with prescribed heat flux. PLoS One. 2014;**9**:1-6

[51] Gudagi Y, Kalyan S. Effect of heat and mass transfer positions on forced convection with porous medium utilizing nanofluid in a vertical channel. Gradiva Review Journal. 2022;**8**:396-410

Chapter 6

# Plane Stress Problem in Isotropic Poroelasticity

*Malla Reddy Perati*

## Abstract

In this chapter, the plane stress problem in isotropic Poroelasticity is presented. The constitutive relations (stress-displacement relations, stress–strain relations, and equations of motion) are developed on the lines of classical Elasticity. It is to be noted that plane stress conditions here involve the condition of fluid pressure. The fluid pressure gradient in the specified direction is assumed to be zero. The generalized plane stress problem is established to overcome the issues of assumptions made in the case of the plane stress problem. Further, the non-dimensionalization of the poroelastic parameters, which are modified in view of plane stress conditions, is presented. The advantages with non-dimensionalization process are also mentioned. From the numerical example, it is clear that the values of parameters of plane stress conditions are less than those of the usual poroelastic parameters.

**Keywords:** isotropic poroelastic solid, plane stress problem, constitutive relations, equations of motion, non-dimensionalization

## 1. Introduction

In the classification of Elasticity, a branch of Solid Mechanics, it is well known that there are two types of problems. In one type, forces are distributed such that the constitutive relations are independent of one coordinate. An example for this type is the one pertaining to the cylindrical solids where external forces are distributed in such a way that the component of deformation in the direction of the axis of the cylinder vanishes, and the remaining components do not vary along the length of the cylinder. Mathematically, if the axis of the cylinder is along the z-axis of the Cartesian system xyz, then, in the strain matrix, the last row elements are zeros, and other elements are functions of x, y only. In the second type, the thickness of a plate-type solid is so small that stress components with respect to the direction of thickness vanish. The first type is called plain strain problem, and the second type is called plane stress problem. These two situations are exploited in the research of various domains of Elasticity, such as Poroelasticity to make three-dimensional problems two-dimensional, thereby reducing the complexity of the solution method.

This chapter is concerned with the second type (plane stress problem) in isotropic Poroelasticity. The significance of the plane stress problems has been increasing, particularly in the destructive research areas. For example, plates in the aircraft-wing spoilers are made of light-weight poroelastic alloy materials, and body impact resistance study for their design is very important for the stability of the spoilers. The thickness of the plate is small compared to its length. Therefore, it comes under plane stress problems. Moreover, if a human bone is fractured, it will be implanted with some functional metal plate which has favorable characteristics, such as immunity to corrosion, biocompatability, and osseointegrability. The selection of the metal plate depends upon the strength of the bone, which in turn depends upon its dimensions. If the bone is like plate type with small thickness, then the problem comes under plane stress. The usual constitutive relations cannot be employed, and the relations of the plane stress problem are to be used. In the case of isotropic poroelastic solids, the constitutive relations are revised under the plane stress conditions [1]. Recently, the constitutive relations are revised under plane stress conditions in the case of double porosity solids [2]. The double porosity solids contain both pores and fractures that may be filled with different fluids. This chapter is confined only to the plane stress problem in solids of single porosity.

The rest of the chapter is organized as follows: In Section 2, the plane stress problem of isotropic poroelastic solids and the related relations are presented. In Section 3, the generalized plane stress problem is established. In Section 4, the non-dimensionalization process is presented. A numerical example is illustrated in Section 5. In Section 6, the conclusion is presented, and the references are placed at the end.

## 2. Plane stress problem in isotropic poroelastic solids

Plane stress conditions apply to plate-type solids, bounded by two parallel planes (bases), where the distance between the bases is small when compared to other dimensions. A state of stress is said to be in plane if the stress vector in a plane parallel to the bases is zero throughout its volume. Consider an element in an isotropic poroelastic solid (**Figure 1(a)**). If the thickness of the element is comparable to its length and breadth, the problem is three-dimensional (**Figure 1(b)**), and if the thickness of the element is small compared to that of other dimensions, then the problem is plane stress (**Figure 1(c)**).

The stress components $\sigma_{xy}, \sigma_{yy}, \sigma_{yz}$, and $\frac{\partial s}{\partial y}$ on the plane parallel to xz plane are approximately zeros, that is

$$\sigma_{xy} = \sigma_{yy} = \sigma_{yz} = \frac{\partial s}{\partial y} = 0 \tag{1}$$

throughout its volume [1]. The last condition indicates that there is no change of fluid pressure '$s$' in the direction of thickness, that is, there is no leakage of the fluid in the said direction. The governing equations of the plane stress problems in poroelastic solids were not readily found in the literature. Therefore, by following the lines of a purely elastic solid [3], the related equations in isotropic poroelastic solids are derived from the corresponding equations of the general three-dimensional problems [4], and the modified equations are presented in the following:

Let $\overleftarrow{u} = (u, v, w)$ and $\overleftarrow{U} = (U, V, W)$ be the displacement vectors of the solid skeleton and the fluid, respectively, in the plate element of small thickness $2\delta$ (**Figure 1 (c)**). In view of Eq. (1), the stress–strain relations in this case become

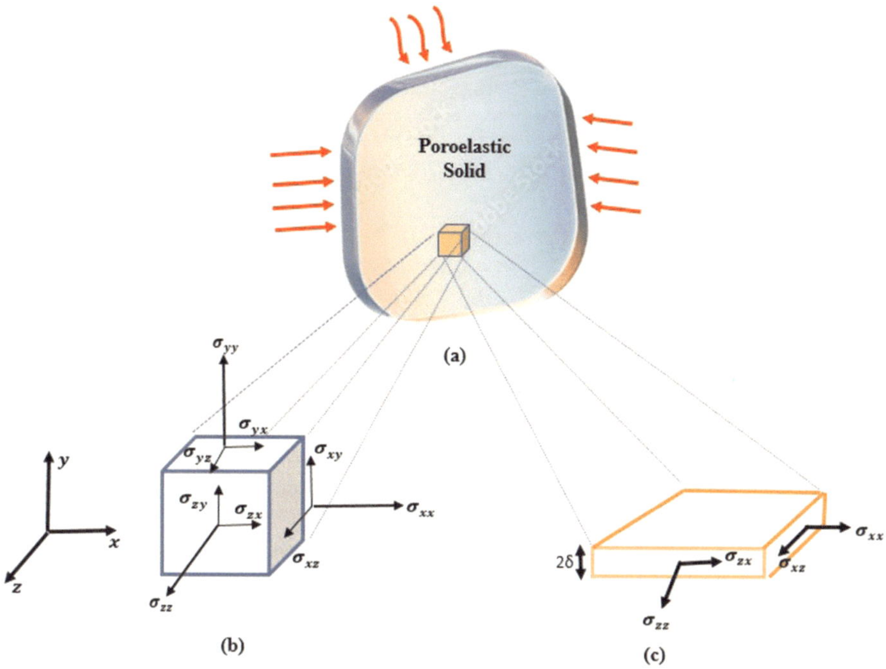


**Figure 1.**
*(a). Plate-type isotropic poroelastic solid, (b). Three-dimensional element (if the thickness is large), and (c). Plane stress problem (if the thickness is small).*

$$\begin{aligned}
\sigma_{xx} &= 2Ne_{xx} + A^*e + Q^*\varepsilon, \\
\sigma_{xz} &= Ne_{xz}, \\
\sigma_{zx} &= 2Ne_{zz} + A^*e + Q^*\varepsilon, \\
s &= Q^*e + R^*\varepsilon, \\
\frac{\partial u}{\partial y} + \frac{\partial v}{\partial x} &= 0, \frac{\partial v}{\partial z} + \frac{\partial w}{\partial y} = 0.
\end{aligned} \tag{2}$$

In the above, $\sigma_{ij}$ are the stress components, and $e_{ij}$ are the strain components, and the symbols $e$ and $\varepsilon$ are the dilatations of the solid and the fluid, respectively, and $A^*, Q^*, R^*$ are the modified poroelastic parameters in a plane stress problem given by

$$\begin{aligned}
A^* &= \frac{A(R+2N) - Q^2}{A + 2Q + R + 2N}, \\
Q^* &= \frac{Q(Q+2N) - AR}{A + 2Q + R + 2N}, \\
R^* &= \frac{R(A+2N) - Q^2}{A + 2Q + R + 2N}.
\end{aligned} \tag{3}$$

In Eq. (3), $A, N, Q, R$ are the poroelastic constants of general isotropic Poroelasticity. $A^*, Q^*, R^*$, and $N$ are all positive in sign, and one can infer from Eq. (3) that $Q^* < Q, R^* < R$; hence, from Eq. (3), it is clear that the liquid pressure $s$ in the plane stress problem, in general, is less than that of the plane strain and the general three-dimensional problems. The strain components $e_{ij}$ in terms of displacements are given below:

$$\begin{aligned} e_{xx} &= \frac{\partial u}{\partial x}, \\ e_{zz} &= \frac{\partial w}{\partial z}, \\ e_{xz} &= \frac{1}{2}\left(\frac{\partial u}{\partial z} + \frac{\partial w}{\partial x}\right), \\ e &= \frac{\partial u}{\partial x} + \frac{\partial w}{\partial z}, \\ \varepsilon &= \frac{\partial u}{\partial x} + \frac{\partial w}{\partial z} \end{aligned} \tag{4}$$

From Eqs. (2) and (4), it is clear that there are three equations in two unknown solid displacement components $u, w$. Therefore, for unique solutions, strain components must satisfy certain conditions called strain compatibility conditions. These conditions can be expressed in terms of stress components and the resultant ones are called stress compatibility conditions [5], For isotropic poroelastic solids, there are three distinct equations which are given below

$$\begin{aligned} \nabla^2\sigma_{xx} + \frac{P^* + A^*}{P^* + 2A^*}\frac{\partial^2\theta}{\partial x^2} &= -\frac{A^*}{P^*} divF^{-1} - 2\frac{\partial F_x}{\partial x} \\ \nabla^2\sigma_{zz} + \frac{P^* + A^*}{P^* + 2A^*}\frac{\partial^2\theta}{\partial z^2} &= -\frac{A^*}{P^*} divF^{-1} - 2\frac{\partial F_z}{\partial z} \\ \nabla^2\sigma_{zx} + \frac{P^* + A^*}{P^* + 2A^*}\frac{\partial^2\theta}{\partial x \partial z} &= -\left(\frac{\partial F_x}{\partial x} + \frac{\partial F_z}{\partial z}\right) \end{aligned} \tag{5}$$

where $\theta = \sigma_{xx} + \sigma_{zz}$ (stress dilatation), and $\overleftarrow{F} = (F_x, 0, F_z)$ is the body force vector and $P^* = A^* + 2N$.

On the other hand, for equilibrium, the resultant force acting on the matter must vanish. The resultant ones are the equations of equilibrium, in this case, they reduce to [5].

$$\begin{aligned} \frac{\partial \sigma_{xx}}{\partial x} + \frac{\partial \sigma_{xz}}{\partial z} &= -F_x \\ \frac{\partial \sigma_{xz}}{\partial x} + \frac{\partial \sigma_{zz}}{\partial z} &= -F_z \end{aligned} \tag{6}$$

From the constitutive relations (2) and (4) and the equilibrium Eq. (6), one can have the following dynamical equations in absense of body forces [1]

$$\begin{aligned} N\nabla^2 u + (A^* + N)\frac{\partial e}{\partial x} + Q^*\frac{\partial \varepsilon}{\partial x} &= \frac{\partial^2}{\partial t^2}(\rho_{11}u + \rho_{12}U) + b\frac{\partial}{\partial t}(u - U), \\ N\nabla^2 w + (A^* + N)\frac{\partial e}{\partial x} + Q^*\frac{\partial \varepsilon}{\partial x} &= \frac{\partial^2}{\partial t^2}(\rho_{11}w + \rho_{12}W) + b\frac{\partial}{\partial t}(w - W), \\ Q^*\frac{\partial e}{\partial x} + R^*\frac{\partial \varepsilon}{\partial x} &= \frac{\partial^2}{\partial t^2}(\rho_{12}u + \rho_{22}U) - b\frac{\partial}{\partial t}(u - U), \\ Q^*\frac{\partial e}{\partial z} + R^*\frac{\partial \varepsilon}{\partial z} &= \frac{\partial^2}{\partial t^2}(\rho_{12}w + \rho_{22}W) - b\frac{\partial}{\partial t}(w - W). \end{aligned} \tag{7}$$

In Eq. (7), $b$ is a dissipative coefficient and $\rho_{ij}$ are the mass coefficients [3] such that the sums $\rho_{11} + \rho_{12}$ and $\rho_{12} + \rho_{22}$ represent the mass of the solid and the fluid per unit volume of the bulk material respectively. The coefficient $\rho_{12}$ is the mass coupling parameter between fluid and solid phases. Furthermore, the mass coefficients obey the inequalities [4].

$$\rho_{11} > 0, \rho_{12} \leq 0, \rho_{22} \geq 0, \left(\rho_{11}\rho_{22} - \rho_{12}^2\right) > 0. \tag{8}$$

Eq. (6) together with Eq. (2) have to be satisfied at every interior point of the body, and on the surface of the body, the stresses must satisfy the boundary conditions. Since the equations of motion (7) are a system of partial differential equations, a set of well-posed boundary conditions must be prescribed to ensure the uniqueness of the solutions. In general, the boundary conditions are of two types: Dirichlet and Neumann. The first type involves solid displacements and fluid pressure, while the second type involves stress components. These conditions can also be alternated to form mixed boundary conditions. In the case of single poroelastic solid with traction free outer surface, all the solid stress components and fluid pressure will be zero. In the case of composite layered solids, the stresses and displacements of the individual materials will be assumed to be intact. Thus, the solutions of a dynamical problem in isotropic poroelastic solids can be completely determined in the case of plane stress problem.

## 3. Generalized plane stress problem

From Eqs. (2) and (3), it is clear that the displacement components and stress components depend on the y-coordinate, which would appear as a parameter in pertinent equations, and the problem is not truly two-dimensional. However, as given in textbooks [5, 6], it is possible to modify the equations such that the problem becomes two-dimensional, as given below.

From **Figure 1**, it is clear that the generators of the plate are parallel to $x$-axis, and the bases are in the planes $y = \pm\delta$, and the plane $y = 0$ is the centre plane of the plate. The tractions on the surfaces $y = \pm\delta$ are taken to be zero, while the loadings on the edge of the plate have no $y$-component and are either independent of $y$ or are symmetrically distributed through the thickness. Under these assumptions, the out-of--the-plane displacement will be an odd function of $y$, that is, $v(x, y, z) = -v(x, -y, z)$ and the points on the middle plane will have no $y$-displacement components, that is $v(x, 0, z) = 0$. These conditions imply that the average value of $v$ denoted by $\overline{v}$ will be zero. That is

$$\overline{v} = \frac{1}{2\delta}\int_{-\delta}^{\delta} v(x, y, z)dy = 0. \tag{9}$$

The stresses on $y = \pm\delta$ can be expressed as

$$\sigma_{xy}(x, \pm\delta, z) = \sigma_{yy}(x, \pm\delta, z) = \sigma_{yz}(x, \pm\delta, z) = 0 \tag{10}$$

The equilibrium equation in the $y$-direction becomes

$$\frac{\partial\sigma_{xy}}{\partial x} + \frac{\partial\sigma_{yy}}{\partial y} + \frac{\partial\sigma_{yz}}{\partial z} + F_y = 0 \tag{11}$$

Eq. (11) is valid throughout the plate, in particular, on the faces of the plate. Therefore, from Eqs. (10) and (11), one can have

$$\frac{\partial \sigma_{xy}}{\partial x}(x, \pm\delta, z) = \frac{\partial \sigma_{yz}}{\partial z}(x, \pm\delta, z) = 0,$$

and consequently

$$\frac{\partial \sigma_{yy}}{\partial y}(x, \pm\delta, z) = 0. \tag{12}$$

For any $\delta_1 > 0$, expanding $\sigma_{yy}$ in Taylor's series, in the neighborhood of $(x, \pm\delta, z)$, one obtains

$$\sigma_{yy}(x, \pm\delta \pm \delta_1, z) = \sigma_{yy}(x, \pm\delta, z) + \delta_1 \sigma_{yy,y}(x, \pm\delta, z) + \frac{\delta_1^2}{2!}\sigma_{yy,yy}(x, \pm\delta, z) + - - - \tag{13}$$

From Eqs. (10), (11), and (12), one can see that $\sigma_{yy}(x, \pm\delta \pm \delta_1, z)$ is of order $\delta_1^2$, and this further justifies that the $\sigma_{yy}$ vanishes throughout the interior of the plate, when $\delta$ is small. The remaining equilibrium equations are

$$\frac{\partial \sigma_{ix}}{\partial x} + \frac{\partial \sigma_{iy}}{\partial y} + \frac{\partial \sigma_{iz}}{\partial z} + F_i = 0, \qquad i = x, z \tag{14}$$

The integration of Eq. (14) with respect to $y$ between the limits $-\delta$ and $\delta$ yields

$$\frac{1}{2\delta}\int_{-\delta}^{\delta}\left(\frac{\partial \sigma_{ix}}{\partial x} + \frac{\partial \sigma_{iy}}{\partial y} + \frac{\partial \sigma_{iz}}{\partial z} + F_i\right)dy = 0, \tag{15}$$

Since

$$\int_{-\delta}^{\delta}\frac{\partial \sigma_{iy}}{\partial y}dy = \sigma_{iy}(x, \pm\delta, z) = 0 \qquad i = x, z, \tag{16}$$

Eqs. (14) and (15) yield

$$\begin{aligned} &\frac{1}{2\delta}\int_{-\delta}^{\delta}\left(\frac{\partial \sigma_{ix}}{\partial x} + \frac{\partial \sigma_{iz}}{\partial z} + F_i\right)dy = 0 \\ \Rightarrow \quad &\frac{\partial \overline{\sigma}_{ix}}{\partial x} + \frac{\partial \overline{\sigma}_{iz}}{\partial z} + \overline{F}_i = 0 \end{aligned} \tag{17}$$

where

$$\begin{aligned} \overline{\sigma}_{ij}(x, z) &= \frac{1}{2\delta}\int_{-\delta}^{\delta}\sigma_{ij}(x, y, z)dy, \quad i, j = x, z \\ \overline{F}_i(x, z) &= \frac{1}{2\delta}\int_{-\delta}^{\delta}F_{ij}(x, y, z)dy, \end{aligned} \tag{18}$$

are the average values of $\sigma_i$ and $F_i$, respectively. Eq. (17) is the equilibrium equation.

From Eqs. (16) and (17), it is pertinent that the average values of the relevant stress components satisfy that the equations of equilibrium without any constraints, as discussed in the first paragraph of this section. Now, repeat the process of the plane stress problem presented in the second section with these average values. The constituent relations (strain displacement relations and stress–strain relations) and equations of motion will be obtained with these average values. In this way, the problem becomes truly two-dimensional, and the inconvenience of the plane stress problem is eliminated.

## 4. Non-dimensionalization

Non-dimensionalization of material parameters is the process in which one parameter is multiplied or divided by some reference value of the same type as the parameter. The significance and advantage of non-dimensionalization are clear from the following:

### 4.1 Buckingham theorem

If there are $n$ variables in a problem, and these $n$ variables contain $l$ primary dimensions, such as $M, L, T$, then the equations relating all the variables will have $n - l$ dimensionless groups.

i. For some poroelastic solid, suppose values of some parameters are available, and values of one or two parameters are not available by any means. Then, the non-dimensionalization results in dimensionless parameters that follow certain equations, so that the dimensionless parameters for which values are not known would be determined.

ii. The non-dimensionalized parameter can be neglected if the denominator value is large, This way, the complexity of the problem will be reduced.

iii. In the case of confidential research, if the researcher does not want to reveal explicit information, then they use non-dimensionalization so that others do not, in general, clearly understand which actual parameters are involved.

In the case of usual problem of isotropic Poroelasticity, the non-dimensionalization is as follows

$$a_1 = \frac{P}{H}, a_2 = \frac{Q}{H}, a_3 = \frac{R}{H}, \tag{19}$$

where $P = A + 2N$, $H = P + 2Q + R$, one can see that $a_1 + 2a_2 + a_3 = 1$. The modified poroelastic constants $A^*, Q^*, R^*$ take the following forms in terms of non-dimensional parameters

$$\begin{aligned}
A^* &= \frac{A(R+2N) - Q^2}{A + 2Q + R + 2N} = \frac{\left(\frac{P-2N}{H}\right)\left(\frac{R}{H} + \frac{2N}{H}\right) - \frac{Q^2}{H}}{\left(\frac{P-2N}{H}\right) + 2\frac{Q}{H} + \frac{R}{H} + 2\frac{N}{H}} \\
A^* &= \frac{(a_1 - 2a_4)(a_3 + 2a_4) - a_2^2}{(a_1 - 2a_4) + 2a_2 + a_3 + 2a_4}, \\
Q^* &= \frac{Q(Q+2N) - AR}{A + 2Q + R + 2N} = \frac{a_2(a_2 + 2a_4) - (a_1 - 2a_4)a_3}{(a_1 - 2a - 4) + 2a_2 + a_3 + 2a_4}, \\
R^* &= \frac{R(A+2N) - Q^2}{A + 2Q + R + 2N} = \frac{a_3 a_1 - a_2^2}{(a_1 - 2a_4) + 2a_2 + a_3 + 2a_4}.
\end{aligned} \tag{20}$$

## 5. Numerical example

The values of isotropic poroelastic parameters $A, N, Q, R$ for bone element are computed from the values of its porosity, jacketed compressibility, unjacketed compressibility, and fluid content, which are experimentally determined [7]. The values are as follows

$$\begin{aligned} A &= 3.126 \times 10^6 lb/inch^2, \quad N = 1.17 \times 10^6 lb/inch^2 \\ Q &= 0.636 \times 10^6 inch^2/lb, \quad R = 0.247 \times 10^6 inch^2/lb \end{aligned} \tag{21}$$

If the thickness of a particular plate-type bone is small, then the plane stress conditions are applicable. Using relation (3), the values of the modified poroelastic parameters are computed, which are given below

$$\begin{aligned} A^* &= 1.0999 \times 10^6 lb/inch^2, \quad N = 1.17 \times 10^6 lb/inch^2 \\ Q^* &= 1.1604 \times 10^6 inch^2/lb, \quad R = 0.1356 \times 10^6 inch^2/lb \end{aligned} \tag{22}$$

From the parameter values of (21) and (22), it is very clear that the values pertaining to a thin plate-type bone are less than that of any general bone, except the case of the shear modulus, as there is no change in it. This kind of information is very useful in the implantation of fractured bone.

## 6. Conclusion

This chapter is concerned with the plane stress problem in isotropic poroelastic solids. The modified constitutive relations, namely, the strain–displacement relation, stress–strain relations, and the equations of motion are presented. The average values of stress and body force components are introduced to overcome issues due to approximations in the plane stress conditions. The compatibility conditions are also derived for the uniqueness of solutions in the case of isotropic poroelastic solids. Further, the non-dimensionalization is discussed. It is hoped that this information will help advance research in the domain of Poroelasticity.

## Author details

Malla Reddy Perati
Department of Mathematics, Kakatiya University, Warangal, TG, India

*Address all correspondence to: mperati@kakatiya.ac.in

## References

[1] Malla Reddy P, Tajuddin M. Edge waves in Poroelastic plate under plane stress conditions. The Journal of the Acoustical Society of America. 2003; **114**(1):185-193. DOI: 10.1121/1.1569258

[2] Venkanna D, Malla RP. Study of symmetric and symmetrical edge waves in double porosity semi infinite thin plate plane stress problem. Mechanics of Solids. 2022;**57**(3):629-643

[3] Kumar S. Edge waves in plates. In: Proceedings of International Symposium Stress Wave Propagation in Materials. New York: Inderscience; 1960

[4] Biot MA. Theory of propagation of elastic waves in a fluid saturated porous solid. I. Low frequency range, reprinted from the. The Journal of the Acoustical Society of America. 1956;**28**(2):168-178

[5] Sokolnikoff IS. Mathematical Theory of Elasticity. 2nd ed. New York: Mc Graw-Hill Book Company Ltd; 1956

[6] Sadd M, H. Elasticity: Theory, Applications and Numbers. 1st ed. New Delhi, India: Academic Press (An Imprint of Elsevier); 2005

[7] Nowinski JL, Davis CF. Propagation of longitudinal waves in circularity cylindrical bone element, transactions of ASME. Journal of Applied Mechanics. 1971;**38**(3):578-584

Section 2

# Engineering Applications

Chapter 7

# Flow Boiling Heat Transfer Characteristics and Enhancement Mechanisms in Gradient Porous Media

*Chuang-Yao Zhao, Li-Yuan Yang, Yu-Ming Guo and Chen-Min Zheng*

## Abstract

The escalating heat flux in modern high-power electronics and energy systems poses major challenges for thermal management solutions. Among various cooling techniques, flow boiling is regarded as one of the most effective methods due to its ability to dissipate large amounts of heat through phase change. In this context, porous media have drawn increasing attention for their potential to significantly enhance flow boiling heat transfer, owing to their high surface area, capillary action, and ability to promote bubble nucleation and liquid replenishment. Compared to experimental approaches, numerical simulations offer advantages such as lower cost and higher efficiency, making them an effective tool for analysing the macroscopic heat transfer behaviour in porous structures. This chapter presents a systematic analysis of flow and heat transfer within porous media. First, the mathematical models and calculation methods applicable to porous media are introduced, including the Darcy-Forchheimer model and the mass transfer model. Next, the influence of porous media placement on heat transfer performance and the underlying mechanisms of boiling enhancement in gradient porous media (GPM) are highlighted. The results indicate that axially filled negative GPMs exhibit superior flow boiling heat transfer performance compared to homogeneous porous media (HPMs) and positive GPM structures. Furthermore, by considering both heat transfer augmentation and flow resistance, a comprehensive evaluation of thermal–hydraulic performance was conducted, revealing that an optimal filling ratio is achieved near $H_p^* = 0.6$.

**Keywords:** gradient porous media, homogeneous porous media, enhanced heat transfer, flow boiling, numerical simulation

## 1. Introduction

### 1.1 Porous media

Porous media consist of a solid framework with interconnected voids, forming a heterogeneous structure that facilitates fluid flow and heat transfer. They are typically characterised by a multiphase composite structure, containing both solid and pore phases [1]. Porous media can also be described as solid materials with a multitude of pores, which may either be interconnected to form a network or isolated from each other. The unique internal pore structure endows these materials with excellent capabilities for heat and mass transfer, rendering them highly versatile for a wide range of industrial applications. In the energy and power sectors [2–4], porous media are widely employed as a high-efficiency heat dissipation element, significantly enhancing the boiling heat transfer performance in high-power electronic devices and thermal systems.

## 1.2 Advantages of heat transfer in porous media

The structural diversity of porous media directly influences their heat transfer characteristics. Generally, porous media can be categorised into two distinct classifications: HPM and GPM. Heat transfer in porous material occurs through a combination of mechanisms, including heat conduction in the solid matrix, convective heat transfer in the pore spaces, radiative heat transfer, and latent heat transfer associated with phase change. The contribution of each mechanism varies depending on the specific application scenario.

In the context of boiling heat transfer, the enhancement mechanisms of porous media are mainly reflected in three aspects [5, 6]: (1) a substantial increase in the effective heat transfer surface area; (2) the provision of abundant nucleation sites for vapourization; and (3) the enhancement of capillary-driven liquid replenishment.

Compared to HPM, GPM offers significant advantages in both heat transfer and flow regulation, due to their tailored pore size and distribution. Gradient porous structures can be designed with spatially varying porosities and pore dimensions, leading to improved overall performance. This design not only provides a larger specific surface area to facilitate efficient heat transfer but also helps reduce flow resistance through optimised structural arrangement.

## 1.3 Research status of flow and heat transfer in porous media

In recent years, GPMs have become a research hotspot in the field of thermal fluids, due to their tuneable pore structures and superior heat transfer performance. With the maturity of 3D manufacturing technologies [7], the precise fabrication of GPM has become feasible, enabling superior performance in both single-phase and boiling heat transfer.

Chen et al. [8] investigated the differences in hydrodynamic and heat transfer behaviours of channels with variable and fixed gradients of porous structures. Their results demonstrated that the variable gradient configuration exhibits higher internal flow velocities and more pronounced temperature variations compared to the fixed gradient case. Yang et al. [9] proposed a simplified phase change model for GPM to address the slow heat transfer rate and low computational efficiency associated with latent thermal energy storage systems. The results showed that increasing the quantity of gradient layers improves the heat transfer performance of the latent heat storage unit, with optimal enhancement achieved at four layers—reducing solidification time by 7.62% and enhancing heat transfer efficiency by 8.25% compared to homogeneous structures.

Shi et al. [10] studied the combined effects of porosity and pore size gradients under various filling ratios and found that, at a filling ratio of 0.7, structures with radially decreasing pore size outperformed homogeneous media in thermal performance. Asiaei et al. [11] numerically investigated mixed convection in a square chamber with a rotating heater and cooler, and observed that GPM effectively suppressed vortex formation near the walls, thereby enhancing heat transfer. Compared to high-permeability HPM, the heat transfer efficiency was improved by 17%. Zheng et al. [12] divided a circular tube into concentric layers filled with porous media of varying porosities and found that the number of porous layers significantly influenced thermal–hydraulic performance, with continuous improvement as layering increased. Wang et al. [13, 14] analysed the impact of axial and radial porosity gradients on the thermal performance of heat exchangers. Their findings indicated that, in contrast to HPM, GPM configurations reduced both the friction factor and overall thermal resistance. Mo et al. [15] experimentally compared HPM with radially graded GPM in pool boiling experiments, showing that the latter offered superior heat transfer performance.

In summary, due to their pore gradient distribution, GPM are more effective in balancing flow resistance and heat transfer enhancement, leading to superior performance in both single-phase and boiling heat transfer scenarios. However, challenges remain in fully understanding the phase-change mechanisms, critical heat flux limits, and flow pattern stability under flow boiling conditions. In this chapter, the heat transfer behaviour of GPM partially filled in the centre of a channel is investigated through numerical simulation, using the Volume of Fluid (VOF) method and the Tanasawa phase-change model. Both qualitative and quantitative analyses are conducted to elucidate the mechanisms of enhanced heat transfer during flow boiling.

## 2. Numerical model

### 2.1 Governing equations

In order to simplify model, the following assumptions are made: (1) The fluid is assumed to be incompressible and in a steady laminar flow; (2) The Darcy-Forchheimer model [16] is employed to simulate the flow in the porous medium; (3) Local thermal equilibrium exists between the solid and liquid phases within the porous medium; (4) Both the fluid and the porous medium possess constant thermophysical properties. Depending on the given conditions, the flow dynamics and heat transfer are modelled by the following equations:

The equation is:

$$\nabla \cdot \mathbf{v} = \dot{m}\left(\frac{1}{\rho_{\mathrm{v}}} - \frac{1}{\rho_{\mathrm{l}}}\right), \tag{1}$$

$$\frac{\partial \rho \mathbf{v}}{\partial t} + \nabla \cdot (\rho \mathbf{v}\mathbf{v}) = -\nabla p + \nabla \cdot (\mu \nabla \mathbf{v}) + \rho \mathbf{g} + \sigma \kappa \nabla \cdot \alpha_{\mathrm{l}} + S_{\mathrm{p}} m, \tag{2}$$

$$\frac{\partial T}{\partial t} + D_{\mathrm{c}} \nabla \cdot (\mathbf{v} T) = \nabla \cdot (D_{\mathrm{k}} \nabla T) - D_{\mathrm{c}} \dot{m}\left(h_{\mathrm{fg}} + \left(c_{p\mathrm{l}} - c_{p\mathrm{v}}\right) T_{\mathrm{sat}}\right), \tag{3}$$

where $\mathbf{v}$ is the velocity vector, $\dot{m}$ is the mass transfer rate, $\rho_{\mathrm{v}}$ and $\rho_{\mathrm{l}}$ are the vapour phase density and liquid phase density, respectively. p is pressure, $\rho\mathbf{g}$ is gravitational body forces, $\sigma$, $\kappa$ and $\alpha_{\mathrm{l}}$ are surface tension, surface curvature and liquid volume fraction, $S_{\mathrm{p}}$ is the source term caused by porous medium. $h_{\mathrm{fg}}$ is the latent heat during boiling, and $c_p$ is the specific heat. Generally, compared with the latent heat, the value of $(c_{pl} - c_{pv})\, T_{sat}$ can be neglected.

According to the VOF model [17], the thermal properties of the fluid in the above equation are calculated as follows:

$$y = \alpha_l y_l + (1.0 - \alpha_l) y_v \quad y \in [\rho, \mu, c_p], \tag{4}$$

where $\rho$ is material density, $\mu$ is viscosity, and $c_p$ is specific heat capacity.

### 2.2 Darcy-Forchheimer model

The momentum sink term $S_p$ in the porous region combines viscous drag (Darcy term) and inertial losses (Forchheimer term):

$$S_p = -\frac{\mu}{K}\mathbf{v} - \frac{C_E}{\sqrt{K}}\rho|\mathbf{v}|\mathbf{v}, \tag{5}$$

where $C_E$ and $K$ are respectively the Ergun coefficient of the porous medium and the permeability of the porous medium [18], which are defined as:

$$C_E = \frac{1.75}{\sqrt{150\varepsilon^3}}, \tag{6}$$

$$K = \frac{d^2\varepsilon^3}{122(1-\varepsilon)^2}. \tag{7}$$

In Eq. (2), the expression for the surface curvature $\kappa$ is:

$$\kappa = -\nabla \cdot \left(\frac{\nabla \alpha_l}{|\nabla \alpha_l|}\right). \tag{8}$$

In Eq. (3), $D_k$ and $D_c$ are defined as:

$$D_c = \frac{(\rho c_p)_f}{(\rho c_p)_m}, \tag{9}$$

$$D_k = \frac{\lambda_{eff}}{(\rho c_p)_m}, \tag{10}$$

where $(\rho c_p)_m$ and $(\rho c_p)_f$ are the volumetric heat capacities of the porous medium and the fluid, respectively [19], and they are defined as:

$$(\rho c_p)_m = (1-\varepsilon)(\rho c_p)_s + \varepsilon(\rho c_p)_f, \tag{11}$$

$$(\rho c_p)_f = \rho_l c_{pl}\alpha_l + \rho_v c_{pv}(1-\alpha_l). \tag{12}$$

The subscripts f and s represent the liquid phase and solid phase, respectively; v and l represent the vapour phase and liquid phase, respectively; m is the average value.

The effective thermal conductivity coefficient ($\lambda_{eff}$) of a porous medium is defined by Ref. [19]:

$$\lambda_{\text{eff}} = \lambda_{\text{f}} \left[ \frac{(\lambda_{\text{f}} + \lambda_{\text{s}}) - (1-\varepsilon)(\lambda_{\text{f}} - \lambda_{\text{s}})}{(\lambda_{\text{f}} + \lambda_{\text{s}}) + (1-\varepsilon)(\lambda_{\text{f}} - \lambda_{\text{s}})} \right]. \tag{13}$$

The fluid thermal conductivity($\lambda_{\text{f}}$) is defined as:

$$\lambda_{\text{f}} = \alpha_{\text{l}}\lambda_{\text{l}} + (1-\alpha_{\text{l}})\lambda_{\text{v}}. \tag{14}$$

### 2.3 Mass transfer model

The transport equation for volume fraction is:

$$\frac{\partial \alpha_{\text{l}}}{\partial t} + \mathbf{v}\nabla \cdot (\alpha_{\text{l}}) = -\dot{m}\left(\frac{1}{\rho_{\text{l}}} - \alpha_{\text{l}}\left(\frac{1}{\rho_{\text{l}}} - \frac{1}{\rho_{\text{v}}}\right)\right), \tag{15}$$

where $\dot{m}$ is the mass source term.

The Tanasawa model [20], derived from the Schrage theory, is used to simulate phase transition in this study. This model has been successfully used in reproducing the evaporative falling film outside a horizontal tube in our previous study [21]. The volumetric mass transfer rate at the liquid-vapour interface can be calculated by:

$$\dot{m} = \frac{2\gamma}{2-\gamma}\sqrt{\frac{1}{2\pi R}}\frac{\rho_{\text{v}}h_{\text{fg}}(T - T_{\text{sat}})}{T_{\text{sat}}^{3/2}}|\nabla\alpha_{\text{l}}|, \tag{16}$$

where $\gamma$ represents the molecular weight of the liquid to the vapour during the phase change process, and its value ranges between 0.1 and 1.0 [22]. The present analysis suggests a value of 0.62.

By simplifying the Clausius-Clapeyron equation, the relationship between the change in the local saturation temperature ($T_{\text{sat, local}}$) and the local pressure ($p$) can be derived.

A simplified form of the Clausius-Clapeyron equation allows for the determination of how local pressure ($p$) influences the saturation temperature ($T_{\text{sat, local}}$),

$$T_{\text{sat,local}} = \frac{1}{\frac{1}{T_{\text{sat}}} - \frac{R}{h_{\text{fg}}}\ln\left(\frac{p}{p_{\text{sat}}}\right)}, \tag{17}$$

where $R$ denotes the gas constant.

## 3. Analysis and discussion

### 3.1 Porous media filling position inside the channel

In the field of fluid flow and heat transfer, the spatial arrangement of porous media within a channel plays a critical role in determining thermohydraulic performance. For channels partially filled with porous media, traditional investigations have primarily focused on two simplified configurations: parallel-plate flow and axisymmetric flow. These models are well suited for cylindrical channels with porous inserts but are clearly obvious limited when applied to square channels. In a circular tube, heat transfer characteristics under uniform thermal boundary conditions typically exhibit symmetrical distributions around the periphery. However, in square channels, non-uniform heat transfer

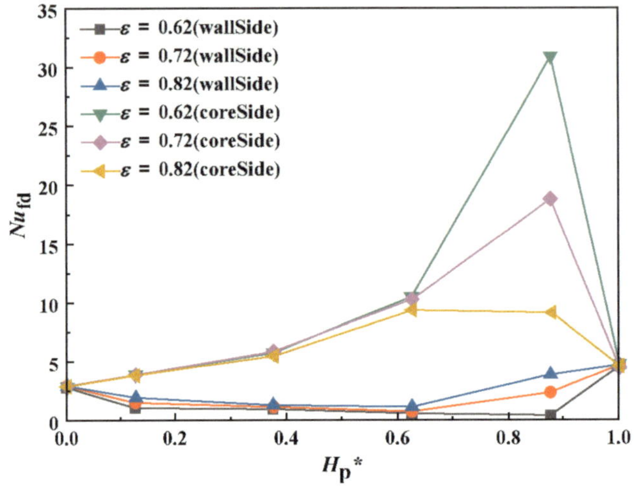


**Figure 1.**
*Fully developed Nusselt number comparison.*

behaviour is observed even under the same boundary conditions. Based on this understanding, Bibin et al. [23, 24] emphasised that special attention must be paid to the insertion position of the porous medium when analysing flow in square channels.

**Figure 1** compares the fully developed Nusselt number ($Nu_{fd}$) for filling ratios $H_p^*$ = 0.2–1.0. It is evident that the configuration with the porous medium located at the channel centre yields higher $Nu_{fd}$ values than when it is placed near the wall. This improvement is attributed to the enhanced convective heat transfer near the heated wall due to core-side filling. Additionally, results show that for wall-side fillings, $Nu_{fd}$ initially decreases with increasing porous layer thickness, reaching a minimum before increasing again with further increase in $H_p^*$. **Figure 2** shows that the pressure drop in the core-side configuration is consistently lower than that in the wall-side

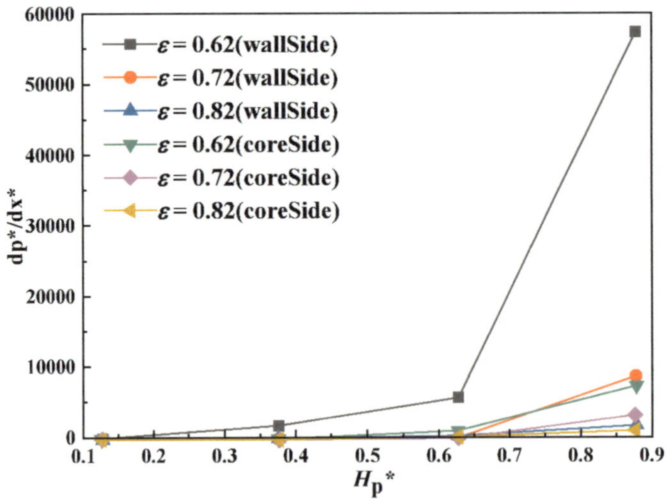


**Figure 2.**
*Pressure drop comparison.*

configuration for the same $H_p^*$. Therefore, the core-side placement is generally recommended to enhance heat transfer performance while minimising pressure loss.

## 3.2 Characterisation of boiling heat transfer in GPM

### *3.2.1 Model parameters and solver validation*

Building upon the square channel model in Section 3.1, this section investigates how GPM influences heat transfer characteristics when the porous insert is centrally aligned within the channel. The computational domain comprises a vertical square channel ($H \times H \times L$ = 10 mm × 10 mm × 200 mm), as illustrated in **Figure 3**. The porous insert has dimensions $H_p \times H_p \times L$ and the dimensionless filling ratio is defined as $H_p^* = H_p^*/H$ [25]. Copper is selected as the porous material because of its high thermal conductivity. Propane (R290) is chosen as the working fluid for its widespread use in heat pumps and heat pipes [23, 26]. Thermophysical properties are obtained from Ref. [27], as shown in **Table 1**.

To quantify spatial variation in porosity, a porosity gradient Δ? is introduced and defined as:

$$\Delta\varepsilon = \varepsilon_{\text{out}} - \varepsilon_{\text{in}}. \tag{18}$$

where $\varepsilon_{\text{in}}$ and $\varepsilon_{\text{out}}$ represent the intlet and outlet porosity, respectively. The porosity distribution is illustrated in **Figure 4**.

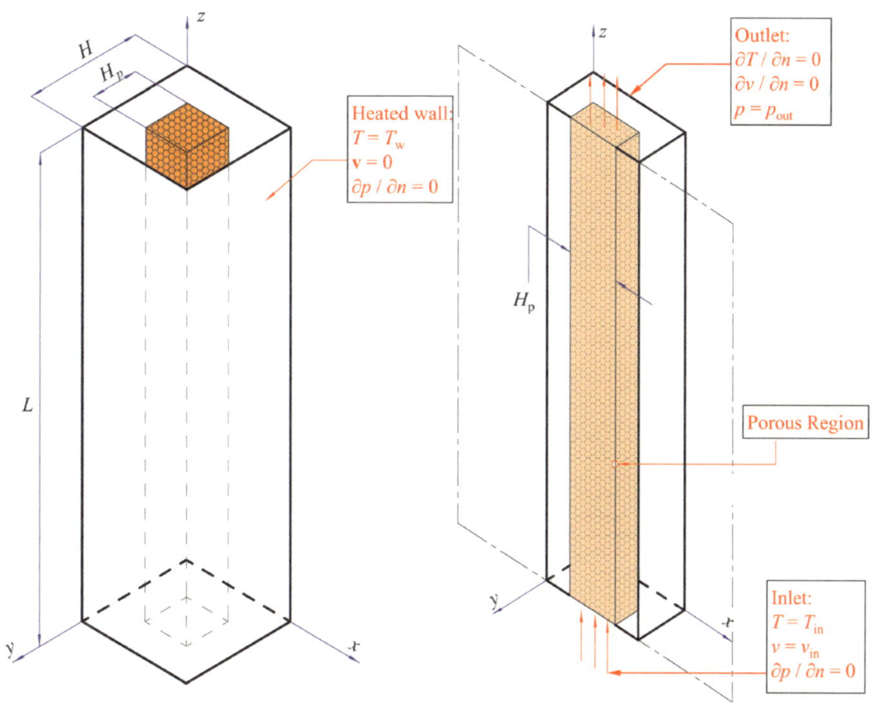


**Figure 3.**
*Square channel partially central filled by gradient porous media physical models.*

| Property | Liquid | Vapour | Porous |
|---|---|---|---|
| $\rho$/kg·m$^{-3}$ | 446.55 | 39.7755 | 8978.0 |
| $c_p$/J·kg$^{-1}$·K$^{-1}$ | 3067.37 | 2431.02 | 381.0 |
| $\lambda$/W·m$^{-1}$·K$^{-1}$ | 0.081 | 0.024 | 387.6 |
| $\nu$/m$^2$·s$^{-1}$ | 1.5725e$^{-7}$ | 2.5067e$^{-7}$ | / |
| $\sigma$/N·m$^{-1}$ | 0.0589 | / | / |
| $h_{fg}$/kJ·kg$^{-1}$ | 420 | / | / |

**Table 1.**
*Thermophysical properties of porous medium and fluid (R290 at saturation* $p_{sat}$ *= 1.75 MPa).*

In this study, OpenFOAM open source CFD software is used, which allows users to freely access and modify its source code and to develop derivative products. Based on this platform Zhao et al. [25] developed a multiphase solver called "GPMPhase-ChangeFoam" to simulate phase change heat transfer in GPM.

The solver was first validated for the permeability properties of porous media using the dam-break model [28]. Next, its phase-change capabilities were verified through the Stefan problem [26]. Finally, the model's ability to simultaneously handle hydrodynamics and phase-change heat transfer in porous media was confirmed using the heat-pipe evaporation problem [29], ensuring overall reliability. The errors of all three validation models are within 7%, which provides a robust framework to fully test the capabilities of the new solver, thus ensuring the generalizability of the model.

### *3.2.2 Temperature distribution in a square channel*

**Figure 5** depicts dimensionless temperature ($T^*$) contours along the channel's diagonal plane for porosity gradients ($\Delta\varepsilon$ = 0 to 0.3) at varying filling ratios ($H_p^*$).

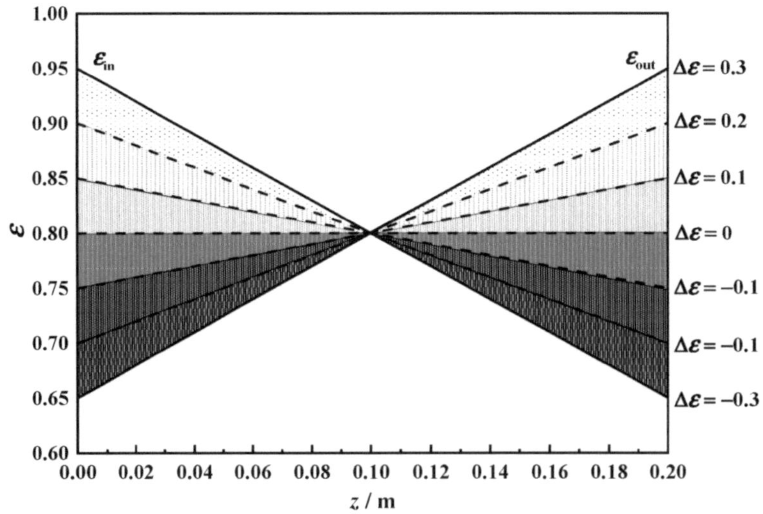


**Figure 4.**
*The porosity arrangement of gradient porous media.*

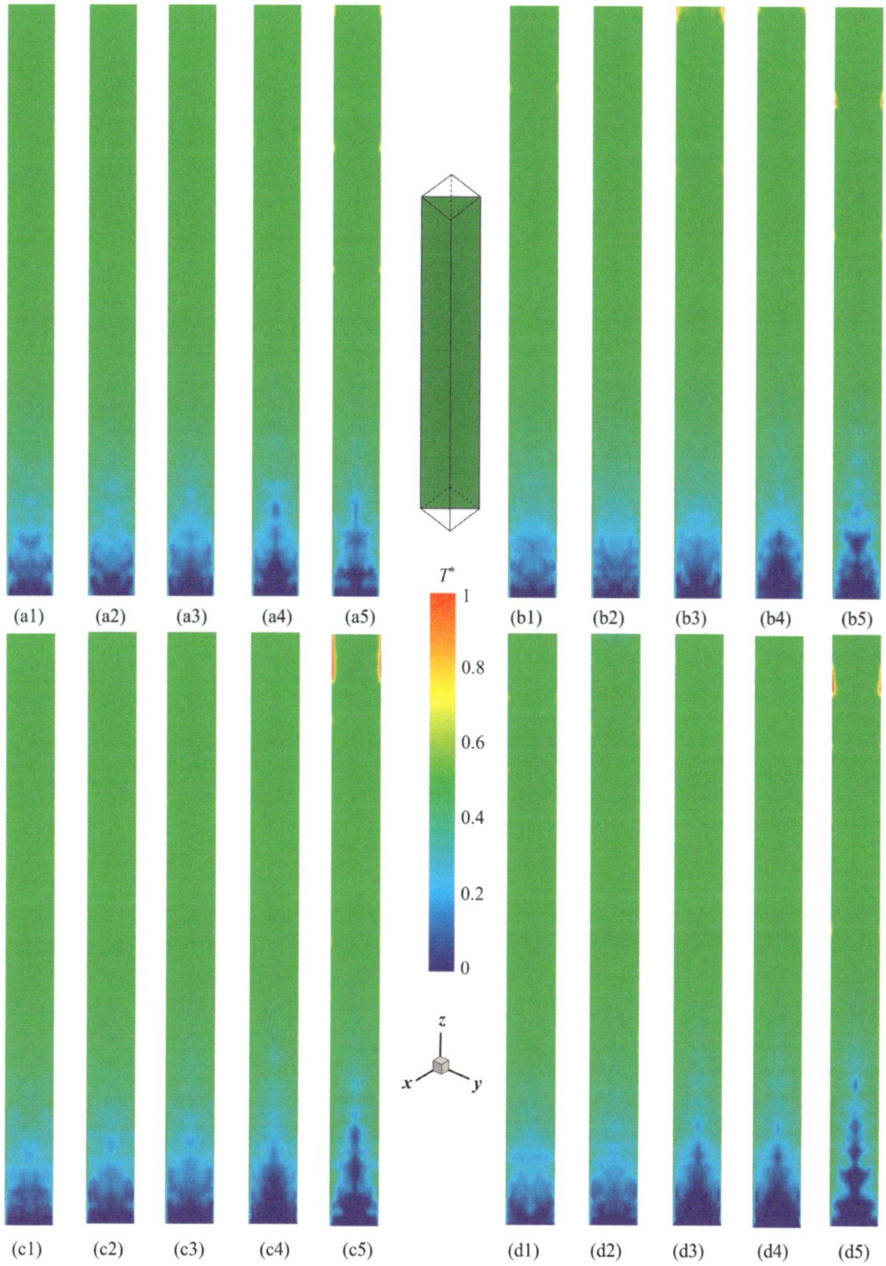


**Figure 5.**
*Contours of dimensionless temperature* ($T^*$) *at different filling ratios for* $\Delta\varepsilon$ *at diagonal sections along the channel, (a)* $\Delta\varepsilon = 0$, *(b)* $\Delta\varepsilon = 0.1$, *(c)* $\Delta\varepsilon = 0.2$, *and (d)* $\Delta\varepsilon = 0.3$; *(1)* $H_p^* = 0.2$, *(2)* $H_p^* = 0.4$, *(3)* $H_p^* = 0.6$, *(4)* $H_p^* = 0.8$, *and (5)* $H_p^* = 1.0$.

When $\Delta\varepsilon = 0$, increasing $H_p^*$ from 0.2 to 1.0 leads to a notable rise in fluid temperature near the heated wall at the inlet, as shown in **Figure 5(a)**. This can be attributed to two coupled mechanisms: (1) the increased thickness of the porous layer enhances local heat transfer; (2) increased flow resistance causes lateral fluid migration from

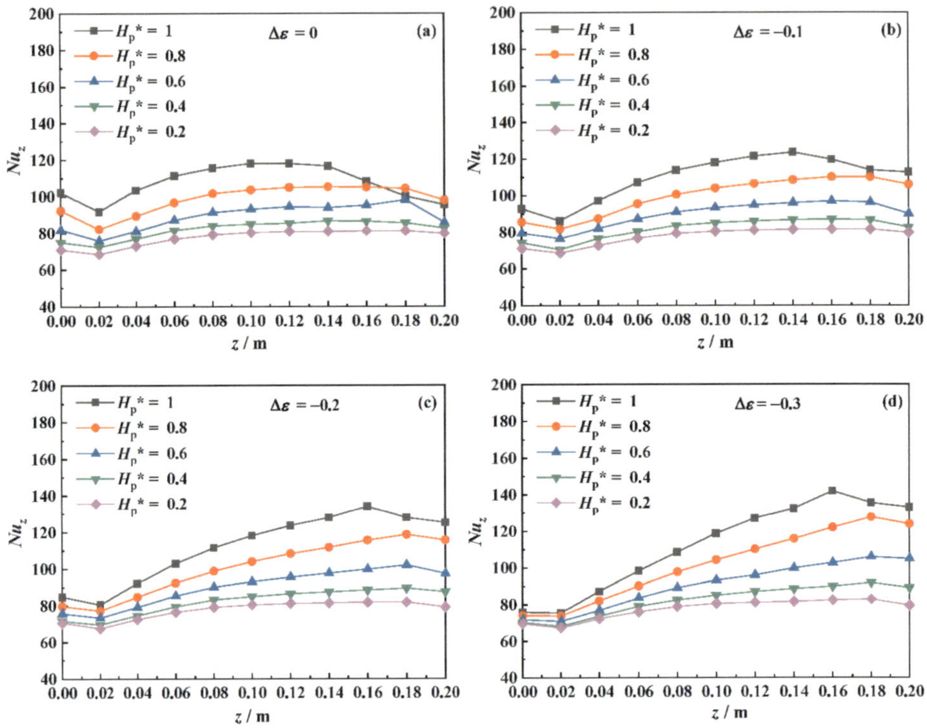


**Figure 6.**
*Plots of local Nusselt number* ($Nu_z$) *over the channel length for (a)* $\Delta\varepsilon = 0$, *(b)* $\Delta\varepsilon = -0.1$, *(c)* $\Delta\varepsilon = -0.2$, *and (d)* $\Delta\varepsilon = -0.3$.

the high-resistance porous region to the low-resistance region, thereby thinning the thermal boundary layer and accelerating the heating of subcooled liquid.

As $\Delta\varepsilon$ increases, this effect becomes more pronounced (**Figure 5c–d**). Additionally, with higher $H_p^*$, the low-temperature zone at the inlet (blue region) expands, particularly at higher $\Delta\varepsilon$. In contrast, a high-temperature zone appears near the outlet when $\Delta\varepsilon \neq 0$, particularly when $H_p^* = 1.0$.

**Figure 7(c–d)** presents the temperature field for negative GPM ($\Delta\varepsilon = -0.1$ to $-0.3$). These results differ significantly from those under positive gradients due to the higher initial porosity at the inlet, which reduces Darcy resistance and suppresses lateral fluid migration from the porous zone to the unfilled zone.

### *3.2.3 Localised Nussle number comparison*

The thermodynamic properties of liquid flow depend ultimately on the heat transfer properties. **Figure 8(a–d)** shows the local Nusselt number ($Nu_z$) along the channel length ($z$) for $H_p^*$ between 0.2 and 1.0, and $\Delta\varepsilon$ between 0 and 0.3. When $\Delta\varepsilon = 0$, as shown in **Figure 8(a)**, $Nu_z$ initially decreases in the range of z = 0 to 0.2 in the inlet region, mainly due to the fact that the single-phase fluid is dominated by the sensible heat transfer, and the $Nu_z$ follows the pattern of the classical heat development region; when the fluid temperature rises up to the saturation point in the flow direction, the high specific surface area of the porous structure and the microscale effect significantly promotes the vapour nucleation, and the porosity gradient further accelerates

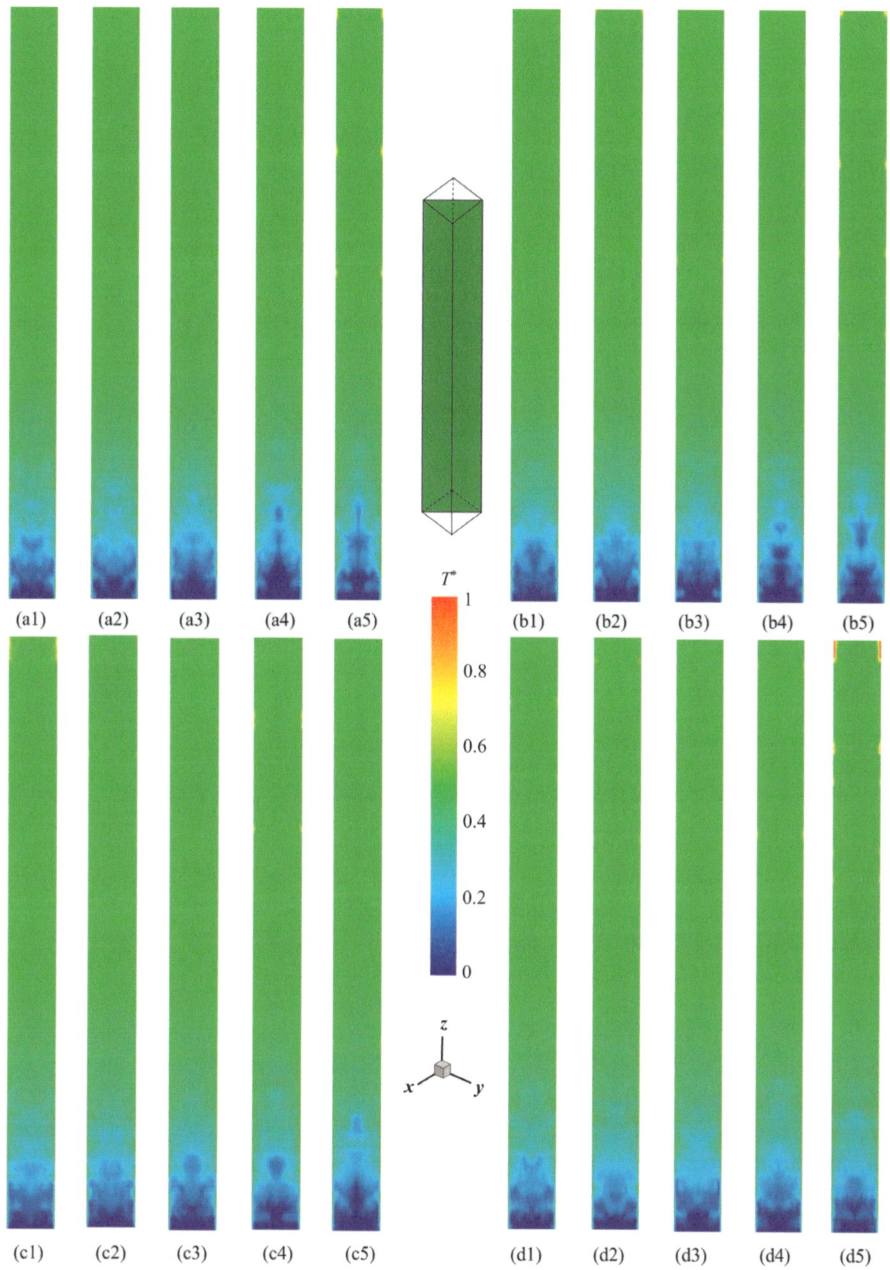


**Figure 7.**
*Contour plots of dimensionless temperature ($T^*$) at different filling ratios for Δε at diagonal slices along the channel space, (a) Δε = 0, (b) Δε = −0.1, (c) Δε = −0.2, and (d) Δε = −0.3; (1) $H_p^*$ = 0.2, (2) $H_p^*$ = 0.4, (3) $H_p^*$ = 0.6, (4) $H_p^*$ = 0.8, and (5) $H_p^*$ = 1.0.*

the phase change process, so that the latent heat transfer gradually dominates. At this time, the turbulence enhancement generated by the porous medium and the local heat balance effect work together to lead to a sharp increase in the $Nu_z$ number. However, in the outlet region of the channel, due to the aggregation of bubbles to form an air film covering the thermal surface, leading to a significant increase in the thermal

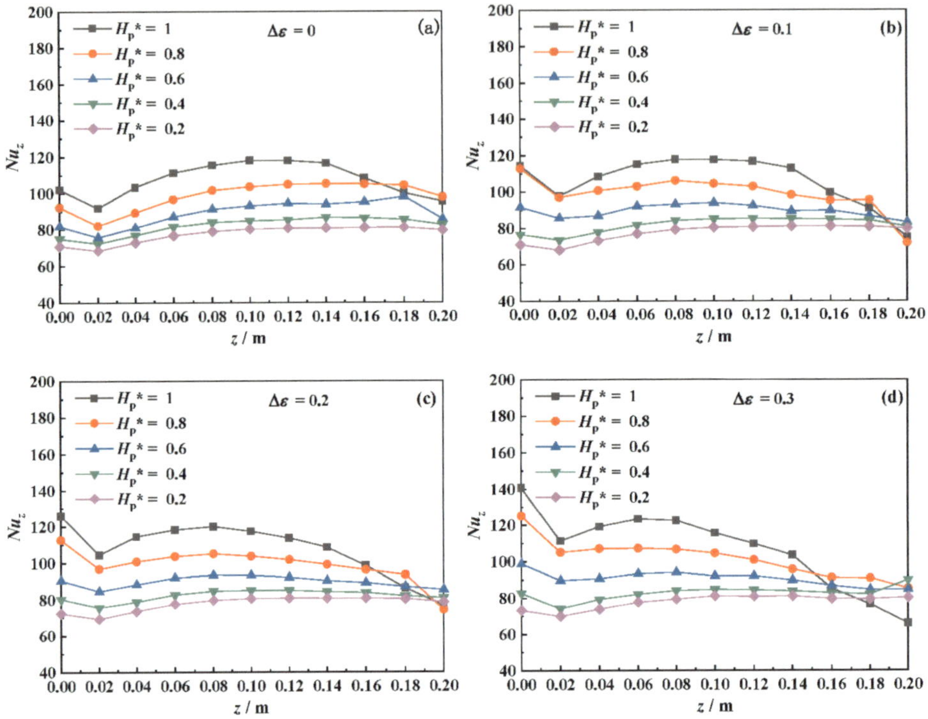


**Figure 8.**
*Plots of local Nusselt number* ($Nu_z$) *over the channel length for (a)* $\Delta\varepsilon = 0$, *(b)* $\Delta\varepsilon = 0.1$, *(c)* $\Delta\varepsilon = 0.2$, *and (d)* $\Delta\varepsilon = 0.3$.

resistance to heat transfer, the $Nu_z$ number decreases, which results in the formation of an obvious peak heat transfer characteristic. However, when $H_p^* = 1.0$, $Nu_z$ decreases sharply upstream of the channel because as the fluid advances along the channel, the dispersed bubbles gradually gather and coalescence, which reduces heat transfer efficiency due to local dryout. This change in trend is consistent with the pattern of change in Ref. [23].

A similar trend to **Figure 8(a)** is observed in **Figure 6** for negative porosity gradient, though distinct differences emerge. As $\Delta\varepsilon$ becomes more negative, $Nu_z$ curves converge near the inlet and diverge near the outlet, with these trends becoming more pronounced when $\Delta\varepsilon$ decreases from $-0.1$ to $-0.3$. This is attributed to the higher porosity in the inlet zone, which reduces the flow resistance, decreases the contact area between the fluid and the solid surface, and reduces the heat transfer performance. On the contrary, lower porosity in the outlet zone increases the flow resistance and improves the heat transfer performance. In addition, changes in $H_p^*$ amplify these trends.

### *3.2.4 Comparison of Nussle numbers in the fully developed segment*

**Figure 9** compares fully developed Nusselt numbers ($Nu_{fd}$) versus the filling ratio ($H_p^*$) for different GPM configurations. In all cases, negative porosity gradients yield higher $Nu_{fd}$ than positive ones. This is because flow from the high-porosity region at the inlet to the low-porosity region at the outlet increases fluid velocity (due to mass conservation), enhancing turbulence and promoting vapour generation. In addition,

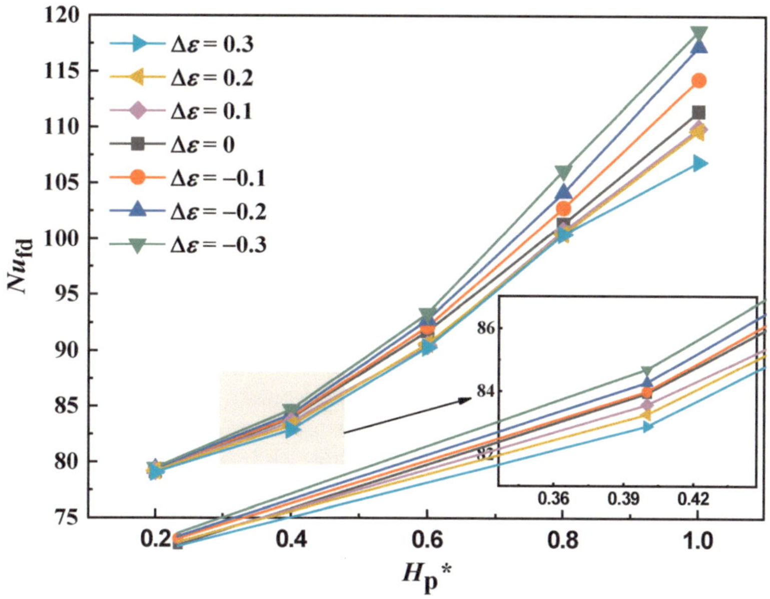


**Figure 9.**
*Fully developed Nusselt number comparison.*

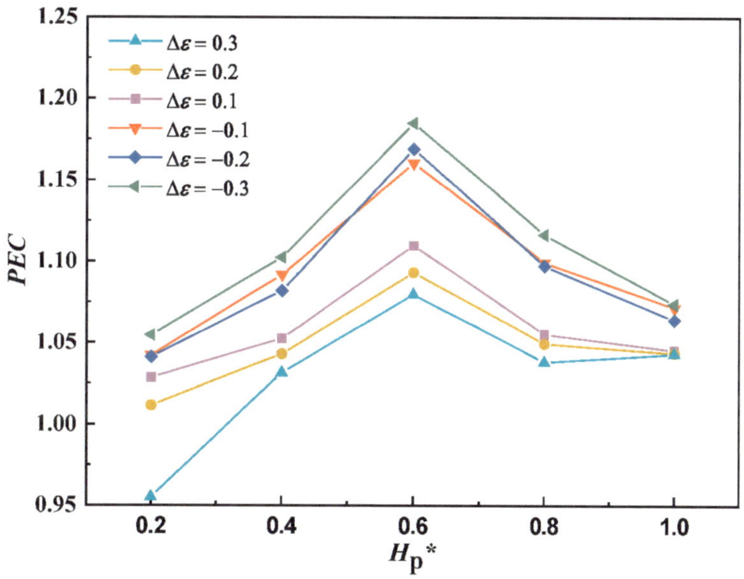


**Figure 10.**
*Heat transfer characterisation evaluation.*

the negative GPM provides more room for vapour generation in the inlet region. Reduced porosity in the outlet region reduces the accumulation of vapour, which lowers the local pressure and reduces the risk of local dryout. Conversely, for positive GPM, the fluid flows from the smaller porosity at the inlet to the larger porosity at the outlet, with a gradual decrease in velocity. This significantly affects the development of the flow boundary layer, increasing its thickness and reducing convective heat transfer, explaining why negative GPM leads to higher $Nu_{fd}$ than positive GPM.

In conclusion, the channel filled with negative GPM has higher superior heat transfer performance compared to the channel filled with positive GPM, which produces 13.4% higher $Nu_{fd}$ compared to the channel with $\Delta\varepsilon = 0.3$ and $H_p^* = 1.0$.

### *3.2.5 Performance evaluation criteria*

Flow boiling involves complex interactions between phase change and multiphase flow. Its unique porosity structure significantly influences the boiling heat transfer and hydrodynamics *via* two primary mechanisms: (1) the gradient porous provides numerous nucleation sites, reducing the superheat required for boiling onset; (2) the porous network promotes bubble detachment and enhances fluid mixing. These multi-scale interaction makes flow boiling in gradient porous media present unique heat transfer and flow characteristics that distinguish it from a HPM channel, including higher critical heat flux, higher heat transfer coefficients, and distinctive pressure drop behaviour.

In order to systematically evaluate the comprehensive performance of porous media under flow boiling conditions, researchers proposed the performance evaluation criterion (PEC), which is a scientific quantitative analysis method, and its calculation method is shown in Eq. (19). The innovation of this standard is that, on the one hand, it breaks through the limitation of traditional single index evaluation, and on the other hand, it realises the comprehensive analysis of gradient by establishing the coupling relation between the heat transfer enhancement effect and the flow resistance.

$$PEC = \frac{Nu/Nu_{ave}}{(f/f_{ave})^{1/3}}, \tag{19}$$

where $Nu$ indicates the actual heat transfer intensity of flow boiling in porous media, the larger the $Nu$ indicates the higher the heat transfer capacity. $Nu_{ave}$ indicates the Nusselt number of the HPM channel under the same conditions as the base heat transfer performance. $f$ indicates the flow resistance coefficient in the porous media, reflecting the flow pressure drop characteristics. $f_{ave}$ indicates the flow resistance coefficient of the HPM channel as the base flow resistance.

From the results of evaluating the effect of axial filling porosity on the heat transfer characteristics of PEC shown in **Figure 10**, it can be seen that with the gradual increase of the filling ratio $H_p^*$, the evaluation index of the integrated heat transfer performance of the PEC shows a non-monotonous tendency of first increasing and then decreasing. When the filling ratio is low ($H_p^* < 0.6$), the increase in pore structure significantly enhances the fluid perturbation effect and heat transfer area, so that the PEC value continues to improve with the increase in the filling ratio; when the filling ratio exceeds the critical value ($H_p^* = 0.6$), the situation is transformed overly high filling ratio of the porous medium makes the flow space in the flow channel significantly narrower, the fluid flow in which the fluid needs to overcome greater resistance, resulting in a sharp increase in flow resistance, and its negative impact on heat transfer performance began to dominate. This competitive mechanism makes the PEC index peak at $H_p^* = 0.6$, in which the filling ratio, the system in the heat transfer performance, and flow resistance achieve an optimal balance between the higher heat transfer efficiency but also to control the flow resistance in a relatively reasonable range.

This phenomenon further validates the existence of an optimal value for the porous material filling thickness at 0.6. However, it is worth attention that, by comparing the PEC values under different gradient conditions, under the same gradient change ratios, the comprehensive performance indexes of the arrangement scheme using negative gradient (the porosity gradually decreases along the flow direction) are significantly better than those of the positive gradient (the porosity gradually increases) scheme.

## 4. Summarise

This chapter investigates the flow boiling heat transfer performance of channels incorporating GPM, with a focus on the influence of filling ratio ($H_p^*$) and porosity gradient ($\Delta\varepsilon$). Key findings reveal that axially filled negative GPM configurations outperform both HPM and positive GPM channels in flow boiling heat transfer. Furthermore, a comprehensive evaluation of thermohydraulic performance—considering both heat transfer enhancement and hydraulic resistance—determines an optimal filling ratio of $H_p^* = 0.6$, providing a theoretical foundation for optimising porous media channel designs.

## Acknowledgements

This work was supported by the Natural Science Foundation of Shaanxi Province (2025JC-YBMS-439) and the National Natural Science Foundation of China (51976144). The authors also gratefully acknowledge the support of the Youth Innovation Team of Shaanxi Universities, the International Joint Research Centre for Building Service Science and the Underground Space Environment, Shaanxi (China).

## Conflict of interest

The authors declare no conflict of interest.

## Nomenclature

| | |
|---|---|
| $C_E$ | Ergun's coefficient |
| $c_p$ | specific heat at constant pressure, J·kg$^{-1}$·K$^{-1}$ |
| $D_h$ | hydraulic diameter, m |
| $H$ | length of square channel, m |
| $h$ | heat transfer coefficient, W·m$^{-2}$·K$^{-1}$ |
| $h_{fg}$ | latent heat, kJ·kg$^{-1}$ |
| $H_p$ | thickness of porous media, m |
| $H_p^*$ | thickness ratio, $H_p^* = H_p/H$ |
| $K$ | permeability of porous media, m$^2$ |
| $L$ | height of the channel, m |
| $\dot{m}$ | volumetric mass transfer rate, kg·m$^{-3}$·s$^{-1}$ |
| $Nu$ | Nusselt number, $Nu = q/(T_w - T_b) \cdot D_h/\lambda_f$ |
| $p$ | pressure, N·m$^{-2}$ |

$q$ heat flux, $W \cdot m^{-2}$
$R$ gas constant, $J \cdot kg^{-1} \cdot K^{-1}$
$S$ source term, $N \cdot m^{-3}$
$T^*$ dimensionless temperature, $T^* = (T - T_{in})/(T_w - T_{in})$
$T_b$ bulk fluid temperature, $T_b = [\int(\rho c_p)_{in} vTdA]/[\int(\rho c_p)_{in} v dA]$, K

## Greek

$\alpha$ volume fraction
$\varepsilon$ porosity
$\eta$ thermal diffusivity, $m^2 \cdot s^{-1}$
$\kappa$ curvature, $m^{-1}$
$\lambda$ thermal conductivity, $W \cdot m^{-1} \cdot K^{-1}$
$\mu$ dynamic viscosity, $m^2 \cdot s^{-1}$
$\nu$ kinematic viscosity, $m^2 \cdot s^{-1}$

## Subscript

eff effective
f fluid
fd fully developed
in inlet
l liquid
out outlet
sat saturation
s solid
v vapour

## Abbreviation

GPM gradient porous medium
HPM homogeneous porous media
PEC performance evaluation criteria

## Author details

Chuang-Yao Zhao[1,2]*, Li-Yuan Yang[1], Yu-Ming Guo[1] and Chen-Min Zheng[1]

1 School of Building Services Science and Engineering, Xi'an University of Architecture and Technology, Xi'an, P. R. China

2 International Joint Laboratory on Low Carbon Built Environment, Ministry of Education of China, P. R. China

*Address all correspondence to: cyzhao@xauat.edu.cn

## References

[1] Bernabé Y, Maineult A. Physics of porous media: Fluid flow through porous media. Treatise on Geophysics. 2015;**11**: 19-41

[2] Shamberger PJ, Bruno NM. Review of metallic phase change materials for high heat flux transient thermal management applications. Applied Energy. 2020;**258**: 113955

[3] Xiaojun W, Hengyun Z, Zhaozang Y, Xinghua H, Hui G, Shunliang Z. Thermal characteristics of refrigerant flow boiling in two mini-channel heat sinks of different aspect ratios for battery thermal management. Applied Thermal Engineering. 2022;**217**:119173

[4] Ali R, Shinichi O, Mahmoud A. Thermal management of concentrator photovoltaic systems using two-phase flow boiling in double-layer microchannel heat sinks. Applied Energy. 2019;**241**:404-419

[5] Saman R, Javad Abolfazli E, Abbas R. A review on the applications of porous materials in solar energy systems. Renewable and Sustainable Energy Reviews. 2017;**73**:1198-1210

[6] Mao-Yu W, Kuang-Jang J, Ching-Yen H. The characteristics of boiling heat transfer and pressure drop of R-600a in a circular tube with porous inserts. Applied Thermal Engineering. 2014; **64**(1):348-357

[7] Lawrence EM, Sara MG, Diana AR, Edwin M, Jennifer H, Krista NA, et al. Metal fabrication by additive manufacturing using laser and electron beam melting technologies. Journal of Materials Science and Technology. 2012; **28**(1):1-14

[8] Chen X, Liu J, Fan Y. Hydrodynamics and heat transfer characteristics of a double gradient microchannel porous model: Fixed gradients and variable gradients structures. Advanced Theory and Simulations. 2020;**3**(4):1900238

[9] Yang C, Xu Y, Cai X, Zheng ZJ. A simplified phase change model and performance optimization of plate latent heat storage unit with graded porous. International Journal of Heat and Mass Transfer. 2022;**196**:123309

[10] Shi C, Wang M, Yang J, Liu W, Liu Z. Performance analysis and multi-objective optimization for tubes partially filled with gradient porous media. Applied Thermal Engineering. 2021;**188**: 116530

[11] Asiaei S, Zadehkafi A, Siavashi M. Multi-layered porous foam effects on heat transfer and entropy generation of Nanofluid mixed convection inside a two-sided lid-driven enclosure with internal heating. Transport in Porous Media. 2019;**126**:223-247

[12] Zheng Z-J, Li M-J, He Y-L. Optimization of porous insert configurations for heat transfer enhancement in tubes based on genetic algorithm and CFD. International Journal of Heat and Mass Transfer. 2015; **87**:376-379

[13] Wang B, Hong Y, Wang L, Fang X, Wang P, Xu Z. Development and numerical investigation of novel gradient-porous heat sinks. Energy Conversion and Management. 2015;**106**: 1370-1378

[14] Wang B, Hong Y, Hou X, Xu Z, Wang P, Fang X, et al. Numerical configuration design and investigation of heat transfer enhancement in pipes filled with gradient porous materials. Energy Conversion and Management. 2015;**105**: 206-215

[15] Mo D-C, Yang S, Luo J-L, Wang Y-Q, Lyu S-S. Enhanced pool boiling performance of a porous honeycomb copper surface with radial diameter gradient. International Journal of Heat and Mass Transfer. 2020;**157**:119867

[16] Vafai K. Handbook of Porous Media. Boca Raton: CRC Press Taylor and Francis Group; 2015

[17] Hirt CW, Nichols BD. Volume of fluid (VOF) method for the dynamics of free boundaries. Journal of Computational Physics. 1981;**39**:201-225

[18] Ergun S. ed. Fluid Flow Through Packed Columns. 1952

[19] Solomon AB, Ramachandran K, Asirvatham LG, Pillai BC. Numerical analysis of a screen mesh wick heat pipe with Cu/water nanofluid. International Journal of Heat and Mass Transfer. 2014; **75**:523-533

[20] Chuang-Yao Z, Di Q, Wen-Tao J, Pu-Hang J, Wen-Quan T. A comprehensive review on computational studies of falling film hydrodynamics and heat transfer on the horizontal tube and tube bundle. Applied Thermal Engineering. 2022;**202**:117869

[21] Chuang-Yao Z, Li-Wen L, Li-Yuan Y, Fang-Fang Z, Avijit K, Di Q, et al. Modelling of evaporative falling-film heat transfer over a horizontal tube. Applied Thermal Engineering. 2024;**255**: 123997

[22] Marek R, Straub J. Analysis of the evaporation coefficient and the condensation coefficient of water. International Journal of Heat and Mass Transfer. 2001;**44**(1):39-53

[23] Bibin KS, Jayakumar JS. Modelling of boiling in square channels partially filled with porous medium. International Communications in Heat and Mass Transfer. 2022;**131**:105835

[24] Bibin KS, Jayakumar JS. Thermal hydraulic characteristics of square ducts having porous material inserts near the duct wall or along the duct Centre. International Journal of Heat and Mass Transfer. 2020;**148**:119079

[25] Zhao CY, Yang LY, Zhao XS, Liang LW, Zhang FF, Qi D, et al. Numerical study on flow-boiling heat transfer in square channels partially centrally filled with gradient porous media. International Journal of Heat and Mass Transfer. 2024;**235**:126170

[26] James AW, Majid H, Ranganathan K. Phase-change mechanism for evaporation in porous media using volume of fluid: Implicit formulation of interfacial temperature. International Communications in Heat and Mass Transfer. 2019;**103**:90-99

[27] Younglove BA, Ely JF. Thermophysical properties of fluids. II. Methane, ethane, propane, Isobutane, and Normal butane. Journal of Physical and Chemical Reference Data. 1989;**53**: 577-789

[28] Manuel JLL, Inigo JL. Three-dimensional interaction of waves and porous coastal structures: Part I: Numerical model formulation. Coastal Engineering. 2012;**64**:57-72

[29] Tomar G. Two-phase flow with phase change in porous channels. Journal of Thermal Science and Engineering Applications. 2015;7(2):021006

Chapter 8

# Impact of Porous Media on Combustion Processes and Waste Heat Recovery

*Pablo Donoso-García and Luis Henríquez-Vargas*

## Abstract

Due to the demand for energy efficiency and sustainability, porous media has emerged as a promising solution due to its high surface area, structural versatility, and enhanced heat transfer performance. The application of porous media in combustion and waste heat recovery offers a wide range of fields for developing new efficient thermal technologies. Typically, these processes operate under turbulent regimes, where flow instabilities significantly influence chemical reactions and thermal transport; therefore, modeling turbulence is essential to capture the physical phenomena governing combustion behavior, flame stability, and heat exchange efficiency. Approaches based on Reynolds-Averaged Navier–Stokes equations and the Local thermal non-equilibrium model allow the description of the phenomenology governing porous media. This chapter provides a concise overview of the fundamentals of turbulent transport phenomena in porous media and highlights their role in advanced thermal systems. In addition, it reviews the current status of porous media combustion and waste heat recovery technologies, identifying their applications, challenges, and future trends.

**Keywords:** porous media, turbulence model, heat transfer, porous media combustion, waste heat recovery, thermoelectric generation

## 1. Introduction

The energy demand grows yearly, and fossil fuels predominate in total energy consumption. However, the excessive use of these fuels promotes severe energetic crises and environmental troubles such as global warming and acid rain, among others [1]. This overview has generated a significant boost within the scientific community in finding new energetic technologies to reduce the use of fossil fuels.

Some measures that can increase the energy efficiency of processes can include proper planning of the productive process, investment in energy-efficient equipment, recycling of energy in the industrial production process, and recovery of excess energy and subsequent utilization in other processes [2]. This search for more efficient and sustainable energy technologies has allowed the development of advanced thermal systems capable of producing cleaner energy and reusing the heat that is typically wasted. In this context, porous media (PM) have emerged as a promising solution, offering enhanced heat transfer capabilities and enabling both energy conversion and recovery in a compact and effective manner [3]. Including PM as metals foams, ceramics, or polymers in combustion systems or waste heat recovery (WHR) devices improves heat transfer due to the higher surface area and interconnected pores [4].

The porous media combustion (PMC) consists of the propagation of a reaction zone inside a PM. This approach offers several advantages compared with conventional systems, such as better heat distribution, more flame stability, and lower emission of pollutants [5]. The high thermal conductivity of the solids favors a thermally efficient thermal feedback mechanism, preheating the reactive mixture and increasing the overall efficiency of the process [6]. On the other hand, WHR refers to the heat typically wasted in exhaust flows, combustion gases, or heat surfaces, transforming it into valuable energy or electricity through technologies such as thermoelectric generators (TEGs) [7]. The PM offers a structure that can be designed to optimize heat transfer by conduction and convection [8].

In several thermal applications, the PMC and WHR occur in turbulent conditions. This is because the systems usually operate at high Reynolds numbers, generating chaotic flows. Turbulence plays a fundamental role in directly affecting the momentum, energy, and mass transport, accelerating the mixture among the reactive and increasing the heat transfer coefficients [9]. The combination of PMC and WHR enables the development of integrated energy systems that are more thermally efficient, consume less fuel, and significantly reduce emissions.

## 2. Theoretical principles

Porous materials are found vastly in nature. With the exception of metals, some dense rocks, and some plastics, virtually all solid and semisolid materials are porous to varying degrees [10]. Characterizing the porous structure are the average pore size $d$ and porosity $\phi$, which can be found in a wide range of values. These pores are usually filled with one or more fluids and could be interconnected, allowing the flow of fluids through the material [11]. With such frequent occurrence, the phenomena of flow in PM take place in many scientific and engineering applications. Some examples from various areas are as follows: fluidized bed combustion, enhanced oil reservoir recovery, underground spreading of chemical waste, chemical catalytic reactors, absorption/desorption columns, mass transfer through membranes, packed bed chromatography, salt water intrusion into coastal aquifers, transpiration cooling, dehumidifying, and enhanced combustion within porous matrix [12, 13]. In general, the transport in porous media can easily fall in the turbulent regime [14] and can be characterized by the presence of eddies and associated with a pore $\mathrm{Re}_d$ greater than 120 [15]. Motivated by the above, the research community has made efforts to study the transport characteristics in PM [14, 16–21].

To numerically simulate the turbulent flow, it is necessary to calculate the motion of all eddies. However, the difference in scale between eddies increases as Reynolds numbers increase, becoming too computationally expensive at high Reynolds numbers. In this sense, adopting some approaches to solve the Navier–Stokes equations in PM is necessary.

### 2.1 The double decomposition

Derivation of mathematical macroscopic models of turbulence in PM requires a double decomposition (time and space) for every property $\xi$ within the porous system and subsequent corresponding average operations over local conservation equations [13]. Following Pedras et al. [22], the intrinsic $\langle\cdot\rangle^{\mathrm{i}}$ and volumetric $\langle\cdot\rangle$ averages are related through the porosity $\phi$:

$$\langle\xi\rangle^{\mathrm{i}} = \frac{1}{\Delta V_i} \int\limits_{\Delta V_i} \xi dV \tag{1}$$

$$\langle\xi\rangle = \phi\langle\xi\rangle^{\mathrm{i}} \tag{2}$$

$$\phi = \frac{\Delta V_i}{\Delta V} \tag{3}$$

The property $\xi$ can then be defined as the sum of $\langle\xi\rangle^{\mathrm{i}}$ and a spatial deviation ${}^{\mathrm{i}}\xi$ within the Representative Elementary Volume (REV) for which by definition:

$$\langle{}^{\mathrm{i}}\xi\rangle^{\mathrm{i}} = 0 \tag{4}$$

$$\xi = \langle\xi\rangle^{\mathrm{i}} + {}^{\mathrm{i}}\xi \tag{5}$$

Statistical analysis is required in the classical (clear fluid) treatment of turbulent flow. For that, the time average operator $\bar{\cdot}$ allows to obtain the time averaged value of property $\xi$:

$$\overline{\xi} = \frac{1}{\Delta t}\int_t^{t+\Delta t} \xi dt \tag{6}$$

**Figure 1** shows a representative elementary volume (REV) where the instantaneous property $\xi$ can be defined as the sum of $\overline{\xi}$ and a time-fluctuating component $\xi'$ for which by definition:

$$\overline{\xi'} = 0 \tag{7}$$

$$\xi = \overline{\xi} + \xi' \tag{8}$$

For a rigid medium, the volume of fluid (phase) $\Delta V_i$ will be dependent only on space and not on time [23]. If the time interval chosen for temporal averaging in 6 is the same for all REVs, then the volumetric average 1 commutes with the time average because both integrations domains are independent of each other [22]. Necessary to the development of macroscopic models are the relationships between volumetric average of derivatives and the derivatives of volumetric averages, which can be found in the literature as the Theorems of Local Volumetric Averaging [13, 23].

$$\langle\nabla\xi\rangle = \nabla\left(\phi\langle\xi\rangle^{\mathrm{i}}\right) + \frac{1}{\Delta V}\int_{A_i} \mathbf{n}_i \xi dS, \tag{9}$$

$$\langle\nabla\cdot\boldsymbol{\xi}\rangle = \nabla\cdot\left(\phi\langle\boldsymbol{\xi}\rangle^{\mathrm{i}}\right) + \frac{1}{\Delta V}\int_{A_i} \mathbf{n}_i\cdot\boldsymbol{\xi} dS, \tag{10}$$

$$\left\langle\frac{\partial\xi}{\partial t}\right\rangle = \frac{\partial}{\partial t}\left(\phi\langle\xi\rangle^{\mathrm{i}}\right) - \frac{1}{\Delta V}\int_{A_i} \mathbf{n}_i\cdot(\mathbf{u}_{A_i}\xi) dS \tag{11}$$

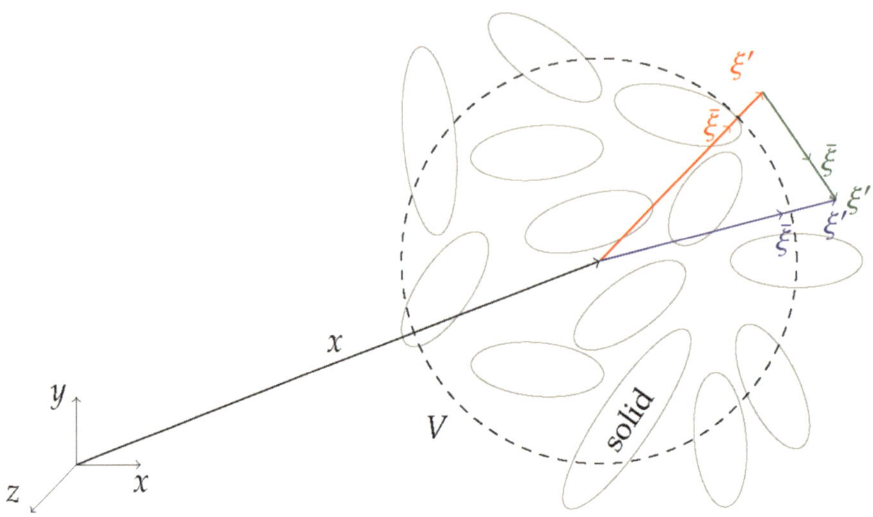


**Figure 1.**
*REV shown as a circle (sphere) with dashed line, centered at* $\mathbf{x}$*.*

## 2.2 $k$ - $\varepsilon$ model

Applying the decomposition approach leads to the inclusion of the Reynolds stress term in the governing equations. One way to model the Reynolds stress is by introducing the turbulent viscosity ($\mu_t$). Several models have been developed, which are classified depending on how many equations are solved to estimate $\mu_t$. This viscosity is related to the turbulent kinetic energy ($k$) and the turbulent kinetic energy dissipation rate $\varepsilon$ that are necessary to solve the Reynolds Averaged Navier-Stoke (RANS) system equations. One of the simplest and widely used models is the $k$ - $\varepsilon$ model. The model, originally presented by Jones and Launder [24] to be applied in PM, requires introducing the porosity, obtaining the following transport equations [25]:

$$\rho\left[\frac{\partial}{\partial t}\left(\phi\langle k\rangle^{i}\right)+\nabla\cdot\left(\overline{\mathbf{u}}_{D}\langle k\rangle^{i}\right)\right]=\nabla\cdot\left[\left(\mu_{t}+\frac{\mu_{t\phi}}{\sigma_{k}}\right)\nabla\left(\phi\langle k\rangle^{i}\right)\right] \\ -\rho\left\langle\overline{\mathbf{u}'\mathbf{u}'}\right\rangle^{i}:\nabla\overline{\mathbf{u}}_{D}+C_{k}\rho\phi\frac{\langle k\rangle^{i}|\overline{\mathbf{u}}_{D}|}{\sqrt{K}}-\rho\phi\langle\varepsilon\rangle^{i} \tag{12}$$

$$\rho\left[\frac{\partial}{\partial t}\left(\phi\langle\varepsilon\rangle^{i}\right)+\nabla\cdot\left(\overline{\mathbf{u}}_{D}\langle\varepsilon\rangle^{i}\right)\right]=\nabla\cdot\left[\left(\mu_{t}+\frac{\mu_{t\phi}}{\sigma_{k}}\right)\nabla\left(\phi\langle\varepsilon\rangle^{i}\right)\right] \\ +C_{1,\varepsilon}\left(-\rho\left\langle\overline{\mathbf{u}'\mathbf{u}'}\right\rangle^{i}:\nabla\overline{\mathbf{u}}_{D}\right)\frac{\langle\varepsilon\rangle^{i}}{\langle k\rangle^{i}}+C_{2,\varepsilon}\rho\phi\left[C_{k}\frac{\langle\varepsilon\rangle^{i}|\overline{\mathbf{u}}_{D}|}{\sqrt{K}}-\frac{\langle\varepsilon\rangle^{i^{2}}}{\langle k\rangle^{i}}\right] \tag{13}$$

where the turbulent viscosity is defined by:

$$\mu_{t\phi}=\rho C_{\mu}\frac{\langle k\rangle^{i^{2}}}{\langle\varepsilon\rangle^{i}} \tag{14}$$

Solving these two transport equations together with the momentum equations makes it feasible to solve the mean flow without using empirical relations.

### 2.3 The thermal non-equilibrium model

The RANS equation for the heat transport is defined as follows:

$$\rho C_p \left( \frac{\partial \overline{T}}{\partial t} + \overline{u}_j \frac{\partial \overline{T}}{\partial x_j} \right) = \frac{\partial}{\partial x_j} \left[ \left( k + \frac{k_t}{Pr_t} \right) \frac{\partial \overline{T}}{\partial x_j} \right] \tag{15}$$

However, this only works for a clear flow. When the flow includes a PM, it is necessary to consider another phenomenon, such as thermal dispersion, fluid–solid inter action, and tortuosity. To get the equations of the turbulent heat transport in PM, thermal local equilibrium can be assumed, which considers that the fluid in a solid matrix has the same temperature but is not successful when internal heat sources are non-uniformly distributed or high thermal gradients are produced. In this sense, a two-energy equation model considering thermal local non-equilibrium has been proposed [26]. This model distinguishes between fluid and solid temperature by the following equations:

$$\frac{\partial}{\partial t} \left( (1-\phi) \rho_s c_s \langle \overline{T_s} \rangle^i \right) = \nabla \cdot \left( \mathbf{K}_{\mathrm{s,eff}} \cdot \nabla \langle \overline{T_s} \rangle^i \right) - h a_i \left( \langle \overline{T_s} \rangle^i - \langle \overline{T_f} \rangle^i \right) \tag{16}$$

$$\frac{\partial}{\partial t} \left( \phi \rho_f c_f \langle \overline{T_f} \rangle^i \right) + \nabla \cdot \left( \rho c_f \overline{\mathbf{u}}_D \langle \overline{T_f} \rangle^i \right) = \nabla \cdot \left( \mathbf{K}_{\mathrm{f,eff}} \cdot \nabla \langle \overline{T_f} \rangle^i \right) + h a_i \left( \langle \overline{T_s} \rangle^i - \langle \overline{T_f} \rangle^i \right) \tag{17}$$

where the effective conductivity tensor for the fluid and solid phases is given by:

$$\mathbf{K}_{\mathrm{s,eff}} = [(1-\phi)\lambda_s]\,\mathbf{I} + \mathbf{K}_{s,f} \tag{18}$$

$$\mathbf{K}_{\mathrm{f,eff}} = [\phi \lambda_s]\,\mathbf{I} + \mathbf{K}_{f,s} + \mathbf{K}_{disp} + \mathbf{K}_t + \mathbf{K}_{disp,t} \tag{19}$$

Each coefficient is defined as follows:

$$\text{Thermal dispersion}: \quad -(\rho c_f) \left( \phi \langle {}^i\overline{\mathbf{u}}\, {}^i\overline{T}_f \rangle^i \right) = \mathbf{K}_{disp} \cdot \nabla \langle \overline{T_f} \rangle^i \tag{20}$$

$$\text{Turbulent thermal dispersion}: \quad -(\rho c_f) \left( \phi \overline{\langle {}^i\mathbf{u}'\, {}^iT'_f \rangle^i} \right) = \mathbf{K}_{disp,t} \cdot \nabla \langle \overline{T_f} \rangle^i \tag{21}$$

$$\text{Turbulent heat flux}: \quad -(\rho c_f) \left( \phi \overline{\langle \mathbf{u}' \rangle^i \langle T'_f \rangle^i} \right) = \mathbf{K}_t \cdot \nabla \langle \overline{T_f} \rangle^i \tag{22}$$

$$\begin{aligned} \text{Local conduction}: \quad \nabla \cdot \left[ \frac{1}{\Delta V} \int_{A_i} \mathbf{n}_i k_f \overline{T}_f dA \right] &= \mathbf{K}_{s,f} \cdot \nabla \langle \overline{T_f} \rangle^i \\ &= \mathbf{K}_{f,s} \cdot \nabla \langle \overline{T_s} \rangle^i - \nabla \cdot \left[ \frac{1}{\Delta V} \int_{A_i} \mathbf{n}_i k_s \overline{T}_s dA \right] \end{aligned} \tag{23}$$

## 3. Applications

### 3.1 Combustion in porous media

The concept of enhanced combustion within a porous matrix or PMC originates in the idea of recirculating the energy from the reaction zone to the incoming reagents, aiming to reach higher temperatures and thermodynamic efficiencies in the process. In contrast to traditional combustion systems that take place in open spaces, PMC occurs within a highly porous solid medium, such as metal foams, refractory ceramics, or fibrous structures, which facilitates heat transfer between solid and gas phases.

In PMC, the propagation of the flame can be understanding as the interaction of two coupled waves: a thermal and combustion waves. Similar propagation speeds for these waves can produce the superadiabatic combustion temperatures (**Figure 2**), also known as the matching condition [28, 29]. Propagation of the combustion wave takes place at subsonic velocities ($\sim$ 0.1 (mm/s)) in the so-called low velocity regime [30], thus a method to detain or stabilize the combustion wave needs to be applied for continuous operation of the burner.

A dimensionless criterion for stabilization is the modified Péclet number, which is measured where the combustion wave propagates through two zones. When the $Pe_m \geq 65$, the flame can propagate, but if $Pe_m < 65$, the propagation is inhibited [31].

A dimensionless criterion for the stabilization is the modified Péclet number (Eq. (24)), which is measured where the combustion wave propagates through two zones. When the $Pe_m \geq 65$, the flame is capable to propagate, but if $Pe_m < 65$, the propagation is inhibited [31]. Adjusting the Péclet number to these values, by using two porous bodies with different thermophysical properties (commonly different $d_p$), the combustion wave is stabilized at the interface of the porous bodies due to local flame quenching.

$$Pe_m = \frac{S_L d_p \rho_f c_f}{\lambda_f} \tag{24}$$

Bubnovich et al. [32] conducted an experimental study on flame stabilization in a two-layer PM burner using reticulated alumina foam, honeycomb alumina foam, and SiC foam as porous ceramics. Their results indicated that the combustion wave propagation rate for all three ceramic materials increased with increased inlet velocities and decreased equivalence ratios of the premixed gases. Combustion temperatures ranged between 1160 and 1380 K, with higher flow rates leading to higher temperatures. Additionally, CO and NOx emissions remained below a few ppm, indicating that combustion was almost complete.

Micro-combustors are portable power devices that provide energy efficiently, where heat recirculation is a key factor influencing the combustion

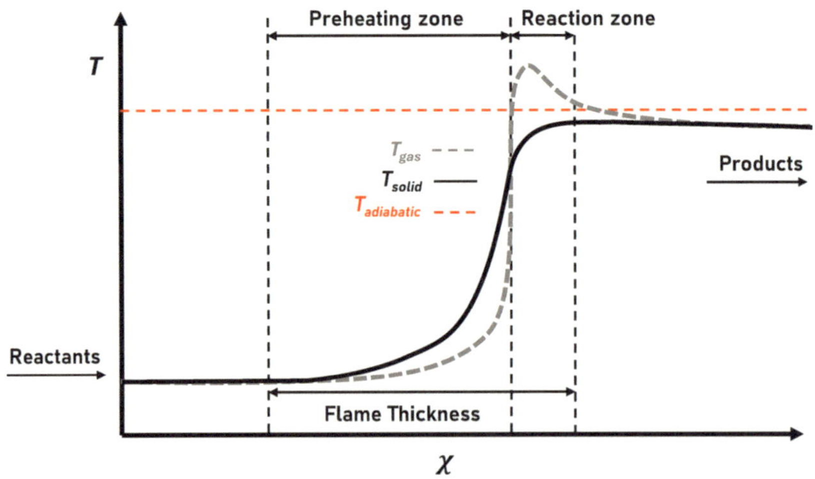


**Figure 2.**
*Gas-and solid-phase temperature profiles in PM-based combustion. Adapted from Ref. [27].*

process. Yan et al. [33] proposed a heat-recirculating micro-combustor embedded with PM to enhance heat recirculation and improve combustion stability. The results showed that micro-combustor embedded with porous media (MCEPM) improved the radial temperature uniformity and enhanced the preheating capability. When the embedding thickness was set to 15 mm, the combustor achieved an optimal balance between stable combustion and efficient heat recirculation.

Vignat et al. [34] evaluated the stabilization of $NH_3/H_2$/air flames using PM burners. The results showed that when operating the burner near its extinction limit, the NOx emission reaches a minimum level and tends to increase with increasing mass flux rates. In addition, due to the unburned $NH_3$ emissions remaining around 1400 ppm, this regime offers similar or better performance compared with the common fuel-rich operation.

Huang et al. [35] investigated numerically the methane-hydrogen combustion on a cylindrical double-layer porous burner geometry considering the effect of the temperature, combustion rate, and diffusion characteristics of the methane-hydrogen-precipitated gas flame within the PM. The authors noticed an accelerated flame diffusion within the porous combustion zone with an increase in the hydrogen blending ratio, leading to an enhanced combustion rate. Besides, the results showed that in the porous combustion section are located the high-temperature regions.

A swirler combustor is a combustion chamber that employs a swirler to induce a strong swirling motion in the flow. This action creates a low-pressure region near the burner, causing hot combustion gases to recirculate back toward the burner. This promotes ignition and flame stability [36]. Tolouei and Gharehghani [37] conducted a numerical study to evaluate the use of catalytic platinum PM in $NH_3/H_2$ fuel blends to improve the flame characteristics and reduce NOx emissions in a microscale swirl combustor. The results showed that a reduction in NOx emissions leads to 69% when the platinum catalytic is used compared to non-catalytic combustion. Moreover, increasing the porosity of the platinum PM from 0.4 to 0.8 led to a 50% decrease in NOx emissions during pure ammonia combustion, highlighting the effectiveness of catalytic PM in emission control.

In general, the PM has a homogeneous porous structure. However, in some cases, a gradual change of either porosity or pore size can exist along a defined axis called gradient PM [38]. Chen et al. [39]. proposed a gradient porous structural burner to investigate numerically the effects of pore size and porosity gradient on the combustion characteristics and stability limits of hydrogen-enriched natural gas flames. The results showed that the axial pore size increment structure improved the stability limit by approximately 117%, while the radial pore size increment structure increased the stability limit by 47% compared to the conventional two-stage uniform structure.

Unlike conventional burners, a porous radiant burner involves a fuel-air mix into the PM, stabilizing the combustion very close to the porous exit plane. The hot gases heated the PM, releasing energy as radiation [40]. Deb and Muthukumar [41] studied the air entrainment and flame stability through the geometrical dimensions of a Self-Aspirated Porous Radiant Burner of 5 kW capacity. The authors analyzed the effect of Orifice Position, Orifice Diameter, and Mixing Tube Diameter. The results showed a stable operation in a partially submerged combustion mode with an optimized setup featuring a 0.3 mm Orifice Diameter positioned 30 mm from the datum, coupled with a 29 mm Mixing Tube Diameter.

### 3.2 Enhanced heat transfer and waste heat recovery

Industrial waste heat (WH) is defined as heat rejected from industrial processes, in which energy is used to produce high-added value products. The most common ones are gaseous streams (e.g. exhaust gas, flaring gas, low-quality steam, cooling air), liquid streams (e.g., hot oil, cooling water), and solids (e.g., commodities and products, such as hot steel) [42]. The amount of WH as a fraction of energy consumption greatly varies among the various industrial sectors [43]. The temperature at which heat is available can vary over a wide range depending on the industrial sector and process, from around 50°C up to 1000°C or higher. This allows for exploring innovative thermal management strategies to maximize energy recovery efficiency. A promising approach involves the use of PM, such as metals, ceramics, or polymers, offering unique thermo-physical properties that make them highly suitable for enhancing heat transfer processes. Its high superficial surface and interconnected porous structure allow an efficient heat exchange through conduction and convection [44].

There are several heat recovery technologies, and one of them consists of heat exchangers (HX) that take heat from exhaust gases to transfer it to another fluid. Hossein et al. [45] considered a porous-filling technique to improve the heat transfer rate in a shell and tube HX. The researchers performed a numerical study to achieve a homogeneous thermal distribution by identifying a mechanism that manipulated the flow structures and heat transfer. The results showed a uniform thermal distribution in both porous-filled designs. In addition, they improved the heat transfer efficiency to 60% compared to the porous filled with a conventional type but with a high-pressure drop.

Bogdan and Abdulmajeed [46] investigated experimentally and numerically the effect of metallic PM, inserted in a pipe, on the rate of heat transfer. The pipe was subjected to a constant and uniform heat flux. The effects of $\phi$, $d_p$, and thermal conductivity as well as $\mathrm{Re}_d$ number on the heat transfer rate and pressure drop were investigated. The results showed that higher heat transfer rates can be achieved using porous inserts at the expense of a reasonable pressure drop. But, the effective thermal conductivity is necessary for accurate simulation.

Scott et al. [47] addressed a transient analysis to solve the convective and conductive heat transfers based on a finite-element scheme. The model considered phase change aspects of a frozen medium through the latent heat effect. The model was validated against six test cases with known analytical solutions, such as (a) single-phase fluid flow through PM, (b) radial conduction with and without phase change, (c) conductive and convective heat transfer in an aquifer, and (d) two-phase immiscible flow in PM showing, in all cases, a strong agreement with analytical solutions.

Gao and Hodgson [48] studied computationally regenerator optimization and its operating parameters using unstable 3D flow and heat transfer. The heat exchange process in the regenerators and the effects of the regenerator structure and operating parameters such as gas mass flow, inversion time, regenerator height, sphere diameter, and thermophysical properties of the spheres were studied with the model to determine the efficiency of heat recovery.

Franklin and Ramesh [49] analyzed a Pebble Bed HX integrated with an internal combustion (IC) engine system, designed to recover heat from exhaust gases. In their study, a thermal energy storage (TES) unit was employed to store the energy generated during operation. It was found that nearly 9–12% of fuel power stored as heat energy in the combined storage system at a higher temperature can be used for suitable applications.

Xu et al. [50] presented a 2D steady model to investigate WHR in a sinter vertical tank based on the PM theory and the local thermal non-equilibrium (LTNE) model. The influences of the air flow rate, sinter flow rate, and sinter particle diameter on the gas–solid heat transfer process were investigated numerically. The results indicated that both heat dissipation and power consumption gradually increase with an increase in air mass flow rate. In contrast, increasing the sinter flow rate reduces heat dissipation, causing an increase in power consumption.

Lee et al. [51] conducted a numerical-experimental study of HX equipped with finned heat pipes, modeling the finned regions as PM using the LTNE approach. A computational framework that integrated Computational Fluid Dynamics (CFD) simulations with experimental validation achieve a WHR of 30.2 kW and a thermal effectiveness of 0.78. The model exhibited high accuracy in predicting heat transfer rates, with a maximum error of 6.2%, outperforming conventional Local Thermal Equilibrium (LTE) approaches, which tended to overestimate efficiency.

Another way to recover waste heat is through TES systems, which can store thermal energy at high or low temperatures. It is categorized into three kinds of systems depending on the state of energy: sensible, latent heat storage, or thermochemical heat storage [52]. Sensible heat storage represents the most economical form of thermal storage and has been performed using solid materials such as rocks, ceramics, and metals, among others [53]. Amiri et al. [54] conducted the feasibility of WHR by TES systems. The researchers used an exhaust gas inside a packed rock bed as a TES. The findings of this study reveal that, although the total thermal energy storage capacity of the system remains largely unaffected by variations in mass flow rate, operating at a lower mass flow rate extends the working period of the thermal energy storage system. On the other hand, a PCM is used to store latent heat. Yang et al. [55] considered a shell-and-tube thermal storage unit to simulate a two-dimensional axisymmetric natural convection employing an open-cell metal foam with a porosity of 0.94 and a pore density of 15 PPI in heat transfer fluid and PCM domains. The results showed that heat transfer can be enhanced if a metal foam is considered due to the reduction of thermal resistance in heat transfer fluid. In addition, the temperature uniformity of PCM can be improved if a metal foam is embedded.

Thermoelectric generators (TEG) are devices that convert the heat differential in voltage potential through the Seebeck effect [56]. **Figure 3** shows a representation of these thermoelectric generators. Donoso and Henríquez [58] used a PM burner to simulate the adiabatic recovery coupled with thermoelectric elements. Two semiconductors, n and p type, combined in series, form a thermoelectric unit. The authors observed that the increase in energy content of the mixture and duplicating the thermal conductivity contributes to reaching high values of electric potential, thermoelectric efficiency, and flame temperature.

Nithyanandam and Mahajan [59] studied exhaust systems using thermoelectric generators and investigated the performance of metal foam-based HX to reduce the thermal resistance of the hot side in thermoelectric generators. The authors found that the maximum net electric power is 5.7 to 7.8 times higher when a metal foam is used than when it is not. Choi et al. [60] investigated thermoelectric modules based on customized bismuth-telluride applied to the WHR of a diesel engine. A maximum efficiency of 2.83% was achieved using an optimal porosity of 0.416 at engine rotation of 1400 rpm. In contrast, at 1000 rpm, the conversion efficiency was optimized at 10.1% compared with a base case without PM. Donoso et al. [61] investigated numerically the WHR using hot air through a square section duct with spherical PM. A global efficiency was obtained in the range of 0.11–4.51%. Furthermore, increased fluid inlet

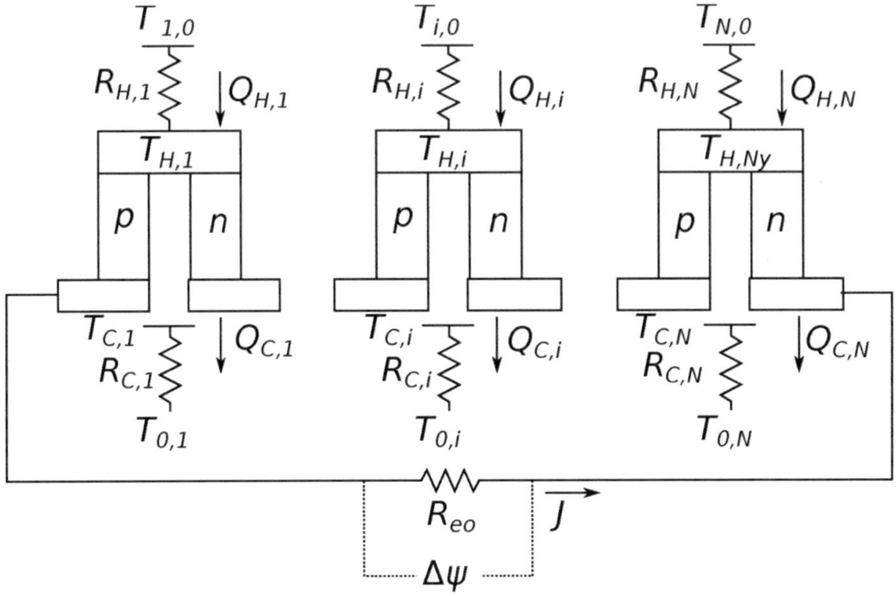


**Figure 3.**
*Schematic representation of thermoelectric elements [57].*

temperature and speed or a higher external heat transfer coefficient improved the system's performance.

Li et al. [62] investigated the impact of TDs and porous copper foam on TEGs. Through experiments and CFD simulations, they analyzed the insertion of foam into the central flow region with varying pore densities and volumetric ratios. The results showed that TDs with low thermal conductivity and high electrical conductivity increased power output by 230% compared to thermogenerators without foam. The model was validated with errors below 5.2%.

Li et al. [63] conducted a numerical and experimental study of a TD designed to recover WH from cables, evaluating the effect of fins filled with porous metal foam. Through CFD simulations and thermal analysis, they demonstrated that foams with high porosity (91.9%) and low pore density (10 PPI) enhance heat transfer by 20%, increasing the generated electrical power. They concluded that this design optimizes $\eta_E$ by compensating for pressure drop losses and is feasible for practical implementation in energy systems.

### 3.3 Porous media combustion and thermoelectric generators

Micro-power generators are small-scale devices that capture energy from surrounding environmental sources and convert it into electrical energy [64]. These systems incorporate micro-combustion chambers, which encounter issues related to flame stability and energy efficiency [65]. Fu et al. [66] investigated the optimization of combustion performance and energy conversion efficiency of micro-power generator. The researchers evaluated the $H/NH_3$ blending ratio found that a small blended fraction of $NH_3$ significantly influences the flame regime of $H_2/NH_3$/air combustion. Besides, PM enhances flame stability, improving heat transfer within the combustion chamber.

Katsuki et al. [67] considered a reciprocating flow combustion for thermoelectric power generation to evaluate its performance. A reciprocatory flow of dilute fuel gas was introduced, and a 200 K/cm gradient was obtained. Despite using fuel mixtures normally not flammables, the system reached an energy density of up to 7 kW/cm$^3$.

Weinberg et al. [68] studied the maximum theoretical efficiency in regenerative burners. A novel configuration is proposed, which includes a two-stage heat rejection system: The first transfers heat to the incoming reactants, enhancing efficiency, while the second expels residual heat to the ambient environment. These devices significantly exceed the maximum efficiency of current thermoelectric conversion systems.

Hanamura [69] proposed a novel power generation system based on reciprocating-flow super-adiabatic combustion within a catalytic and thermoelectric porous element. The authors established a trapezoidal temperature profile, which result in a temperature gradient in the thermoelectric porous element. Through numerical evaluation, the heat transferred reached 94% through the thermoelectric element by conduction.

Bubnovich et al. [70], based on super-adiabatic combustion, proposed a conversion combustion heat system into electric power. The numerical results showed a total thermal efficiency of 4.6%, equaling the thermoelectric element's conversion efficiency.

Bubnovich et al. [71] proposed a combustion-based system for generating electricity using a thermoelectric porous element. Numerical simulations showed that using leaner mixtures and higher combustion wave propagation velocities increases the overall thermal efficiency, approaching the thermoelectric element's conversion efficiency.

Henríquez-Vargas et al. [72] studied numerically a continuous flow PM burner for the direct conversion of thermal to electric energy. The results showed that there is an optimal length and placement region for the thermoelectric elements within the step thermal gradients, which maximizes power output.

Ding et al. [73] developed a thermoelectric system based on PM microcombustors to enhance energy conversion in Micro-Electro-Mechanical Systems. The results showed that, under low equivalence ratio mixtures and high flow rates, a maximum $\eta_E$ of 5.51% was achieved, while mixed-fuel combustion increased the power output to 57.8 W. The optimization used segmented materials ($Bi_2Te_3$/PbTe) and enhanced thermal conductivity, emphasizing the importance of thermal uniformity.

Yan et al. [74] investigated a two-stage thermophotovoltaic-thermoelectric system incorporating PM to enhance energy efficiency in microcombustion. Through numerical simulations was observed a power output of 14.88 W and an $\eta_E$ of 7.33%. The study highlights the importance of thermal uniformity and parameter selection in energy recovery systems.

## 4. Challenges and trends

### 4.1 Combustion in porous media

PMC offers several advantages that highlights its capacity to improve the thermal efficiency due to the unique structure of PM, allowing better fuel utilization and lower operational losses. Another key aspect is the reduction of pollutant emissions due to the stabilization of the flame within the PM and its subsequent heat distribution, which promotes a more complete and homogeneous combustion virtually free from cold/hot spots that generate CO/NOx contaminants. Moreover, it is compatible with WHR technologies, such as TEGs, expanding its potential in advanced thermal systems. However, the transition of the PMC to industrial or commercial scales faces several technical and physical challenges. Modeling such systems involves complex multiphase flows across multiple scales (from the pore level to the macroscopic level), significantly increasing computational demands. Additionally, the Péclet number becomes a critical parameter in determining whether combustion can be sustained, relating to flame stabilization and a minimum pore size threshold. Poor fuel mixtures or high inert gas contents can further complicate ignition and propagation. Furthermore, due the higher temperatures reached, the porous material is degraded and/or its physical properties changed (such as thermal expansion giving high mechanical stresses, others), affecting the long-term stability and performance of the system. Complementary, in real implementations of PMC burners, it may be necessary to perform maintenance of the equipment—and porous bed—which may be troublesome. In particular applications for household users (stoves), there can be cultural barriers presented by the use of (diffusive) free flame conventional systems that need to be overcome. To address these challenges, continued development of accurate, scalable models, systems and advanced materials is essential to consolidate PMC as a key tool in the energy transition.

### 4.2 Waste heat recovery

WHR technologies have gained prominence in recent years, achieving integration into a wide range of applications, from HXs, power generation, industrial furnaces, and internal combustion engines. Combined with the advanced materials, WHR systems can significantly increase heat transfer rates and overall energy recovery. However, despite their potential, they still present several challenges that can be addressed by designing materials that are resistant to degradation and offer high chemical stability and high mechanical strength, and that generate reasonable pressure drops, resulting in lower costs per fluid drive. One limitation of the use of PM is the variability of WH conditions in terms of temperature and chemical composition of the fluids. A decrease in temperature can generate undesirable conditions such as moisture condensation and sulfuric acid production in flue gases from diesel-powered industrial boilers. This makes it difficult to develop a comprehensive system where each thermal solution is tailored to each specific case, and the initial implementation cost and the return-on investment may pose barriers for small- and medium-sized industries, which limits large-scale adoption. The development of porous materials such as Metal–Organic Frameworks (MOFs), Covalent Organic Frameworks (COFs), and Polymers of Intrinsic Microporosity (POPs) that offer high selectivity, recyclability, and thermal and chemical stability, can be explored as specific alternatives in WHR processes through neural network-assisted modeling, the generation of porous structures with machine learning or other forms of artificial intelligence (AI) assistance. Continuous advances in functional materials, such as metallic foams or porous ceramics, and advances in direct conversion technologies, such as high-performance TEGs, are expanding the application possibilities which can be extended to non-conventional renewable energy systems, promoting energy sustainability and the circular economy.

## Acknowledgements

The authors thank University of Santiago of Chile for the financial support to this research.

## Nomenclature

| | |
|---|---|
| $a_i$ | specific surface ($m^2/m^3$) |
| $A_i$ | interface area between the fluid and the solid phases ($m^2$) |
| $C_\mu$ | turbulent model constant |
| $C_{1,\varepsilon}$ | turbulent model constant |
| $C_{2,\varepsilon}$ | turbulent model constant |
| $C_k$ | turbulent model constant |
| $c_f$ | fluid specific heat ($J/kg^{-K}$) |
| $c_s$ | solid specific heat ($J/kg^{-K}$) |
| $h$ | convective heat transfer coefficient ($W/m^2K$) |
| $k$ | turbulent kinetic energy ($m^2/s^2$) |
| $\mathbf{K}_{eff}$ | effective conductivity tensor (W/m-K) |
| $\mathbf{K}_{disp}$ | conductivity tensor due to dispersion (W/mK) |
| $\mathbf{K}_{f,s}$ | local conduction tensor (W/m-K) |
| $\mathbf{K}_{disp,t}$ | conductivity tensor due to turbulent dispersion (W/mK) |
| $\mathbf{K}_t$ | conductivity tensor due to turbulent heat flux (W/mK) |
| $\mathbf{I}$ | tensor unit |
| $\mathbf{n}_i$ | unit vector |
| $p$ | pressure (Pa) |
| $\mathrm{Pr}_t$ | Prandtl turbulent number |
| $\mathrm{Pe}_m$ | modified Peclet number |
| $\mathrm{Re}_d$ | pore-based Reynolds number |
| $S_L$ | laminar stoichiometric flame speed (m/s) |
| $T$ | temperature (K) |
| $\mathbf{u}$ | velocity (m/s) |
| $\bar{\mathbf{u}}_D$ | Darcy velocity (m/s) |

### Greeks

| | |
|---|---|
| $\varepsilon$ | dissipation rate of k |
| $\Delta V$ | elementary volume ($m^3$) |
| $\Delta V_i$ | fluid volume in $\Delta V$ ($m^3$) |
| $\phi$ | porosity |
| $\mu$ | viscosity (kg/ms) |
| $\mu_{t\phi}$ | macroscopic coefficient of exchange for porous media (kg/ms) |
| $\xi$ | generic property within porous media |
| $\lambda$ | thermal conductivity (W/mK) |
| $\rho$ | density ($kg/m^3$) |
| $\sigma$ | turbulent model constant |

### Subindex

| | |
|---|---|
| $i$ | phase index |
| $f$ | fluid phase |
| $s$ | solid phase |

## Operators

| | |
|---|---|
| $\langle \cdot \rangle$ | volume average |
| $\langle \cdot \rangle^i$ | intrinsic volume average |
| $^i\cdot$ | spatial perturbation |
| $\bar{\cdot}$ | time average |
| $\cdot'$ | time perturbation |

## Abbreviations

| | |
|---|---|
| CFD | computational fluid dynamics |
| HX | heat exchanger |
| PCM | phase change material |
| PM | porous media |
| PMC | porous media combustion |
| PPI | porous per inch |
| RANS | Reynolds averaged Navier-Stoke |
| REV | representative elementary volume |
| TEG | thermoelectric generators |
| TES | thermal energy storage |
| WH | waste heat |
| WHR | waste heat recovery |

## Author details

Pablo Donoso-García*† and Luis Henríquez-Vargas†
Department of Chemical Engineering and Bioprocesses, University of Santiago of Chile, Santiago, Chile

*Address all correspondence to: pablo.donosoga@usach.cl

† These authors contributed equally.

## References

[1] Yanyan X, Xue Y, Cai W, Qi H, Li Q. Experimental study on performances of flat-plate pulsating heat pipes without and with thermoelectric generators for low-grade waste heat recovery. Applied Thermal Engineering. 2023;**225**:120156

[2] Johansson MT, Söderström M. Electricity generation from low temperature industrial excess heat an opportunity for the steel industry. Energy Efficiency. 2014;7:203-215

[3] Banerjee A, Paul D. Developments and applications of porous medium combustion: A recent review. Energy. 2021;**221**:119868

[4] Aboujafari M, Valipour MS, Hajialimohammadi A, Honnery D. Porous medium applications in internal combustion engines: A review. Transport in Porous Media. 2022;**141**(3): 799-824

[5] Abdul Mujeebu M, Abdullah MZ, Abu Bakar MZ, Mohamad AA, Muhad RMN, Abdullah MK. Combustion in porous media and its applications – A comprehensive survey. Journal of Environmental Management. 2009;**90** (8):2287-2312

[6] Wood S, Harris AT. Porous burners for lean-burn applications. Progress in Energy and Combustion Science. 2008; **34**(5):667-684

[7] Olabi AG, Al-Murisi M, Maghrabie HM, Yousef BAA, Sayed ET, Alami AH, et al. Potential applications of thermoelectric generators (tegs) in various waste heat recovery systems. International Journal of Thermofluids. 2022;**16**:100249

[8] Xuan Z-H, Fang W-Z, Yu-Hao L, Yang C, Tao W-Q. Significance of the natural convection to the heat transfer of porous media: A pore-scale study. International Journal of Heat and Mass Transfer. 2024;**222**:125163

[9] Wood BD, He X, Apte SV. Modeling turbulent flows in porous media. Annual Review of Fluid Mechanics. 2020;**52**(1): 171-203

[10] Dullien FAL. Porous Media: Fluid Transport and Pore Structure. New York: Academic Press; 1979

[11] Ganji DD, Kachapi SHH. Nanofluid Flow in Porous Medium. Norwich, New York, United States: Elsevier; 2015. pp. 271-316

[12] Kaviany M. Principles of Heat Transfer in Porous Media. second ed. New York: Springer; 1995

[13] de Lemos M. Turbulence In Porous Media: Modeling and Applications. second ed. Oxford: Elsevier; 2012

[14] Jouybari NF, Maerefat M, Nimvari ME. A macroscopic turbulence model for reacting flow in porous media. Transport in Porous Media. 2015;**106**:355-381

[15] Yarahmadi A, Nobari M, Hosseini R. A numerical investigation of laminar and turbulent premixed flames in porous media. Combustion Science and Technology. 2011;**183**(11):1164-1182

[16] Masuoka T, Takatsu Y. Turbulence model for flow through porous media. International Journal of Heat and Mass Transfer. 1996;**39**(13):2803-2809

[17] Antohe B, Lage J. A general two-equation macroscopic turbulence model for imcompressible flow in porous media. International Journal of Heat and Mass Transfer. 1997;**40**(13):3013-3024

[18] Nakayama A, Kuwahara F. A macroscopic turbulence model for flow in a porous medium. Journal of Fluids Engineering. 1999;**121**:427-433

[19] Teruel F, Uddin R. Characterization of a porous medium employing numerical tools: Permeability and pressure-drop from Darcy to turbulence. International Journal of Heat and Mass Transfer. 2009;**52**:5878-5888

[20] Soulaine C, Quintard M. On the use of a Darcy-Forchheimer like model for a macro-scale description of turbulence in porous media and its application to structured packings. International Journal of Heat and Mass Transfer. 2014; **74**:88-100

[21] Pinson F, Grégoire O, Simonin O. k−ε macro-scale modeling of turbulence based on a two scale analysis in porous media. International Journal of Heat and Fluid Flow. 2006;**27**:955-966

[22] Pedras M, de Lemos M. On the definition of turbulent kinetic energy for flow In porous media. International Communications in Heat and Mass Transfer. 2000;**27**(2):211-220

[23] Gray WG, Lee PC. On the theorems for local volume averaging of multiphase system. International Journal of Multiphase Flow. 1977;**3**:333-340

[24] Jones WP, Launder BE. The prediction of laminarization with a two-equation model of turbulence. International Journal of Heat and Mass Transfer. 1972;**15**(2):301-314

[25] Pedras MHJ, De Lemos MJS. Macroscopic turbulence model adjustment for a porous medium modeled as an infinite array of transversally-displaced elliptic rods. Revista de Engenharia Térmica. 2003;**2** (2):73

[26] Quintard M, Whitaker S. One- and Two-Equation Models for Transient Diffusion Processes in Two-Phase Systems. Cambridge, Massachusetts, United States: Elsevier; 1993. pp. 369-464

[27] Voss S. Porous Media Combustion. 2016

[28] Zhdanok SA, Kennedy LA, Koester G. Superadiabatic combustion of methane air mixtures under filtration in a packed bed. Combustion and Flame. 1995;**100**:221-231

[29] Foutko SI, Shabunya SI, Zhdanok SA, Kennedy LA. Superadiabatic combustion wave in a diluted methane-air mixture under filtration in a packed bed. Symposium (International) on Combustion. 1996;**26**(2):3377-3382

[30] Dobrego KV, Kozlov IM, Bubnovich V, Rosas CE. Dynamics of filtration combustion front perturbation in the tubular porous media burner. International Journal of Heat and Mass Transfer. 2003;**46**:3279-3289

[31] Al-Hamamre Z, Al-Zoubi A. The use of inert porous media based reactors for hydrogen production. International Journal of Hydrogen Energy. 2010;**35**(5): 1971-1986

[32] Bubnovich V, Hernandez H, Toledo M, Flores C. Experimental investigation of flame stability in the premixed propane-air combustion in two-section porous media burner. Fuel. 2021;**291**: 120117

[33] Yan Y, Zhang C, Gange W, Feng S, Yang Z. Numerical study on methane/air combustion characteristics in a heat-recirculating micro combustor embedded with porous media. International Journal of Hydrogen Energy. 2022;**47**(48):20999-21012

[34] Vignat G, Akoush B, Toro ER, Boigné E, Ihme M. Combustion of lean ammonia-hydrogen fuel blends in a porous media burner. Proceedings of the Combustion Institute. 2023;**39**(4):4195-4204

[35] Huang T, Ren X, Chen Y, Ma J, Yi D, Wan Z, et al. Transient combustion characteristics of methane–hydrogen mixtures in porous media burner. ACS Omega. 2024;**9**(17):19525-19535

[36] Pedersen K, Jensen A, Skjothrasmussen M, Damjohansen K. A review of the interference of carbon containing fly ash with air entrainment in concrete. Progress in Energy and Combustion Science. 2008;**34**(2):135-154

[37] Tolouei A, Gharehghani A. Numerical investigation of catalytic effect of platinum porous media on the combustion characteristics of hydrogen/ammonia blend. International Journal of Hydrogen Energy. 2025;**98**:1155-1173

[38] Wang B, Hong Y, Wang L, Fang X, Wang P, Zhongbin X. Development and numerical investigation of novel gradient-porous heat sinks. Energy Conversion and Management. 2015;**106**:1370-1378

[39] Chen Y, Long L, Jie H, Huang T, Chen X, Song X, et al. The combustion characteristics and stable limit of a novel combustor with gradient porous media for hydrogen-enriched natural gas. Journal of the Energy Institute. 2024;**116**:101743

[40] Tong TW, Tarafdar A. Heat Transfer in Porous Radiant Burners. Netherlands: Springer; 1999. pp. 89-100

[41] Deb S, Muthukumar P. Investigations on the primary air entrainment and flame stability in a partially submerged combustion-based porous radiant burner. Energy. 2024;**307**:132623

[42] Zhang H, Wang H, Zhu X, Qiu Y-J, Li K, Chen R, et al. A review of waste heat recovery technologies towards molten slag in steel industry. Applied Energy. 2013;**112**:956-966

[43] Miró L, Bruckner S, Cabeza LF. Mapping and discussing industrial waste heat (iwh) potentials for different countries. Renewable and Sustainable Energy Reviews. 2015;**51**:847-855

[44] Krishan Sharma P, Deepu, and Subrata Kumar. Convective heat transfer in a tube filled with homogeneous and inhomogeneous porous medium. International Communications in Heat and Mass Transfer. 2020;**117**:104791

[45] Zolfagharnasab MH, Pedram MZ, Hoseinzadeh S, Vafai K. Application of porous-embedded shell and tube heat exchangers for the waste heat recovery systems. Applied Thermal Engineering. 2022;**211**:118452

[46] Bogdan I, Abdulmajeed A. An experimental and numerical study on heat transfer enhancement for gas heat exchangers fitted with porous media. International Journal of Heat and Mass Transfer. 2004;**47**(47):4939-4952

[47] Scott D, Debendra K, Vijayagandeeban S, Vidyadhar A. A computational scheme for fluid flow and heat transfer analysis in porous media for recovery of oil and gas. Petroleum Science and Technology. 2006;**23**:843-862

[48] Gao W, Hodgson P. Numerical analisis of heat transfer and the optimization of regenerators. Numerical Heat Transfer, Part A. 2006;**50**:63-78

[49] Franklin B, Ramesh K. Experimental investigation on heat recovery from

diesel engine exhaust using pebble bed heat exchanger and thermal energy storage system. International Journal of Applied Engineering Research. 2015;**10**: 37090-37098

[50] Xu C, Liu Z, Wang S, Liu W. Numerical simulation and optimization of waste heat recovery in a sinter vertical tank. Energies. 2019;**12**:1-19

[51] Lee S, Kang S, Kim Y, Geum G, Kong D, Shin DH, et al. Thermal design framework of heat pipe heat exchanger for efficient waste heat recovery. Energy. 2025;**318**:134731

[52] Kousksou T, Bruel P, Jamil A, El Rhafiki T, Zeraouli Y. Energy storage: Applications and challenges. Solar Energy Materials and Solar Cells. 2014; **120**:59-80

[53] Elouali A, Kousksou T, El Rhafiki T, Hamdaoui S, Mahdaoui M, Allouhi A, et al. Physical models for packed bed: Sensible heat storage systems. Journal of Energy Storage. 2019; **23**:69-78

[54] Amiri L, Ghoreishi-Madiseh SA, Sasmito AP, Hassani FP. A porous medium based heat transfer and fluid flow model for thermal energy storage in packed rock beds. IOP Conference Series: Earth and Environmental Science. 2019;**268**(1):012100

[55] Yang X, Jiabang Y, Guo Z, Jin L, He Y-L. Role of porous metal foam on the heat transfer enhancement for a thermal energy storage tube. Applied Energy. 2019;**239**:142-156

[56] Jaziri N, Boughamoura A, Müller J, Mezghani B, Tounsi F, Ismail M. A comprehensive review of thermoelectric generators: Technologies and common applications. Energy Reports. 2020;**6**: 264-287

[57] Donoso-García P, Henríquez-Vargas L. Numerical study of a waste heat recovery thermogenerator system. Journal of the Brazilian Society of Mechanical Sciences and Engineering. 2019;**41**:356-364

[58] Donoso-García P, Henríquez-Vargas L. Numerical study of turbulent porous media combustion coupled with thermoelectric generation in a recuperative reactor. Energy. 2015;**93**: 1189-1198

[59] Nithyanandam K, Mahajan RL. Evaluation of metal foam based thermoelectric generators for automobile waste heat recovery. International Journal of Heat and Mass Transfer. 2018; **122**:877-883

[60] Choi Y, Negash A, Kim TY. Waste heat recovery of diesel engine using porous medium-assisted thermoelectric generator equipped with customized thermoelectric modules. Energy Conversion and Management. 2019;**197**: 111902

[61] Donoso-García P, Henríquez-Vargas L, Huerta E. Waste heat recovery from air using porous media and conversion to electricity. Energies. 2022;**15**(15):5597

[62] Li Y, Wang S, Zhao Y, Yue L. Effect of thermoelectric modules with different characteristics on the performance of thermoelectric generators inserted in the central flow region with porous foam copper. Applied Energy. 2022;**327**:120041

[63] Li Y, de León F, Zhao Y, Yue L, Zihao H, Wang R. Influence and comprehensive evaluation of porous metal foam filled fins on the performance of power cable surface waste heat recovery thermoelectric power generation device. Thermal Science and Engineering Progress. 2024; **47**:102278

[64] Siddique ARM, Mahmud S, Van Heyst B. A comprehensive review on vibration based micro power generators using electromagnetic and piezoelectric transducer mechanisms. Energy Conversion and Management. 2015;**106**: 728-747

[65] Chou SK, Yang WM, Chua KJ, Li J, Zhang KL. Development of micro power generators - a review. Applied Energy. 2011;**88**(1):1-16

[66] Shuai F, Teng P, Peng Q, Zhang L, Yin R, Yaojie T. Investigation on combustion characteristics and energy efficiency improvement of $H_2/NH_3$ fueled micro power generator with bluff body and porous media. International Journal of Hydrogen Energy. 2024;**80**: 1356-1367

[67] Katsuki F, Tomida T, Nakatani H, Katoh M, Takata A. Development of a thermoelectric power generation system using reciprocating flow combustion in a porous fesi2 element. Review of Scientific Instruments. 2001;**72**(10): 3996-3999

[68] Weinberg FJ, Rowe DM, Min G, Ronney PD. On thermoelectric power conversion from heat recirculating combustion systems. Proceedings of the Combustion Institute. 2002;**29**(1):941-947

[69] Hanamura K, Kumano T, Iida Y. Electric power generation by super-adiabatic combustion in thermoelectric porous element. Energy. 2005;**30**(2–4): 347-357

[70] Bubnovich V, Maiza M, Henriqez L. Modeling of thermoelectric power generation by porous media combustion burner. Chemical Engineering Transactions. 2011;**25**:141-146

[71] Bubnovich VI, Maiza MF, Henríquez-Vargas LA. Analysis of thermal energy conversion into electric power inside a porous media burner. Theoretical Foundations of Chemical Engineering. 2012;**46**(6):666-672

[72] Luis Henriquez-Vargas M, Maiza, Donoso P. Numerical study of thermoelectric generation within a continuous flow porous media burner. Journal of Porous Media. 2013;**16**(10): 933-944

[73] Ding J, Jiaqiang E, Li J, Cai L, Luo B. Effect analysis on energy conversion enhancement of porous medium micro-combustor and thermoelectric system and its optimization. Energy Conversion and Management. 2023;**292**:117441

[74] Yunfei Yan Y, Wei DW, You J, He Z, Zhang C. Numerical study on combustion and energy efficiency characteristics of thermophotovoltaic-thermoelectric two-stage utilization system filled with porous media. Energy. 2024;**308**:132909

Chapter 9

# Pore Formation in Biodegradable Polymers via Supercritical $CO_2$ Foaming: Mechanisms, Characterization, and Applications

*Ignacio García-Casas, Diego Valor, Ludisbel León-Marcos, Antonio Montes and Clara Pereyra*

## Abstract

The development of porous biodegradable polymers via supercritical $CO_2$ foaming represents a promising approach towards sustainable materials engineering. This chapter explores the physical and chemical mechanisms behind pore formation in such systems, focusing on gas-polymer interactions, nucleation dynamics, and pore growth during pressure quenching. The process is placed in the context of transport phenomena in porous media, highlighting mass transfer, solubility, and thermodynamic transitions relevant to the foaming process. Recent advances in pore formation are presented, including multiscale approaches that link molecular interactions with macroscopic pore structures. Experimental methods for characterizing the resulting porous architectures—such as Scanning Electron Microscopy (SEM), Mercury Intrusion Porosimetry (MIP), and X-ray Micro-Computed Tomography (μCT)—are discussed to assess porosity, pore size distribution, and interconnectivity. The chapter also examines how the structure-property relationships of these foamed biodegradable polymers can be optimized for practical applications in biomedical devices, packaging, and filtration systems. Emphasis is placed on the environmental benefits of using $CO_2$ as a green blowing agent and biodegradable polymers as a renewable substrate. This paper aims to bridge fundamental understanding and real-world applicability, in line with current theoretical advances and future opportunities in porous media research.

**Keywords:** $scCO_2$ foaming, biopolymers, porous materials, morphology, scaffolds

## 1. Introduction

The widespread use of conventional plastics has generated growing environmental concern due to their long degradation times and their tendency to accumulate in terrestrial and marine ecosystems. In response, biodegradable polymers have emerged as a promising alternative. These materials can naturally degrade through microbial activity, breaking down into harmless by-products, such as carbon dioxide, water, and biomass. This makes them key components in the transition towards a circular economy based on sustainable materials.

Biodegradable polymers can be classified by origin as natural (e.g., starch, cellulose, and proteins), semi-synthetic or synthetic (e.g., PLA (polylactic acid), PBS (poly(butylene succinate)), and PBAT (poly(butylene adipate-co-terephthalate)) [1]. They have a wide range of applications, including packaging, agriculture, biomedicine, and other environmentally sensitive areas where recyclability is not an option or where contact with living tissues or food requires safety and purity [1, 2].

Despite their environmental advantages, biodegradable polymers often present processing challenges, such as low thermal stability, limited melt strength, or crystallization kinetics, which complicate the fabrication of advanced materials. In this context, polymer foaming has emerged as a technique that can enhance properties while reducing material usage. Foaming introduces a cellular structure into the polymer matrix, resulting in lightweight materials with improved thermal, acoustic, and mechanical properties that are useful in packaging, insulation, filtration, and the biomedical field [1].

Foaming techniques are usually categorized as chemical, mechanical, or physical. Chemical foaming uses blowing agents that thermally decompose to release gas and form cells. However, they often leave behind residual chemicals, which make them less suitable for sensitive applications [3, 4]. Mechanical foaming involves injecting gas through agitation, but this method offers limited control over the final morphology. In contrast, physical foaming with supercritical $CO_2$ ($scCO_2$) provides precise control over the structure and avoids the use of toxic solvents [4].

The $scCO_2$ foaming process involves two main stages: (i) saturation, in which $CO_2$ diffuses into the polymer at an elevated pressure and temperature; and (ii) depressurization, which causes a sudden drop in solubility and leads to the nucleation and growth of gas bubbles [2, 5]. The structure of the resulting foam (cell size, density, and uniformity) is strongly influenced by parameters, such as saturation pressure and temperature, time, and depressurization rate [6, 7]. A key feature of $scCO_2$ is its plasticizing effect, which reduces the glass transition temperature (Tg) and viscosity of polymers, enabling processing at lower temperatures—critical for thermally sensitive biodegradable polymers, such as polylactic acid (PLA) and thermoplastic starch (TPS) [1, 8].

Furthermore, $scCO_2$ is non-toxic, inert, and non-flammable, and it does not leave behind any harmful residues, making it ideal for use in food packaging and biomedical devices [1, 2]. In addition to enabling structural control, $scCO_2$ allows functional additives (e.g., drugs, antioxidants, and antimicrobials) to be impregnated simultaneously during foaming, facilitating the production of multifunctional porous materials [1]. For example, active packaging materials or drug-loaded scaffolds for tissue engineering can be produced in a single step using supercritical $CO_2$.

The foaming efficiency of a polymer depends on its molecular characteristics. PLA, for example, is widely used, but it has low melt strength, which can cause bubbles to collapse [9]. Starch-based polymers, on the other hand, are difficult to foam due to their hydrophilicity and weak mechanical properties [7]. To overcome these issues, researchers have investigated the use of polymer blends, chain extenders, and nanoparticle fillers, such as calcium carbonate ($CaCO_3$) or carbon nanofibers, to improve foam stability, nucleation, and mechanical performance [10, 11].

In summary, $scCO_2$ foaming of biodegradable polymers is a versatile, sustainable, and high-performance method of producing advanced porous materials, with increasing potential in industries that demand environmentally responsible solutions.

## 2. Supercritical carbon dioxide: A smart tool for clean polymer processing

Supercritical carbon dioxide is a fluid that becomes supercritical when the temperature exceeds 31.1°C and the pressure exceeds 7.38 MPa [4, 8]. Under these conditions, it exhibits a unique combination of liquid-like density and gas-like diffusivity, enabling efficient dissolution, swelling, and penetration of polymer matrices—making it an ideal medium for polymer processing and foaming [4].

In polymer foaming, $scCO_2$ acts as a physical blowing agent, plasticizing the polymer and reducing its viscosity and glass transition temperature (Tg) [8]. This enables thermally sensitive biodegradable polymers, such as PLA and starch derivatives, to be processed without compromising their integrity [1]. Its non-toxic, inert, and solvent-free nature also makes it ideal for use in biomedicine and food packaging [2].

A key advantage of supercritical carbon dioxide ($scCO_2$) is that its solvent power can be tuned by adjusting the pressure and temperature. This provides precise control over gas solubility, nucleation, and foam cell morphology, enabling micro- or nanocellular structures to be created that are optimized for specific functions ranging from thermal insulation to biomedical scaffolds [4, 5]. Furthermore, $scCO_2$ can impregnate functional molecules into the polymer matrix during foaming, expanding its use in producing active materials for drug delivery or antimicrobial packaging [1].

Although the potential of $scCO_2$ foaming is promising, its industrial-scale implementation still faces challenges, including the cost of high-pressure equipment, the need to optimize the process for each polymer, and scaling limitations [2]. Nevertheless, advancements in $CO_2$-assisted extrusion, gas co-blowing strategies, and eco-efficient process design are being actively pursued [6, 7]. Overall, $scCO_2$ offers a clean, tunable, and efficient approach to producing biodegradable foam, in line with the principles of green chemistry and modern goals for sustainable manufacturing.

In addition to its technical advantages, $scCO_2$ foaming offers significant environmental benefits. Unlike conventional chemical blowing agents, which may release volatile organic compounds (VOCs) or leave residues, $scCO_2$ is non-toxic, recyclable, and leaves no residue. It enables solvent-free processing and eliminates the need for post-treatment purification, thereby reducing energy and material consumption. From a sustainability perspective, the $CO_2$ used in foaming can be captured from industrial emissions, transforming waste gas into a valuable resource—a process known as carbon capture and utilization (CCU). Furthermore, most industrial systems operate in closed loops, enabling repeated $CO_2$ recovery with minimal losses and low environmental impact. While maintaining supercritical conditions does require energy, this is often offset by the lower processing temperatures enabled by the plasticizing effect of $scCO_2$. This reduces degradation and energy input compared to thermal or solvent-based foaming methods.

Comparative studies show that when biodegradable or bio-based polymers are used, supercritical $CO_2$ foaming achieves one of the lowest environmental footprints among current foaming technologies. The resulting materials are compatible with circular economy models, particularly when polymers such as PLA, poly(ε-caprolactone) (PCL), or starch derivatives are used.

## 3. Biodegradable polymers

This section provides a comprehensive overview of biodegradable polymers, focusing on their definition, classification, and specific types relevant to foam formation using supercritical carbon dioxide ($scCO_2$). Particular attention is paid to the physicochemical properties that determine their suitability for this advanced processing technique, especially thermal transitions and interactions with $scCO_2$.

### 3.1 Fundamental concepts and terminology

A biodegradable polymer is defined as a polymeric material that degrades as a result of the metabolic activity of microorganisms, such as bacteria, fungi, or algae [12]. This biodegradation process leads to the formation of biomass and carbon dioxide ($CO_2$) under aerobic conditions or methane ($CH_4$) under anaerobic conditions, as well as water and mineral salts [12]. This definition highlights the biological mechanism as the main cause of degradation, setting it apart from other degradation pathways, such as photodegradation or oxidative degradation, which do not necessarily involve microbial assimilation or the complete mineralization of substances into harmless end products.

The environmental context and temporal framework are crucial aspects of this definition. Ideally, biodegradable materials should decompose within a reasonable timeframe (e.g., within a year) in the most common disposal environment, transforming into non-toxic end products, such as humus, water, or $CO_2$. However, 'biodegradability' is not an absolute material property, but rather one that depends heavily on specific environmental conditions and time. For example, although polylactic acid (PLA) is compostable under industrial conditions (55–60°C), it may not readily degrade in natural environments such as the ocean or soil, where it may instead fragment into microplastics. Degradation rates are critically influenced by environmental factors, such as temperature, pH, and oxygen availability, as well as polymer formulation [13].

It is important to distinguish between 'biodegradable' and 'degradable' polymers. The latter may decompose via non-biological mechanisms, meaning they are not necessarily microbially assimilated or mineralized. The term 'bioplastics' is also frequently used and can refer to bio-based materials from renewable sources, biodegradable materials, or both [12].

### 3.2 Overview of biodegradation mechanisms

Biodegradation of polymers is a multiphase process that usually starts with biofragmentation. In this stage, polymer chains are broken down into smaller oligomers and monomers by extracellular enzymes released by microorganisms, or through initial abiotic processes that facilitate microbial access. These low-molecular-weight fragments can then be taken up by microbial cells and metabolized.

A polymer's susceptibility to biodegradation is influenced by multiple factors. The chemical structure is paramount in dictating the stability of functional groups, reactivity (especially towards enzymatic hydrolysis), hydrophilicity, and water-swelling behavior, which facilitates microbial attack. Other intrinsic factors include crystallinity (amorphous regions are generally more degradable), molecular weight (shorter chains degrade faster), and morphology. Environmental conditions, such as temperature, humidity, pH, and microbial population composition, are also critical [12].

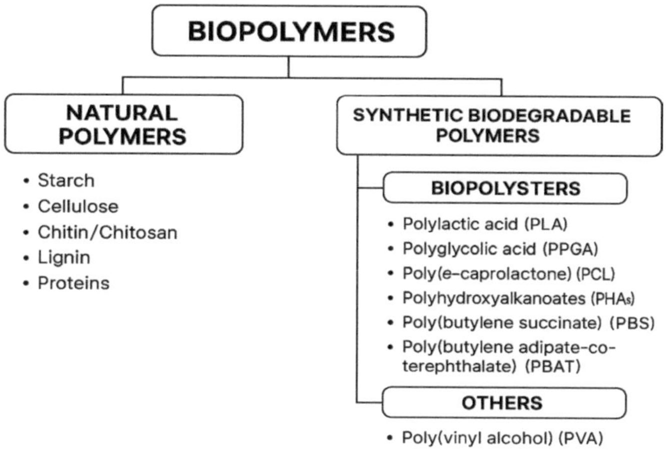


**Figure 1.**
*Scheme of polymers according to their source.*

### 3.3 Classification and types of biodegradable polymers

Biodegradable polymers exhibit considerable diversity. They originate from various sources and their distinct chemical structures determine how they degrade and can be processed (**Figure 1**).

*Natural Polymers*: Derived directly from biomass including plants, animals, or microbial sources. These polymers are inherently biodegradable and renewable but often require modification to meet processing and performance criteria for foam applications.

*Synthetic Biodegradable Polymers*: These are chemically synthesized from either bio-based monomers or fossil-derived precursors. Their biodegradability is engineered into the polymer backbone, often through hydrolysable linkages. They offer tailored properties and controlled degradation rates, making them highly relevant to advanced applications such as scCO2 foaming.

A trade-off exists between source and processability. While natural polymers like starch and cellulose are abundant and renewable, they typically require significant chemical modification (e.g., starch to TPS [14], cellulose to cellulose acetate (CA) [15]) to become thermoplastic and amenable to melt-based foam processing such as $scCO_2$ extrusion. In contrast, synthetic biopolyesters like PLA, PCL, or polyhydroxyalkanoates (PHAs) are often designed for better processability, even though they present challenges such as low melt strength (e.g., PLA). Hence, there is often a balance between the sustainability of raw materials and their direct applicability to advanced processing. Life-cycle considerations, including energy input and chemical modification steps, are essential in assessing the overall sustainability of $scCO_2$-foamed biodegradable products.

## 4. Supercritical foaming process

The supercritical foaming process is a process that uses supercritical fluid, commonly $CO_2$, to generate a porous and interconnected structure in a polymer. The polymer foaming process generally involves three core stages: the creation of a homogeneous polymer/gas system, the nucleation of cells, and the subsequent growth and stabilization of these cells. In this process, $CO_2$ is brought into contact with the polymer to be treated under supercritical conditions. As can be seen in the diagram in **Figure 2**, the $CO_2$ is pumped to the chosen operating conditions (pressure and temperature) inside the chamber where the polymer is located.

The operating temperature must generally be above the glass transition temperature (Tg) for amorphous polymers and above the melting temperature (Tm) for crystalline polymers. This is why amorphous polymers or those with a large amorphous domain are generally used, so that the operating temperature is not too high.

The polymer and $CO_2$ must be able to form a solution where the $CO_2$ needs to diffuse to carry out the nucleation and growth process of the bubbles. The gas's plasticizing effect on the polymer increases the mobility of the polymer molecular chains, which in turn lowers the free energy barrier for cell nucleation. This cell nucleation is driven by thermodynamic instability within the system. Once a contact time has elapsed, the chamber is subjected to controlled depressurization using a micrometric depressurization valve, causing the dissolved $CO_2$ to lose its solubility in the polymer. This reduction in $CO_2$ solubility could also be achieved by increasing the temperature in the chamber.

The supersaturation of the gas (**Figure 3**) within the polymer causes the nucleation of millions of small bubbles (cells). The uniformity and size of these cells are fundamental for the properties of the final product. In this sense, higher saturation pressure and a lower temperature, a priori, favor the formation of smaller and more homogeneous cells. The newly formed bubbles grow as the gas expands. However, only nuclei that attain a critical size can maintain stability. Subsequently, the gas dissolved within the polymer matrix diffuses into the cell nucleus, leading to an increase in its diameter. The viscosity of the polymer and surface tension play an important role in this growth. Precise control of the temperature and decompression rate during this stage is vital to obtain the desired density, cell size, and porous structure of the final porous structure.

This process can be modified by including additives and/or pore generators to achieve greater nucleation and interconnection of the formed pores. The most

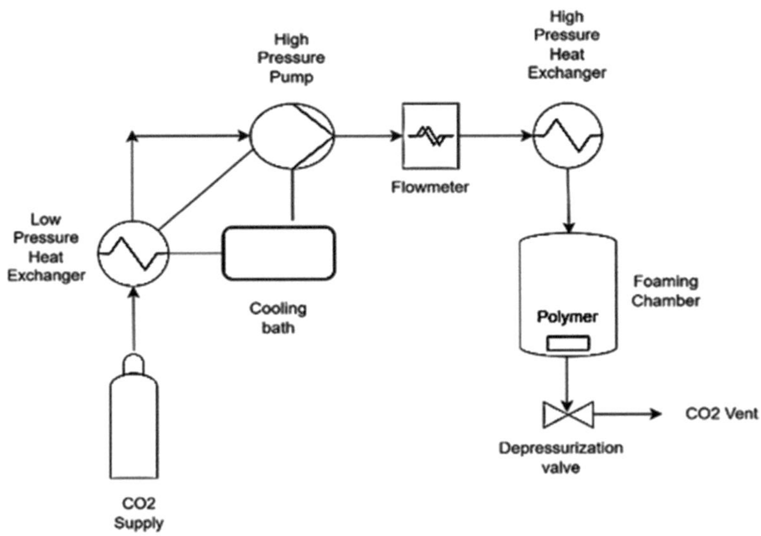


**Figure 2.**
*Supercritical foaming process pilot plant.*

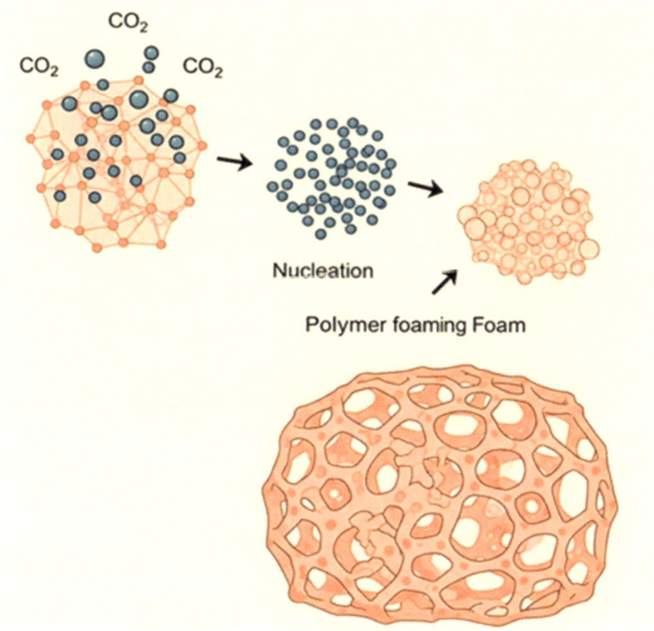


**Figure 3.**
*Supercritical foaming process scheme.*

widespread approach is to use inorganic particles to promote heterogeneous nucleation. These particles provide specific sites in the polymer matrix where the supercritical gas can nucleate more easily when pressure is reduced or temperature is increased. By having a surface with high interfacial energy with the polymer and the gas, they act as seeds for bubble formation. Some examples of particles used are silica nanoparticles [16], titanium dioxide [17], calcium carbonate [18], glass or carbon fibers [19], clays [20], which even improve mechanical properties, or carbon nanotubes [21], which can improve the mechanical, electrical, and thermal properties of the formed polymers. The pore generation effect could also be achieved by adding another polymer with a different glass transition or melting temperature to create heterogeneity in the matrix that favors nucleation, since when one of the polymers reaches its Tg or Tm and the other does not, interfacial stresses are generated that can promote bubble nucleation.

### 4.1 Process optimization strategies and control tools

Optimizing supercritical $CO_2$ ($scCO_2$) foaming conditions requires balancing thermodynamics, gas solubility, polymer rheology, and kinetics. The key variables—pressure, temperature, saturation time, and depressurization rate—must be tailored to each polymer system to achieve desired pore structure, porosity, and interconnectivity. **Table 1** summarizes optimal ranges of pressure and temperature for three commonly used biodegradable polymers.

### 4.2 Guidelines for parameter selection

A simplified decision guide (**Figure 4**) is provided to assist in parameter selection:

| Polymer | Optimal Pressure (bar) | Optimal Temperature (°C) | Notes |
|---|---|---|---|
| PLA | 150–250 | 35–45 | Lower temperatures preserve polymer integrity; high pressure improves cell density |
| PCL | 100–200 | 40–70 | Broader range due to flexibility; higher T favors expansion but risks collapse |
| PBS | 120–160 | 100–115 | Typically foamed near melting point; blends enhance open-cell structure |

**Table 1.**
*Optimal ranges of pressure and temperature.*

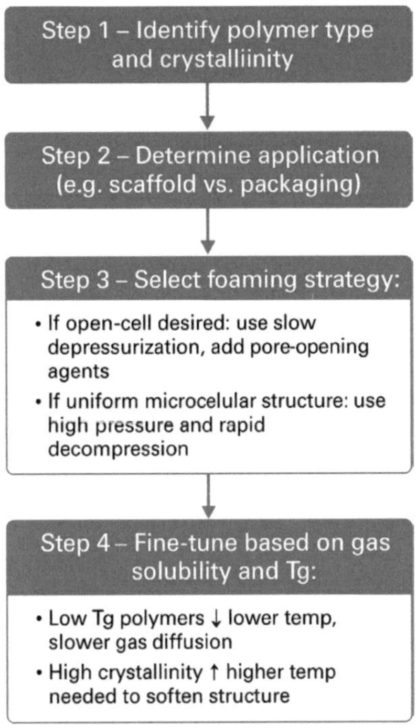


**Figure 4.**
*Parameter selection decision guide.*

## 5. Mechanisms

The processes implied in supercritical foaming are divided into several steps. First, a solution of gas/polymer should be formed. Then, the diffusion and nucleation of bubbles of gas into the polymer structure occur. This nucleation can be homogeneous or heterogeneous, depending on the presence or absence of inorganic particles within the polymer. The subsequent steps include growing, stabilization, and pore formation.

### 5.1 Gas/polymer solution formation

The process starts when a blowing agent, typically carbon dioxide ($CO_2$), is introduced into the polymer. Under high pressure, the $CO_2$ gas directly dissolves into the polymer melt without any chemical reaction. This creates a uniform polymer/gas solution. The more gas that dissolves, the lower the final foam density will be. Factors like temperature, pressure, and the specific chemical attraction between the gas and the polymer all influence how much gas dissolves. Interestingly, the dissolved $CO_2$ acts as a plasticizer for the polymer. This means it significantly reduces the polymer's melt viscosity and lowers its glass transition temperature (Tg). This plasticizing effect is crucial because it makes it easier for the gas to move around and for bubbles to grow [22].

### 5.2 Cell formation

Once the polymer-gas solution is formed, a thermodynamic instability is introduced, usually by rapidly dropping the pressure or increasing the temperature. This makes the dissolved gas supersaturated within the polymer [1]. When the solution becomes supersaturated, the gas wants to escape and clump together, forming tiny gas bubbles called cell nuclei. This process is called nucleation. If the supersaturation is high enough, these nuclei can form spontaneously throughout the solution (this is called homogeneous nucleation). However, if there are particles or impurities in the system, they act as preferred spots for bubble formation, as they lower the energy barrier for nucleation. This is known as heterogeneous nucleation, and it often leads to a more uniform cell structure [23].

Once nuclei are formed, dissolved gas from the surrounding polymer quickly diffuses into these growing bubbles. The main reason why this happens is the difference in gas concentration. The higher the gas diffusivity, the faster the bubbles will grow [22]. The pressure of the gas inside a growing cell is inversely related to its radius, this means smaller bubbles have higher internal pressure. As a result, gas tends to move from areas of higher pressure (smaller bubbles) to areas of lower pressure (larger bubbles). This causes smaller bubbles to shrink and disappear while larger ones continue to grow.

The viscoelastic properties of the polymer melt are also critical for controlling cell growth. The polymer needs enough melt strength to resist the outward pressure from the expanding bubbles, which prevents the bubbles from merging prematurely or rupturing. At the same time, it needs to be fluid enough to allow gas to diffuse and the bubbles to expand.

### 5.3 Cell stabilization

The inherent melt strength of the polymer is a key point for stabilizing the cellular structure. Polymers with higher melt strength are better at holding onto the cellular structure and preventing cells from merging, especially during the growth phase. One common way to stabilize the foam is by rapid cooling [10]. As the temperature drops, the polymer's viscosity increases dramatically, essentially 'freezing' the bubbles in place and preventing them from growing further or collapsing. Alternatively, stabilizing agents like surfactants, specific polymers, or fillers can be added to the polymer solution. These agents increase the viscosity of the cell walls, making them more resistant to drainage and rupture. In some cases, such as with thermoset foams like polyurethane, chemical crosslinking reactions occur during foaming, which creates a rigid network that permanently stabilizes the cellular structure.

## 6. Comprehensive analysis of foam morphologies and properties of key biodegradable polymers

### 6.1 Poly(lactic acid) (PLA) and copolymers (PLGA, PLLA)

Poly(lactic acid) (PLA) and its derivatives, such as poly(L-lactic acid) (PLLA) and poly(lactic-co-glycolic acid) (PLGA), are widely studied biodegradable polymers, primarily due to their predictable resorption rates and well-documented biocompatibility[1]. These polymers have been instrumental in the development of porous scaffolds for tissue engineering, particularly when processed with supercritical carbon dioxide ($scCO_2$).

#### *6.1.1 Porosity, pore size, and connectivity characteristics*

Scaffolds based on PLA and its copolymers typically achieve high porosity values, with multiple reports citing void fractions above 80%. For instance, PLGA-based foams have been shown to achieve porosity levels of up to 94.97% [24], while PLA/PBS blends have demonstrated values as high as 97.7% [25]. Levels of this kind have been recorded in poly(DL-lactide) (P(DL)LA) and poly(DL-lactide-co-glycolide) (P(DL)LGA) systems [26], thus highlighting the suitability of $scCO_2$ processing for the generation of low-density, high-surface-area matrices.

The average pore size in these foams varies significantly depending on both material formulation and processing parameters. For instance, PLGA scaffolds typically exhibit pores with a diameter ranging from 30 to 200 μm, which are conducive to supporting diverse cell types. The addition of chitosan has been demonstrated to result in a reduction of pore size from 190 to 110 μm [27]. The incorporation of magnetic nanoparticles (SPIONs (superparamagnetic iron oxide nanoparticles)) has also been demonstrated to contribute to pore size refinement [24]. In neat PLA, mean pore diameters of approximately 109 μm can be reduced to ~21 μm through the addition of 3 wt% microcellulose fibrils [28]. Ternary blends such as PLA/PBS/PBAT frequently exhibit bimodal pore architectures, with distinct populations of small pores (105–164 μm) and larger macropores (476–889 μm) [29].

Pore interconnectivity is another critical parameter for tissue scaffolds, influencing nutrient diffusion and cell migration. This feature can be significantly enhanced by means of formulation adjustments and precise control of processing conditions [30]. For instance, silica-reinforced PLGA foams exhibit increased network continuity [31], while the addition of chitosan enhances pore interconnections by up to 2.5 times [27]. Blends of PLA/PBS yield highly open structures, with up to 98.2% open cells [25]. The rate of depressurization is also of pivotal significance. Slower rates have been observed to promote pore coalescence and network connectivity, while rapid decompression has been shown to favor smaller and more isolated pores [30].

#### *6.1.2 Effects of processing parameters*

Increased pressure during the process of $scCO_2$ foaming has been shown to reduce average pore size; this trend has been observed to be consistent across a variety of systems, with enhanced homogeneity and finer pore distributions being noted at pressures greater than 250 bar [32]. In composite systems such as PLGA/SPIONs, pressure is a pivotal factor in achieving multimodal morphologies [24].

Temperature exhibits a significant influence on both the rate of diffusion of $CO_2$ and the viscosity of the polymer melt. It has been established that moderate increases in temperature result in the dilatation of pores, while very high temperatures can cause structural collapse due to a reduction in melt strength [33]. However, PLA/PBS/PBAT blends demonstrate an inversion of this trend, with elevated temperatures resulting in diminished pore size. This phenomenon may be attributed to alterations in gas solubility and melt rheology [34]. In the case of more sensitive copolymers such as P(3HB-co-4HB) (poly(3-hydroxybutyrate-co-4-hydroxybutyrate)), lower processing temperatures are required as the proportion of 4-hydroxybutyrate (4HB) increases [35].

Brief periods of saturation may result in the formation of heterogeneous structures, characterized by a dense core and a porous shell. Optimal times (e.g., 10 minutes) have been shown to enhance porosity and average pore size, while prolonged saturation has been observed to decrease pore dimensions by promoting overplasticization [30]. In PLGA-based systems, the application of extended durations of up to 2 hours has been shown to result in the achievement of consistent foaming behavior [36]. Gradual depressurization has been shown to promote interconnectivity and wider pore distributions [30]. Conversely, rapid decompression has been shown to promote homogeneity and smaller pore size, but may increase closed-cell content. In PLGA/SPION systems, it is imperative to meticulously modulate this rate to yield hierarchically and functionally graded pore networks. Nevertheless, it has been demonstrated that a sluggish process of depressurization may impede the process of foam expansion, should gas effusions occur prior to the stabilization of pores [36].

### *6.1.3 Effect of additives and polymer blends*

The blending of PLA with TPU (Thermoplastic Polyurethane) in solvent-free $scCO_2$ processes has been shown to result in the production of foams that exhibit enhanced elasticity and increased open-cell content. Foams containing 50% TPU achieved open porosity levels of up to 96.2%, while blends comprising 30% TPU demonstrated a high expansion ratio of 36.06. Furthermore, TPU has been demonstrated to enhance PLA crystallinity, thereby contributing to enhanced foam stability.

PLA with PBS (poly(butylene succinate)) blends are distinguished by their full biodegradability and highly interconnected structures. The materials demonstrate porosity levels of up to 97.7%, accompanied by 98.2% open-cell content, thus substantiating their efficacy in filtration and environmental applications [25]. Furthermore, PBS has been demonstrated to contribute to morphological tuning, favoring open-cell formation.

PLA/PBS with PBAT (poly(butylene adipate-co-terephthalate)) ternary systems, when reinforced with nanohydroxyapatite (nHA), exhibit a wide porosity range (26.5–94.4%) and diverse open-cell contents (12.7–84.3%), depending on the process conditions to which they are subjected [29]. The presence of bimodal pore populations enhances their suitability for tissue engineering.

PLA with MCFs (Microcellulose Fibrils) functions as a potent nucleating agent, resulting in a substantial reduction in average pore size, from 109 to 21 μm [28]. It has been demonstrated that open porosity increases with low MCF loading (from 71.8 to 76.1% with 1.5% MCF), but may decrease at higher concentrations. This indicates an upper limit beyond which excessive filler impairs porosity.

The incorporation of chitosan has been shown to result in a reduction of the average pore size, accompanied by a substantial enhancement in pore interconnectivity

[27]. However, it is noteworthy that elevated levels of chitosan or the incorporation of phytochemical extracts (e.g., derived from olive pruning) have been observed to impede the formation of fine pores and thereby reduce overall porosity.

The integration of conductive polymers, such as PEDOT:PSS (poly(3,4-ethylenedi oxythiophene):polystyrene sulphonate), into PLGA facilitates the fabrication of electroactive scaffolds. These materials exhibit moderate porosity (approximately 40%) alongside electrical conductivities ranging from $2.2 \times 10^{-7}$ to $1.0 \times 10^{-5}$ S/cm [37]. The uniform distribution of PEDOT:PSS is conducive to both structural integrity and electrical continuity, rendering these scaffolds promising candidates for neural or cardiac regeneration.

## 6.2 Poly(ε-caprolactone) (PCL) and composite systems

Poly(ε-caprolactone) is a semicrystalline aliphatic polyester that has gained a reputation for its biodegradability, biocompatibility, and mechanical flexibility, which render it highly suitable for tissue engineering applications [38]. Its relatively low melting point and ease of processing render it compatible with supercritical $CO_2$ ($scCO_2$) foaming methods.

### *6.2.1 Porosity, pore size, and connectivity characteristics*

Porosity values of up to 89% have been documented in the literature [5], and interconnectivity can attain 96% in certain low-molecular-weight PCL systems (e.g., 27 kDA) [39]. While the incorporation of nanofillers generally results in porosity levels above 80% [38], the inclusion of bulky porogens or ceramic fillers has been observed to reduce porosity due to space competition. For instance, in PCL/HA (hydroxyapatite) composites, porosity typically ranges from 65.6 to 74.4% at 40°C and from 61.9 to 67.1% at 50°C [40]. Analogous trends have been observed in scaffolds impregnated with carvacrol [40].

The mean pore diameter of pure PCL foams is typically found to be within the range of 70–180 μm [39]. However, it is important to note that this range can be significantly altered in the processing stage. For instance, an increase in saturation pressure from 100 to 200 bar (at 40°C) has been shown to result in a reduction of the average pore size from 622 ± 62 to 85 ± 24 μm [33]. PCL/HA composites exhibit smaller pores, attributable to the nucleating effect of HA, with a typical range of 40–250 μm [29]. It is interesting to note that, following impregnation with carvacrol, there was a slight increase in pore size, from 133 to 140 μm [40]. The PCL/nC (nanocellulose) and PCL/nGO (graphene oxide) systems demonstrate notable nucleation behavior. These additives typically result in d10 values under 30 μm, although maximum pore sizes of up to 150 μm have been reported for low nGO loadings (0.2 wt%) [38].

Foams from 27 kDa PCL have been shown to reach interconnectivity values of 96% [39]. The incorporation of modest quantities of ethanol during the $scCO_2$ processing phase has been demonstrated to encourage the formation of more open and uniform pore structures, thereby enhancing the morphology relative to that of neat $CO_2$ [30]. The utilization of plasticizers, such as eugenol, has been demonstrated to promote enhanced throat connectivity and mesenchymal stem cell migration [41]. The employment of advanced processing methodologies, such as two-step depressurization, has resulted in the generation of bimodal, highly interconnected scaffolds, which are highly conducive to applications in regenerative medicine [40].

#### *6.2.2 Effects of processing parameters*

An increase in pressure from 100 to 200 bar at 40°C has been shown to result in a reduction in average pore size of almost sevenfold [33]. Moderate increases in temperature have been shown to enhance $CO_2$ diffusion and reduce melt viscosity, thereby promoting pore expansion [30]. However, at elevated temperatures, structural instability can lead to the formation of collapsed or dense morphologies [39]. In the context of PCL/HA foams, an increase in temperature from 40 to 50°C has been observed to result in a narrowing of the pore size distribution. However, this increase in temperature can also lead to phase coalescence and morphological irregularities [40]. In the context of biomedical scaffolds, a frequently cited compromise between structure and integrity is 70°C [38].

For PCL, durations ranging from seconds to one hour have been the subject of exploration [29]. It has been demonstrated that an increase in duration (e.g., 1 hour) generally results in larger pores and enhanced interconnectivity [38]. It has been demonstrated that rapid decompression (e.g., 20 mmHg/min) results in the uniformization of cell sizes throughout the matrix [38]. It has been demonstrated that a reduction in the rate of depressurization allows for a greater duration in which cell growth and coalescence can occur, thus resulting in the formation of larger, more interconnected pores [30]. Multistep depressurization strategies, including two-step release, have been shown to be especially effective for generating bimodal, interconnected architectures [40].

#### *6.2.3 Effect of additives and composite design*

PCL/HA (hydroxyapatite). It has been demonstrated that both micro- and nano-sized HA function as effective nucleating agents, promoting the formation of smaller pores and increasing their density [38]. The porosities of these composites are typically found to be within the range of 55–85%, with pore sizes varying from 40 to 250 μm [29]. While HA strengthens the structure by integrating into pore walls, micro-HA (μHA) has been observed to produce more heterogeneous morphologies than nano-HA (nHA) [38].

PCL/PANI (polyaniline). The incorporation of polyaniline (PANI) into the composition of the material results in the introduction of electrical conductivity while maintaining the integrity of the foam architecture. Despite the absence of direct foaming, PANI domains are uniformly dispersed within the PCL matrix. The resulting scaffolds demonstrate porosities of approximately 50%, and conductivities that are well suited for bioelectronic applications [42].

### 6.3 Polyhydroxyalkanoates (PHAs), including PHBV

PHAs and PHBV (poly(3-hydroxybutyrate-co-3-hydroxyvalerate)) are microbial-synthesized biodegradable polymers that provide an eco-friendly alternative to fossil-based plastics [35].

#### *6.3.1 Porosity, pore size, and connectivity characteristics*

The porosity levels of PHA foams are found to be contingent on both formulation and processing parameters. In particular, PHBV foams have been found to form open-cell structures with notably high porosity [43]. However, the foaming behavior of PHBV can be restricted due to the limited solubility of $CO_2$ in the polymer matrix [44].

The dimensions of pores in PHAs can be tuned by modifying the conditions under which they are processed. The average cell area of PHBV foams has been shown to be approximately 116.6 μm$^2$ [43], and earlier reports have achieved pore diameters ranging from 6 to 22 μm in microcellular structures [44]. In the context of P(3HB-co-4HB), it has been observed that elevated foaming pressures are associated with the formation of larger cells [35].

The degree of interconnectivity exhibited by PHA foams is subject to variation. PHBV has the capacity to develop open-cell morphologies that permit a certain degree of connectivity, which renders it a promising material for biomedical applications [43]. However, the formation of well-interconnected pore networks may be hindered by the material's mechanical rigidity [43].

### *6.3.2 Effects of processing parameters*

The application of elevated foaming pressure has been demonstrated to enhance $CO_2$ uptake and nucleation density, thereby increasing cell population [45]. Furthermore, elevated pressures have been shown to result in larger cell diameters and higher expansion ratios [35].

The ideal foaming temperature is contingent upon the 4HB ratio and the degree of crystallinity. For instance, P(3HB-co-10% 4HB), with higher crystallinity, forms cell structures around 130°C, while more amorphous versions (e.g., P(3HB-co-30% 4HB)) foam effectively between 50 and 80°C [35]. The utilization of chain extenders has been employed to extend this temperature window [45].

For P(3HB-co-4HB), a 60-minute saturation time followed by rapid depressurization to atmospheric pressure has been applied [35]. It is evident that these variables play a pivotal role in shaping the pore structure, although the precise nature of their influence remains to be elucidated.

## 6.4 Poly(butylene succinate) (PBS) and blends

Poly(butylene succinate) (PBS) is a biodegradable aliphatic polyester that is frequently blended with other polymers to enhance foam morphology, particularly open-cell formation [25].

### *6.4.1 Porosity, pore size, and connectivity characteristics*

Poly(butylene succinate) (PBS)-based foams, particularly those in blend form, have the capacity to attain notably elevated levels of porosity. For instance, the porosity levels of PLA/PBS foams have been reported to reach as high as 97.7% [25]. Pure PBS foam is generally categorized as microcellular, although precise porosity data are not always provided [43].

Foams composed of neat PBS exhibited an average pore area of 673 μm$^2$ [43]. In composite blends such as PLA/PBS/PBAT, dual-size cellular networks are frequently reported, with smaller pores measuring between 105 and164 μm and larger ones between 476 and 889 μm [29]. PBS foams reinforced with cellulose nanocrystals also exhibit bimodal structures, with larger pores measuring approximately 68.9 μm and smaller ones measuring approximately 11.0 μm [25].

PBS is a structural modifier that has been extensively utilized to enhance pore openness. Typically, PLA/PBS foams develop a highly interconnected, lace-like structure with up to 98.2% open-cell content [9]. PLA/PBS/PBAT systems have been shown to achieve up to 84.3% open cells under optimized foaming conditions [44]. In contrast, pure PBS has been observed to form predominantly closed cells with thin walls [43].

### *6.4.2 Effects of processing parameters*

Foaming of PBS is commonly conducted at its melting point (113°C) [43]. In blends that contain PBS, intermediate foaming temperatures have been shown to have a significant effect on macropore formation. A reduction in foaming temperature to approximately 100°C in PLA/PBS/PBAT blends has been shown to result in the formation of larger pores (889.8 ± 66.2 μm) and higher open-cell fractions [34].

Pure PBS foaming was carried out at 15 MPa [43]. In blends, a decrease in intermediate pressure drop (from 120 to 80 bar) has been linked with increased cell size and porosity [34], which is consistent with classical nucleation theory, where greater pressure drops facilitate cell expansion.

PBS foams were subjected to a $CO_2$ saturation time of 30 minutes [43], while PBS/PBAT blends were equilibrated over 90 minutes to ensure complete gas absorption [34].

For neat PBS, the process of depressurization was initiated rapidly (within 2–4 seconds) [43]. Conversely, the utilization of blended systems has been shown to be advantageous due to the stepwise depressurization process, which has been demonstrated to yield bimodal cell structures and enhanced interconnectivity [29].

The incorporation of PBS into PLA has been demonstrated to promote the development of highly interconnected cell architectures [25]. The incorporation of nanoparticles, such as nanohydroxyapatite (nHA), into ternary systems comprising PLA/PBS/PBAT has been demonstrated to enhance hydrophilicity and mechanical performance [34].

## **6.5 Cellulose and chitosan/derivative-based polymers**

Cellulose and chitosan are two naturally abundant biopolymers with promising characteristics for supercritical $CO_2$ ($scCO_2$) foaming. The inherent hydrophilicity, biocompatibility, and renewable nature of these materials render them particularly well suited to applications, such as biomedical scaffolds and the fabrication of porous materials with large surface areas [2].

### *6.5.1 Porosity, pore size, and connectivity characteristics*

The production of cellulose-based aerogels via scCO2 drying processes has been shown to result in the attainment of remarkably elevated levels of porosity, frequently ranging from 80 to 99.9% [46]. For instance, nanocrystalline cellulose (NCC) aerogels demonstrate porosities ranging from 91 to 95%, while ionic-liquid-treated pulp subjected to scCO2 drying reaches approximately 94–96% [46]. Furthermore, microcrystalline cellulose (MCC) processed under analogous conditions has been shown to exhibit porosities of 91–96% [46].

Chitosan scaffolds exhibit a wide range of porosities, commonly reported between 88 and 97 or 75 and 85% when interconnected pore networks are formed [47]. However, the available quantitative data on the foaming of chitosan solely by $scCO_2$ are limited; the majority of findings relate to structures formed through supercritical drying or gelation [47].

Pore dimensions vary considerably depending on the cellulose form and process. In $scCO_2$-dried aerogels, fibrils are typically characterized by a diameter of approximately 40 nm, while mesh distances can extend over several hundred nanometers [46]. Cellulose cryogels have demonstrated an enhancement in average pore size, ranging from 4.3 to 7.5 μm. In foamed cellulose acetate, the range of pore diameters is from 80 to 20 μm, with pressure being a key factor in size reduction [48].

The utilization of supercritical carbon dioxide ($scCO_2$) during the drying process of aerogels typically results in the preservation of their porous networks, which are characterized by significant interconnectedness. This phenomenon can be attributed to the elimination of capillary forces during the drying process, a process that has been shown to help retain the original gel structure and avoid pore collapse [46].

Interconnected architectures are commonly achieved in architectures derived from chitosan, with descriptions including 'polygonal and interconnected pores' or 'spherical pores in communication' [47]. The addition of chitosan to PLGA scaffolds has been demonstrated to augment the number of interpore connections [27]. However, the incorporation of certain additives (e.g., plant extracts) has been observed to impede microchannels, thereby reducing overall connectivity [27].

### *6.5.2 Effects of processing parameters*

In foamed cellulose acetate, an increase in pressure to approximately 250 bar has been observed to result in a reduction of the average pore size from 80 to around 20 μm, accompanied by a more uniform distribution [48]. In contrast, for cellulose aerogels, moderate pressures (~7.39 MPa) during supercritical drying are sufficient to preserve pore morphology while preventing collapse.

It has been demonstrated that foaming cellulose acetate at elevated temperatures (215–245°C) has a propensity to increase pore size [48]. However, for the supercritical drying of aerogels, lower temperatures (e.g., 31.1 or 60°C) are preferable in order to maintain structural integrity. In the case of cellulose acetate, saturation times between 180 and 480 minutes are employed to ensure a homogeneous pore structure [48]. In the process of drying starch gels (which are chemically similar to cellulose), it was determined that a 3-hour $scCO_2$ flow was effective [46].

In the domain of composite foams, the influence of additives on the characteristics of pores is a subject of significant interest. For instance, the incorporation of chitosan into PLGA has been observed to result in a reduction in pore size, while concurrently enhancing interconnectivity [27]. Conversely, the introduction of bioactive extracts (e.g., olive leaf extract (OLE)) has been shown to obstruct smaller pathways, thereby reducing connections [27]. In the field of PCL foams, the utilization of compounds such as μCMC and nanocellulose has been demonstrated to enhance porosity and morphology [38].

## 6.6 Other interesting biodegradable polymers and blends

### *6.6.1 PVA (poly(vinyl alcohol))*

Although direct foaming of PVA using $scCO_2$ is rarely described in the literature, it has been used as a supporting matrix in systems where PMMA (poly(methyl methacrylate)) microparticles are foamed within a PVA film [45]. In such systems, PMMA foaming has been observed to produce nanoporous features (0–1000 nm), with porosity and pore size being amenable to control via process parameters [45].

#### *6.6.2 Starch-based systems*

The process of starch foaming via $scCO_2$ drying results in the formation of aerogels, which exhibit porosity levels ranging from 61 to 76%. These aerogels demonstrate a high degree of interconnectivity and the presence of submicron pores, with a diameter of less than 1 μm, as previously reported in [46]. Fibril diameters of approximately 20 nm have been observed to be prevalent [46]. Typical $scCO_2$ conditions involve saturation at 120 bar and 60°C for a duration of 3 hours. It has been demonstrated that thermoplastic starch undergoes significant microstructural and rheological changes during processing [49].

### 6.7 Long-term stability and degradation of foamed polymers

The long-term stability and degradation behavior of foamed biodegradable polymers are crucial for their effectiveness in applications, such as biomedical scaffolds, environmental filters, and compostable packaging. The stability and degradation behavior of these polymers are influenced by polymer chemistry, foam morphology (e.g., pore size and open-cell content), and environmental conditions, such as humidity, temperature, and microbial activity.

#### *6.7.1 Structural integrity over time*

Foamed PLA and PLGA scaffolds generally maintain structural integrity over short to medium periods (1–3 months), especially under dry or physiological conditions [24, 26]. However, hydrolysis of ester bonds in humid environments gradually reduces mechanical strength and pore wall thickness. For instance, PLGA foams stored at 37°C in a PBS solution undergo mass loss over time, with significant degradation commencing after 6–8 weeks [36].

PCL foams are known for their slow degradation rates. Under physiological conditions (37°C, PBS), structural integrity can be preserved for several months. Kosowska et al. [38] reported that PCL-based foams retained over 80% of their initial compressive strength after 8 weeks of immersion, particularly when reinforced with hydroxyapatite (HA).

PBS-based foams degrade more rapidly when exposed to microbial activity. While PLA/PBS blends are mechanically stable initially, they show signs of crack formation and pore coalescence after ~12 weeks of prolonged composting under industrial conditions [25, 34]. In open-air or marine environments, however, degradation is significantly slower and may lead to microplastic-like fragmentation if the material is not fully mineralized.

#### *6.7.2 Influence of pore structure on degradation*

Highly porous structures degrade faster due to their increased surface area and water uptake. Open-cell foams tend to undergo accelerated hydrolytic degradation, especially in amorphous regions. In contrast, microcellular structures degrade more gradually, but provide longer-lasting mechanical support.

Additives and blends influence long-term behavior.

- Chitosan enhances biodegradation, but it may reduce mechanical resistance over time [27].

- Nanoparticles (e.g., HA and SPIONs) delay degradation and improve mechanical retention [24, 40].
- Plasticizers and surfactants (e.g., eugenol or polyethylene glycol (PEG)) can accelerate degradation by increasing water uptake and chain mobility [41].

### *6.7.3 Outlook and testing recommendations*

Future studies should include:

- Long-term soaking tests (≥3 months) in physiological and composting environments.
- Repeated mechanical testing (compression, Dynamic Mechanical Analysis (DMA)) to assess structural fatigue.
- In situ monitoring (e.g., micro-CT, SEM at intervals) to track pore collapse or coalescence.

It is essential to understand degradation pathways in order to match the lifespan of foam to the application. This is particularly important for temporary implants, biodegradable packaging, and controlled-release systems.

## 6.8 Functional additives in advanced applications

Not only do additives in sc$CO_2$-foamed biodegradable polymers influence pore morphology, they also enable the design of multifunctional materials tailored for biomedical, environmental, and packaging applications. Their functions extend beyond structural reinforcement to include bioactivity, responsiveness, and delivery capabilities.

### *6.8.1 Antimicrobial and antioxidant activity*

Natural bioactive compounds, such as carvacrol and olive pruning extracts, can be successfully impregnated during sc$CO_2$ foaming. This yields porous materials with antimicrobial or antioxidant properties [27, 40]. These materials are particularly promising for use in active packaging and wound dressings, where barrier and therapeutic effects are desired.

### *6.8.2 Bioactivity and osteointegration*

Nanofillers such as HA and chitosan promote osteoconductivity and enhance cell attachment in tissue engineering scaffolds [27, 38]. Their presence also modulates degradation rates and improves pore interconnectivity, making them ideal for bone regeneration and drug-releasing implants.

### *6.8.3 Controlled drug release*

Some additives serve as both structural modifiers and drug carriers. For instance, PLGA foams impregnated with polyphenols or antibiotics via sc$CO_2$ exhibit prolonged release profiles over weeks [36]. These systems enable monolithic delivery scaffolds to be fabricated without the need for post-processing.

*6.8.4 Electrical conductivity*

Incorporating conductive polymers such as PEDOT:PSS or PANI confers electrical conductivity while maintaining foam porosity [37, 42]. These scaffolds are suitable for neural or cardiac tissue engineering, where electroactive environments can enhance cellular responses.

*6.8.5 Thermal and mechanical enhancement*

Carbon-based fillers (e.g., carbon nanotubes and graphene oxide) and microcellulose fibrils can enhance mechanical strength and thermal stability, as well as providing electromagnetic shielding properties.

## 7. A comparative analysis of polymers and emerging trends from 2018 to the present day

A review of recent literature from 2018 to 2025 reveals clear advancements in the use of supercritical $CO_2$ ($scCO_2$) foaming for biodegradable polymers. This encompasses both refinements in processing techniques and a more profound comprehension of how formulation and process variables influence foam morphology.

### 7.1 Porosity trends

The utilization of $scCO_2$ foaming has been demonstrated to be a highly effective method for the production of polymeric foams, with the porosity, pore size, and connectivity of these foams being able to be tuned to meet the requirements of a range of applications. These applications include biomedical scaffolds and packaging materials [5]. **Table 2** has been optimized for visualization.

### 7.2 Processing parameters

In **Table 3**, it can be observed that the role of $scCO_2$ foaming parameters is broadly consistent across different polymers, even though each system may respond uniquely due to differences in crystallinity, viscosity, and gas solubility.

| Parameter | Observations | References |
|---|---|---|
| Porosity | High porosity (>80%) achievable in PLA, PLGA, and PCL. Influenced by $CO_2$ affinity, temperature, and additives like HA. | [5, 40] |
| Pore Size | Range from nanometers to millimeters. Bimodal structures seen in PLA/PBS/PBAT, suitable for bone tissue integration. | [29] |
| Interconnectivity | Interconnected pores up to 96% in PCL and 98.2% in PLA/PBS blends. Affected by depressurization rate and pore-opening additives. | [39] |

**Table 2.**
*Porosity trend parameters.*

| Parameter | Effects | Examples/Notes | References |
|---|---|---|---|
| Pressure | Higher pressure usually reduces pore size (PLA, PLGA, PCL), but may increase in P(3HB-co-4HB). | High nucleation density with elevated pressure. | [30, 35] |
| Temperature | Higher temperatures increase pore size via enhanced diffusivity and lower viscosity, but must be controlled to avoid instability. | Critical for semicrystalline polymers like PCL. | [30, 39] |
| Saturation Time | Longer times lead to uniform nucleation and structure; short times cause inconsistency. | Key for homogeneous $CO_2$ diffusion. | [30] |
| Depressurization Rate | Slow rate improves interconnectivity and larger pores; rapid rate yields smaller, uniform pores but less connectivity. | Staged depressurization effective in PLA/PBS/PBAT. | [29, 30] |

**Table 3.**
*Processing parameter trends.*

| Modification Type | Details | References |
|---|---|---|
| Nanofillers as Nucleating Agents | nHA, MCF, nGO, and SPIONs lower energy barrier for foaming, increase cell density, and reduce pore size. | [38] |
| Blends for Tailored Properties | PLA/PBS blends achieve up to 98.2% interconnectivity. Other blends enhance foam structure. | [25] |
| Crystallinity and Rheological Control | PLA/TPU improves crystallization; PHA + chain extenders boost rheological control. | [35] |
| Hybrid Processing Techniques | $scCO_2$ with two-step depressurization or plasticizers enables bimodal pores, throat tuning, and advanced tissue engineering. | [29, 41] |

**Table 4.**
*Impact of nanofillers and polymer blends.*

### 7.3 Impact of nanofillers and polymer blends

In **Table 4**, it is developed how the employment of nanofillers and polymer blends is evolving into a progressively sophisticated approach to address limitations in foamability, morphology, and functional performance.

## 8. Characterization techniques used

In addition to porosity, pore size, and interconnectivity, studies on supercritical $CO_2$ foaming of biodegradable polymers frequently incorporate a broad range of characterization techniques (**Table 5**). These are essential for understanding the microstructure, chemical composition, thermal and mechanical properties, and even biological performance of the materials. Below is an overview of commonly employed techniques, grouped by characterization objective.

### 8.1 Morphology and pore architecture

Scanning Electron Microscopy (SEM) is widely used to investigate the surface morphology and cellular structure of polymer foams [9, 27, 38, 42, 43]. It provides

| Characterization Category | Common Techniques | Purpose/Measurement |
|---|---|---|
| Morphological Structure | SEM, μCT, Optical Microscopy | Pore size, shape, interconnectivity |
| Chemical Composition | FTIR, XRD | Functional groups, crystallinity |
| Porosity and Surface Area | BET, MIP, Liquid Displacement | Total porosity, surface area, open-cell rate |
| Thermal Properties | DSC, TGA | Tg, Tm, thermal stability |
| Mechanical Properties | Tensile, Compression, DMA, Rheology | Strength, elastic modulus, viscoelasticity |
| Electrical/Biological | Conductivity, Cell Assays | Electroactivity, biocompatibility |
| Image Processing | ImageJ | Quantification from SEM/μCT images |

**Table 5.**
*Techniques used for characterization.*

high-resolution imaging to assess pore size and distribution, cell wall thickness, structural uniformity, and the presence of defects or inclusions.

Transmission Electron Microscopy (TEM) is ideal for studying nanostructures and internal phase morphology, especially in multicomponent systems, such as polymer blends and block copolymers [50]. It offers atomic to nanometer-scale resolution by transmitting electrons through ultrathin samples.

Optical Microscopy is used for general observation of foam structure and cell formation, particularly where high magnification is not required [51].

X-ray Micro-Computed Tomography (μCT/Nano-CT) is a non-destructive 3D (three-dimensional) imaging technique that provides insights into internal structure, pore architecture, and interconnectivity [31]. Nano-CT enables higher resolution (down to ~50 nm voxel size), allowing quantification of porosity, surface area, and tortuosity [52].

## 8.2 Chemical structure and crystallinity

Fourier Transform Infrared Spectroscopy (FTIR) identifies functional groups, additives, and contaminants through characteristic infrared (IR) absorption bands. It provides a molecular fingerprint of the sample based on vibrational transitions.

X-ray Diffraction (XRD) is used to analyze the crystalline structure and degree of crystallinity of polymer materials. It is particularly useful in evaluating structural changes induced by foaming or by the addition of fillers [53].

## 8.3 Porosity and surface area analysis

Nitrogen ($N_2$) Physisorption/Brunauer-Emmett-Teller (BET)/Barrett-Joyner-Halenda (BJH) Methods are used to measure specific surface area (BET method) and pore size distribution (BJH method), particularly in the mesoporous range [54, 55].

Mercury Intrusion Porosimetry (MIP) characterizes pore volume, porosity, and pore size distribution by intruding mercury under increasing pressures. Larger pores fill at lower pressure; pore diameter is calculated via the Washburn equation [36].

Liquid Displacement/Archimedes' Principle determines open porosity and bulk density by measuring the displaced volume of a non-reactive liquid and comparing dry, wet, and submerged sample weights.

### 8.4 Surface and nanoscale properties

Atomic Force Microscopy (AFM) provides true 3D topographical mapping at nanometer resolution and can measure roughness, layer thickness, and domain size. It also offers qualitative mapping of mechanical properties, such as stiffness and adhesion.

### 8.5 Thermal stability and transitions

Differential Scanning Calorimetry (DSC) assesses thermal transitions like the glass transition temperature (Tg), melting point (Tm), and crystallization behavior, supporting material identification and processing optimization.
Thermogravimetric Analysis (TGA) evaluates thermal stability and compositional changes by measuring sample mass loss as a function of temperature or time. It helps determine filler content, moisture levels, and degradation profiles.

### 8.6 Mechanical properties

Tensile Testing measures the tensile strength, elastic modulus, and elongation at break by applying uniaxial tension to the material. Compression Testing is used to evaluate compressive strength and elastic modulus, providing data on how a material behaves under crushing or compressive forces. Shear Testing assesses a material's resistance to sliding deformation, especially useful for adhesives, composites, and soft materials. Dynamic Mechanical Analysis (DMA) characterizes viscoelastic behavior under oscillatory stress, providing data on storage modulus, loss modulus, and Tg across temperature or frequency sweeps. Rheology assesses flow behavior and deformation under different shear rates, temperatures, and time scales. It is crucial for processing studies and optimizing the viscoelastic performance of polymer melts or gels.

### 8.7 Functional and biological performance

For conductive polymer composites, techniques, such as 2-point/4-point probes or impedance spectroscopy, are used to determine electrical resistivity and conductivity. Cellular Assays (WST-1 (Water-soluble tetrazolium salt 1), Fluorostaining, Alkaline phosphatase (ALP)). These methods assess biocompatibility and cell behavior on scaffolds: WST-1: Measures metabolic activity as a proxy for cell viability. Fluorostaining: Uses fluorescent dyes for imaging cell attachment and morphology. ALP Assay: Evaluates osteogenic differentiation via alkaline phosphatase activity, a marker of early stage osteoblast function. Agitation/Cohesiveness Testing: Measures a material's internal cohesion and resistance to breakage during agitation. Techniques include powder flow analysis and unconfined compressive strength tests.

### 8.8 Image analysis tools

ImageJ is an open-source tool extensively used for quantitative analysis of micrographs from SEM or μCT. It allows measurements of cell size, pore density, and structural features through calibration, binarization, and segmentation.

## 9. Current challenges and future research directions

Despite considerable advances in the field of $scCO_2$ foaming of biodegradable polymers, several critical challenges remain to be addressed in order to fully exploit the potential of this technique, particularly with regard to biomedical and environmental applications.

A primary concern is the precise control and reproducibility of pore interconnectivity, which is vital for tissue engineering where nutrient diffusion and cell migration are essential [30]. Despite the noteworthy levels of interconnectivity that have been documented in PCL and PLA/PBS systems, the attainment of uniform, large-scale homogeneity continues to elude researchers. The formation of dense, non-porous surface layers continues to limit the functionality of the resulting foams in practical applications [30].

The intricate interplay of thermodynamic and kinetic phenomena inherent in the process of $scCO_2$ foaming introduces significant challenges to the accurate prediction and control of material morphology [2]. These challenges are further compounded in polymer blends and nanocomposites, where the behaviors of components and the interactions between $CO_2$ and polymers can be challenging to model and replicate.

In terms of scalability, $scCO_2$ foaming offers significant environmental and technical advantages; however, its industrial application is currently limited by several key challenges. These include the high capital investment required for high-pressure systems, the predominance of batch processes, and the need to customize parameters for each polymer formulation.

Several practical strategies have been developed to address these limitations. For instance, $CO_2$-assisted extrusion has emerged as a viable continuous method in which $CO_2$ is injected directly into the molten polymer. This allows for higher throughput and process automation [6]. Hybrid methods combining $scCO_2$ foaming with techniques, such as 3D printing, injection molding, or electrospinning, have also shown promise in modular upscaling, particularly in biomedical applications [56].

Successful pilot implementations include producing PLA/PBS foams for oil adsorption and compostable packaging [25], and manufacturing PCL-based scaffolds using $scCO_2$ that demonstrate clinical applicability [40]. Modular high-pressure systems offered by technology providers such as Thar Technologies and Separex facilitate gradual scale-up, enabling industry validation at different stages. Furthermore, integrating real-time pressure and temperature control systems enhances reproducibility and ensures compliance with regulations.

It is crucial that research efforts prioritize the development of more sophisticated processing methodologies. These methodologies should include multistage depressurization strategies that are precisely tailored to specific pressure-temperature profiles. This approach is expected to improve control over bimodal pore structures and their interconnectivity. Combining $scCO_2$ foaming with complementary fabrication methods such as electrospinning or additive manufacturing (e.g., 3D printing) shows promise in producing complex, hierarchical, and multifunctional materials [56].

Multifunctional Additive Design: Efforts must be concentrated on designing next-generation nucleating agents, plasticizers, and compatibilizers. These agents have a well-documented influence on pore structure, and it is now recognized that they can also impart additional functionalities, such as antibacterial activity, electrical conductivity, and enhanced biocompatibility. The effective dispersion of the additives and their compatibility with the polymer matrix will be pivotal to their success [42].

Modeling and Simulation Tools: Developing robust theoretical and computational models to simulate supercritical $CO_2$ foaming behavior across various polymer systems could significantly reduce the need for trial-and-error experimentation. This would accelerate material design and improve the scalability and predictability of foam properties.

Industrial Scalability and Sustainability: The transition from laboratory-scale to industrial-scale foaming presents a dual challenge, both technical and environmental in nature. The efficient recovery and reuse of carbon dioxide, in addition to process optimization for energy and material sustainability, is imperative for the large-scale manufacturing of $scCO_2$ foaming to be considered viable. Ultimately, the commercialization of $scCO_2$ foaming depends on selecting materials aligned with market needs, retrofitting existing lines for $CO_2$ use, and forming partnerships with end users in medical, packaging, or filtration sectors. The convergence of sustainable materials and scalable clean technology places $scCO_2$ foaming in a favorable position for broader industrial adoption.

## Acknowledgements

We gratefully acknowledge the Spanish Ministry of Science and Innovation (Project PID2020-116229RB-I00) for financial support.

## Conflict of interest

The authors declare no conflict of interest.

## Author details

Ignacio García-Casas*, Diego Valor, Ludisbel León-Marcos, Antonio Montes and Clara Pereyra
Department of Chemical Engineering and Food Technology, Faculty of Sciences, University of Cadiz, International Excellence Agrifood Campus (CeiA3), Wine and Agrifood Research Institute (IVAGRO), Spain

*Address all correspondence to: ignacio.casas@uca.es

## References

[1] Gonçalves LFFF, Reis RL, Fernandes EM. Forefront research of foaming strategies on biodegradable polymers and their composites by thermal or melt-based processing technologies: Advances and perspectives. Polymers. 2024;**16**

[2] Zhou Y, Tian Y, Peng X. Applications and challenges of supercritical foaming technology. Polymers. 2023;**15**

[3] Tapia-Blácido DR, Aguilar GJ, de Andrade MT, Rodrigues-Júnior MF, Guareschi-Martins FC. Trends and challenges of starch-based foams for use as food packaging and food container. Trends in Food Science and Technology. 2022;**119**:257-271

[4] Kemmere M, F, Meyer T. Supercritical Carbon Dioxide. In: Polymer Reaction Engineering. Supercritical Carbon Dioxide: In Polymer Reaction Engineering. Wiley-VCH Verlag GmbH & Co. KGaA; 2005. DOI: 10.1002/3527606726

[5] Liao X, Zhang H, He T. Preparation of porous biodegradable polymer and its nanocomposites by supercritical CO 2 foaming for tissue engineering. Journal of Nanomaterials. 2012;**2012**

[6] Sauceau M, Fages J, Common A, Nikitine C, Rodier E. New challenges in polymer foaming: A review of extrusion processes assisted by supercritical carbon dioxide. Progress in Polymer Science. 2011;**36**(6):749-766

[7] Alavi S, Rizvi SSH. Strategies for enhancing expansion in starch-based microcellular foams produced by supercritical fluid extrusion. International Journal of Food Properties. 2005;**8**(1):23-34

[8] Kazarian SG. Polymer processing with supercritical fluids. Polymer Science. 2000;**42**:78-101

[9] Villamil, Jiménez JA, Le Moigne N, Bénézet JC, Sauceau M, Sescousse R, Fages J. Foaming of PLA composites by supercritical fluid-assisted processes: A review. Molecules. 2020;**25**

[10] Jin FL, Zhao M, Park M, Park SJ. Recent trends of foaming in polymer processing: A review. Polymers. 2019;**11**

[11] Chen Z, Yin X, Chen H, Fu X, Sun Y, Chen Q, et al. Mechanical, crystallization, rheological, and supercritical CO2 foaming properties of polybutylene succinate nanocomposites: Impact of carbon nanofiber content. Polymers (Basel). 2024;**16**(1)

[12] Leja K, Lewandowicz G. Polymer biodegradation and biodegradable polymers - a review. Polish Journal of Environmental Studies. 2010;**19**(2):255-266

[13] Kim S, Chung H. Biodegradable polymers: From synthesis methods to applications of lignin-graft-polyester. Green Chemistry. 2024;**10774-803**

[14] Zhang C, Sun Z, Yang X, Guoyan Duan MW, Qiao Y. Effect of thermoplastic starch content on the properties of poly(butylene adipate-co-terephthalate) (PBAT) composites. Bioresources.cnr.ncsu.edu. 2025;**20**(2):3910-3922

[15] Cardea S, De MI. Cellulose acetate and supercritical carbon dioxide: Membranes, nanoparticles, microparticles and nanostructured filaments. Polymers (Basel). 2020;**12**(1)

[16] Chang HK, Chen PY. Synthesis of silica-based scaffolds with high porosity and controllable microstructure by a sintering-free sol–gel/freeze-casting hybrid method under mild conditions. Journal of Materials Research and Technology. 2020;**9**(6):16167-16178

[17] Haugen HJ, Monjo M, Rubert M, Verket A, Lyngstadaas SP, Ellingsen JE, et al. Porous ceramic titanium dioxide scaffolds promote bone formation in rabbit peri-implant cortical defect model. Acta Biomaterialia. 2013;**9**(2):5390-5399

[18] Olah L, Filipczak K, Jaegermann ZJ, Czigány T. Synthesis, structural and mechanical properties of porous polymeric scaffolds for bone tissue regeneration based on neat poly(ε-caprolactone) and its composites with calcium carbonate. Polymers for Advanced Technologies. 2006;**17**(11-12):889-897

[19] Tejchman A, Znój A, Chlebanowska P, Frączek-Szczypta A, Majka M. Carbon fibers as a new type of scaffold for midbrain organoid development. International Journal of Molecular Sciences. 2020;**21**(17):1-14

[20] Ambre A, Katti KS, Katti D. Nanoclay based composite scaffolds for bone tissue engineering applications. Journal of Nanotechnology in Engineering and Medicine. 2010;**1**

[21] Lalwani G, Gopalan A, D'Agati M, Sankaran JS, Judex S, Qin Y-X, et al. Porous three-dimensional carbon nanotube scaffolds for tissue engineering. Journal of Biomedical Materials Research. Part A. 2015;**103**(10):3212-3225

[22] Chen X, Feng JJ, Bertelo CA. Plasticization effects on bubble growth during polymer foaming. Polymer Engineering and Science. 2006;**46**(1):97-107

[23] Liu S, de Beer S, Batenburg KM, Gojzewski H, Duvigneau J, Vancso GJ. Designer Core–Shell nanoparticles as polymer foam cell nucleating agents: The impact of molecularly engineered interfaces. ACS Applied Materials and Interfaces. 2021;**13**(14):17034-17045

[24] Wang J, Zhang Y, Sun J, Jiao Z. Controllable fabrication of multi-modal porous PLGA scaffolds with different sizes of SPIONs using supercritical CO2 foaming. Journal of Applied Polymer Science. 2022;**139**(23)

[25] Li B, Zhao G, Wang G, Zhang L, Gong J, Shi Z. Biodegradable PLA/PBS open-cell foam fabricated by supercritical CO2 foaming for selective oil-adsorption. Separation and Purification Technology. 2021;**257**:117949

[26] Tayton E, Purcell M, Aarvold A, Smith JO, Kalra S, Briscoe A, et al. Supercritical CO2 fluid-foaming of polymers to increase porosity: A method to improve the mechanical and biocompatibility characteristics for use as a potential alternative to allografts in impaction bone grafting? Acta Biomaterialia. 2012;**8**(5):1918-1927

[27] García-Casas I, Valor D, Elayoubi H, Montes A, Pereyra C. Morphological 3D analysis of PLGA/chitosan blend polymer scaffolds and their impregnation with olive pruning residues via supercritical CO2. Polymers (Basel). 2024;**16**(11)

[28] Oluwabunmi K, D'Souza NA, Zhao W, Choi TY, Theyson T. Compostable, fully biobased foams using PLA and micro cellulose for zero energy buildings. Scientific Reports. 2020;**10**(1)

[29] Salerno A, Di Maio E, Iannace S, Netti PA. Solid-state supercritical CO2 foaming of PCL and PCL-HA nano-composite: Effect of composition,

thermal history and foaming process on foam pore structure. Journal of Supercritical Fluids. 2011;**58**(1):158-167

[30] Bhamidipati M, Scurto AM, Detamore MS. The future of carbon dioxide for polymer processing in tissue engineering. Tissue Engineering - Part B: Reviews. 2013;**19**:221-232

[31] Collins NJ, Bridson RH, Leeke GA, Grover LM. Particle seeding enhances interconnectivity in polymeric scaffolds foamed using supercritical CO2. Acta Biomaterialia. 2010;**6**(3):1055-1060

[32] Léonard A, Calberg C, Kerckhofs G, et al. Characterization of the porous structure of biodegradable scaffolds obtained with supercritical CO2 as foaming agent. Journal of Porous Materials. 2008;**15**:397-403

[33] Baldino L, Cardea S. Generation of biocompatible PCL foams by supercritical foaming. Chemical Engineering Transactions. 2020;**79**:241-246

[34] Chen PH, Chen CW, Chan TH, Lin HY, Tuan KL, Su CS, et al. Foaming of bio-based PLA/PBS/PBAT ternary blends with added nanohydroxyapatite using supercritical CO2: Effect of operating strategies on cell structure. Molecules. 2025;**30**(9)

[35] Zhang T, Jang Y, Lee E, Shin S, Kang HJ. Supercritical CO2 foaming of poly(3-hydroxybutyrate-co-4-hydroxybutyrate). Polymers (Basel). 2022;**14**(10)

[36] Valor D, Montes A, Monteiro M, García-Casas I, Pereyra C, de la Ossa EM. Determining the optimal conditions for the production by supercritical CO2 of biodegradable PLGA foams for the controlled release of rutin as a medical treatment. Polymers (Basel). 2021;**13**(10)

[37] Montes A, Valor D, Penabad Y, Domínguez M, Pereyra C, de la Ossa EM. Formation of PLGA–PEDOT: PSS conductive scaffolds by supercritical foaming. Materials. 2023 Mar;**16**(6)

[38] Kosowska K, Krzysztoforski J, Henczka M. Foaming of PCL-based composites using scCO2: Structure and physical properties. Materials. 2022;**15**(3)

[39] Song C, Luo Y, Liu Y, Li S, Xi Z, Zhao L, et al. Fabrication of PCL scaffolds by supercritical CO2 foaming based on the combined effects of rheological and crystallization properties. Polymers (Basel). 2020;**12**(4)

[40] Satpayeva A, Rojas A, Tyrka M, Ksepko E, Galotto MJ, Zizovic I. Supercritical foaming and impregnation of Polycaprolactone and Polycaprolactone-hydroxyapatite composites with Carvacrol. PRO. 2022;**10**(3)

[41] Tsivintzelis I, Pavlidou E, Panayiotou C. Biodegradable polymer foams prepared with supercritical CO2–ethanol mixtures as blowing agents. Journal of Supercritical Fluids. 2007;**42**(2):265-272

[42] Montes A, Valor D, Delgado L, Pereyra C, de la Ossa EM. An attempt to optimize supercritical CO2 polyaniline-Polycaprolactone foaming processes to produce tissue engineering scaffolds. Polymers (Basel). 2022;**14**(3)

[43] Lima GMR, Mukherjee A, Picchioni F, Bose RK. Characterization of biodegradable polymers for porous structure: Further steps toward sustainable plastics. Polymers (Basel). 2024;**16**(8)

[44] Oluwabunmi KE, Zhao W, D'souza N. Carbon capture utilization for biopolymer foam manufacture: Thermal,

mechanical and acoustic performance of PCL/PHBV CO2 foams. Polymers (Basel). 2021;**13**(15)

[45] Borgmann LM, Johnsen S, Santos de Oliveira C, de Souza M, J e S, Li J, et al. Porous polymeric microparticles foamed with supercritical CO2 as scattering white pigments. Bioinspiration and Biomimetics. 2023;**18**(2)

[46] Baudron V, Gurikov P, Smirnova I, Whitehouse S. Porous starch materials via supercritical-and freeze-drying. Gels. 2019;**5**(1)

[47] Grzybek P, Jakubski Ł, Dudek G. Neat chitosan porous materials: A review of preparation, structure characterization and application. International Journal of Molecular Sciences. 2022;**23**

[48] Reverchon E, Cardea S. Production of controlled polymeric foams by supercritical $CO_2$. Journal of Supercritical Fluids. 2007;**40**(1):144-152

[49] Atiwesh G, Mikhael A, Parrish CC, Banoub J, Le TAT. Environmental impact of bioplastic use: A review. Heliyon. 2021;**7**(9)

[50] Jinnai H. Electron microscopy for polymer structures. Microscopy. 2022;**71**:I148-I164

[51] Li W, Wei F, Xiong C, Ouyang J, Shao L, Dai M, et al. A novel supercritical CO2 foam system stabilized with a mixture of Zwitterionic surfactant and silica nanoparticles for enhanced oil recovery. Frontiers in Chemistry. 2019;7

[52] Scharf J, Chouchane M, Finegan DP, Lu B, Redquest C, Kim MC, et al. Bridging nano- and microscale X-ray tomography for battery research by leveraging artificial intelligence. Nature Nanotechnology. Nature Research. 2022;**17**:446-459

[53] Rinki K, Dutta PK, Hunt AJ, Clark JH, Macquarrie DJ. Preparation of chitosan based scaffolds using supercritical carbon dioxide. Macromolecular Symposia. 2009;**277**(1):36-42

[54] Long LY, Weng YX, Wang YZ. Cellulose aerogels: Synthesis, applications, and prospects. Polymers. 2018;**8**

[55] Rother G, Krukowski E, Wallacher D, Grimm N. Pore size effects on the sorption of supercritical CO2 in mesoporous CPG-10 silica. Journal of Physical Chemistry C. 2012;**116**(1):917-922

[56] Wang Z, Huang H, Wang Y, Zhou M, Zhai W. A review of the preparation of porous fibers and porous parts by a novel micro-extrusion foaming technique. Materials. 2024;**17**

Chapter 10

# Porous Sol-Gel Materials as Electrocatalysts for Sensing Applications

*Mario Sánchez Suárez, Celia Toyos Rodríguez, Ana Arenillas de la Puente, Natalia Rey Raap and Ignacio Cameán Martínez*

## Abstract

The chapter presents a general overview of the fundamentals of electrochemical sensors, describing the materials used and their critical characteristics, such as porosity and electrical conductivity. In this sense, the sol-gel methodology is presented as an effective solution to design a porous structure and other physicochemical properties to meet the needs of electrochemical sensors. This approach allows us to obtain a versatile and customized design of the properties of the materials, improving the sensitivity and durability of these devices.

**Keywords:** porosity, aerogels, carbon materials, electrochemistry, electrocatalyst, sensors

## 1. Introduction

Recent advancements in analytical technologies have enabled the development of efficient and selective sensors for various applications, including environmental and industrial monitoring, food safety, and healthcare [1]. Sensors are devices that allow the detection and quantification of chemical, biological, or physical changes. Among the wide variety of sensor types, electrochemical sensors have emerged as a powerful and effective technology due to their high sensitivity, rapid response time, and cost-effectiveness [2], and they have already reached the market with successful examples such as the glucose meter.

Traditional electrochemical sensors rely on the use of enzymes for the generation of a measurable redox signal, but their stability is influenced by environmental factors [3]. Nonenzymatic alternatives, based on electrocatalysts, offer greater robustness and lower susceptibility to external conditions. The incorporation of electrocatalysts as advanced electrode materials with specific physico-chemical properties is key to facilitating electron transfer, reducing overpotential, and improving sensitivity in nonenzymatic electrochemical sensors [4, 5]. The increasing demand for materials with improved electrocatalytic properties has stimulated research into the field of nanostructured and porous materials. Noble metals, transition metals, carbon materials, and their composites have been widely studied. However, for their application in nonenzymatic electrochemical sensors, it is essential to control the chemical composition as well as their textural and structural properties. To this end, the sol-gel synthesis is presented as a very interesting strategy since it allows operating at low temperatures and atmospheric pressure, offering a simple but effective synthesis process to obtain materials with tailored properties [6].

Therefore, this chapter explores the role of sol-gel porous materials as electrocatalysts in nonenzymatic electrochemical sensors. Thus, Section 2 provides an introduction to electrochemical sensors, explaining their basic principles and highlighting the role of electrocatalysts in nonenzymatic sensing. In Section 3, the sol-gel methodology is presented, including innovative techniques such as sol-gel synthesis assisted by microwave technology and the importance that porosity and electrical conductivity play in the electrocatalytic performance. Section 4 compares sol-gel materials with traditional sensing materials, outlining their pros and cons. Finally, Section 5 presents the conclusions and a future vision of sol-gel material requirements in electrochemical sensors.

## 2. Fundamentals and types of sensors

A sensor is an analytical device that detects changes in physical, chemical, or biological variables, converting this information into a signal that can be processed, measured, and analyzed [7]. Thus, sensors can measure magnitudes and quantities such as temperature, pressure, speed, and concentration of ions, among others, in different applications. Sensors can be classified based on various parameters, such as the transduction mechanism, the stimulus measured, or their specific applications. The most fundamental classification divides the sensors into two main categories based on the type of stimulus measured, which are (i) physical sensors that detect and measure physical properties such as temperature, pressure, changes in light, electrical, magnetic, or optical properties, variations in sound waves or acoustic vibrations, and (ii) chemical sensors that respond by sensing the presence of specific chemical compounds (gases, ions, etc.) and their concentration [8]. This is the case of electrochemical sensors that convert chemical redox reactions into electrical signals. Their versatility, high sensitivity, low cost, and possibility of miniaturization position them as highly relevant analytical tools compared to other types of sensors.

### 2.1 Electrochemical sensors: An effective detection technology

Electrochemical sensors represent a cutting-edge technology in contemporary analytical chemistry that provides an accurate, simple, rapid, and inexpensive detection method [9]. As shown in **Figure 1**, the main components of an electrochemical sensor are (i) the *receptor*, or sensing element, which is in direct contact with the environment and detects changes in the target stimulus; (ii) the *transducer*, which converts the signal from the sensitive element into a processable signal (electrical, optical, mechanical...); and (iii) the *signal conditioner* and processing circuit that amplify, filter, or convert the raw signal from the transducer to a suitable form for further processing and conversion. Finally, the processed signal is sent to an output in order to be interpreted by the user or an automated system. In a traditional electrochemical system, at least two electrodes are required: a working electrode (WE), where the redox reaction takes place, and a reference electrode (RE), commonly Ag/AgCl, that

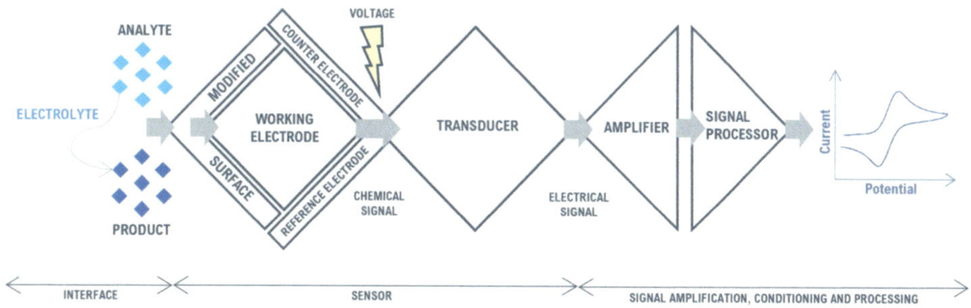


**Figure 1.**
*Scheme of the components of an electrochemical sensor.*

maintains a constant and known potential regardless of the solution composition. A third electrode, the counter electrode (CE), is added in dynamic electrochemical techniques to connect the electrolytic solution and complete the circuit [10, 11].

The operating mechanism is based on the interpretation of electrical signals produced by chemical or biochemical oxidation-reduction reactions in the interface between the electrode surface and the solution containing the analyte or analytes to be studied [12]. In redox reactions, two complementary processes occur simultaneously: oxidation and reduction. Oxidation involves chemical species, called reducing agents, that lose electrons, thus increasing their oxidation state, while reduction involves other species, known as oxidizing agents, that gain electrons from reducing agents, resulting in a decrease in their oxidation state.

These reactions involve electron transfer in a complex physicochemical process, producing detectable electrical variations that can be directly proportional to the concentration of the analyte [13].

## 2.2 Types of electrochemical sensors

Electrochemical sensors can be classified based on how electrical signals are generated and measured in response to the target analyte, which means according to the operating principle [14]. Each category uses measurement techniques and sensor configurations optimized for the specific analytical application. In this sense, it is possible to distinguish between *potentiometric sensors,* which measure the difference between the potential working and reference electrodes under minimal current flow [15]; *amperometric sensors,* in which a fixed potential induces redox reactions generating measurable currents [16]; *impedimetric sensors,* which analyze electrical impedance variations [12]; *conductimetric sensors,* which detect changes in solution conductivity during chemical interactions [14]; and *voltammetric sensors,* which analyze current responses by applying a variation in potential [17]. All of them provide information about electron transfer kinetics, reaction mechanisms, and the concentration of the analytes. Each technique offers different advantages whose selection will depend on the specific application and the analyte properties.

An important part of electrochemical sensors is the receptor. If the sensor includes a biological receptor, the device is called a biosensor. Biological recognition elements such as enzymes, proteins, nucleic acids, or whole cells are used to specifically interact with target analytes, producing an associated chemical signal. Electrochemical sensors can also be classified according to the recognition element used. Enzymatic sensors use immobilized enzymes as a biological recognition element to catalyze specific reactions with target molecules to generate an electrochemical signal [10]. The most extended enzymatic sensor is the glucose sensor, based on the enzyme glucose oxidase [18], but there are also other detection devices focused on compounds such as lactate [19] or neurotransmitters [20] and even herbicides [21]. The catalytic activity of enzymes leads to the formation of an electrochemically active compound that can be detected. This is the reason why enzymes are not just used as receptors in biosensors but also used as signal generators in combination with other biological recognition elements such as nucleic acids or antibodies.

Nonenzymatic electrochemical sensors work through direct electrochemical reactions between analytes and electrocatalysts, avoiding the intrinsic limitations of enzymes. These sensors use electrocatalysts, such as metals or carbon nanomaterials, which allow them to overcome the instability and high cost of traditional enzymatic sensors [22]. Their greater durability and resistance to adverse environmental conditions [23] facilitate their use in industrial, environmental, and biomedical applications [24–26]. However, their main challenge remains in selectivity, as they can respond to interferents present in complex samples. Therefore, current research focuses on the design of new materials and strategies to improve both the sensitivity and selectivity of these devices.

### 2.3 Electrocatalysts in nonenzymatic electrochemical sensors

Nonenzymatic electrochemical sensors use tailored electrode materials that directly participate in the electron transfer processes with the target analytes. The composition and structure of the electrocatalyst influence the sensitivity, selectivity, detection limit, and stability of the sensor. A wide range of materials has been explored for nonenzymatic sensors, including noble metals [27], transition metals and their oxides [28], carbon-based nanomaterials [29], and composite structures [30]. Each category of materials offers unique advantages and addresses specific challenges in electrochemical sensing applications. All these materials have been used as electrocatalysts both in bulk and at the nanoscale. Nanomaterials have been extensively applied in electrochemical sensors in the past years as they enhance the electrocatalytic activity by increasing the surface area, electrical conductivity, and electron transfer kinetics, and improving surface chemistry.

Noble metals, particularly gold, have been extensively studied as electrocatalysts for nonenzymatic sensors due to their excellent conductivity and catalytic properties [31]. These metals demonstrate a high electron transfer capacity and exhibit good biocompatibility, making them suitable for biomedical applications. However, their high cost and susceptibility to surface poisoning by intermediates have led researchers to explore alternative materials to improve their efficiency and durability [32]. Some effective approaches involve alloying noble metals with other more abundant elements or even creating core-shell structures that maintain catalytic performance while reducing material costs. Similarly, the use of transition metals and their oxides/hydroxides represents a promising alternative to replace noble metals, offering comparable or superior performance with lower costs [33]. Metals such as nickel, copper, cobalt, manganese, and iron, and their oxides, have shown remarkable electrocatalytic activity for a wide variety of analytes [30], for example, by using metal-organic frameworks [34].

Carbon-based nanomaterials, including carbon nanofibers [35], carbon nanotubes [36], graphene, and reduced graphene oxide [31, 37], have been used as electrocatalysts, improving the performance of the sensor. The combination of carbon nanomaterials with metal catalysts to produce composite and hybrid materials creates synergistic effects that enhance electron transfer and electrocatalytic properties [38], turning out to be one of the approaches that has gained the most attention recently in this field. Besides, depending on the type of analyte, carbon-based materials can be used without metal catalysts. Thus, for example, graphene-based materials represent a significant category of electrocatalysts for nonenzymatic sensors due to their exceptional electrical conductivity and large surface area, enhancing direct electron transfer and increasing sensor sensitivity [39].

### 2.4 Carbon-based electrocatalysts in sensing applications

Carbon materials are becoming one of the most widely used electrocatalysts in nonenzymatic electrochemical sensors due to their structural versatility, chemical and thermal stability, exceptional electrocatalytic properties, large surface area, excellent electrical conductivity, and ability to be functionalized. As a result, these materials improve the analytical performance of the sensors, operating at wider electrochemical potential windows, with lower background currents, and reducing nonspecific adsorption.

Carbon-based electrochemical sensors in environmental and industrial monitoring fields focus on, for example, monitoring toxic chemicals in water such as hydrazine [40], phenolic compounds [41], organic compounds like bisphenol A [42], antibiotics, or pharmaceutical compounds [43]. It is also possible to highlight the detection of pesticides using nitrogen-doped carbon materials [44] or the monitoring of hydrogen peroxide using carbon materials modified with an aluminum-based metallic agent [45]. Regarding the food safety field, food freshness monitoring by histamine quantification [46], lactose detection to monitor milk products [47], cholesterol detection in meat samples [48], and detection of heavy metals in foods [49] are some of the examples of the applicability of carbon-based materials.

The diverse applications of carbon-based materials in electrochemical sensing highlight their versatility, with healthcare being the most prominent field. In this sense, glucose is the most extensively studied analyte in nonenzymatic electrochemical sensors for biomedical applications, as it is critical in diabetes monitoring [50]. Carbon materials, such as porous carbon materials [51], carbon nanotubes [52], carbon layers [53], or other types of carbonaceous structures [32], have been widely studied for this application, all incorporating metallic agents in different forms as active phases. Although glucose monitoring dominates this field, nonenzymatic electrochemical carbon-based sensors have successfully achieved the detection of other clinically relevant analytes, such as uric acid [54], dopamine [55], lactate [56], or ions such as calcium [57].

In conclusion, electrochemical sensors stand out for their high sensitivity, rapid measurement, and low cost, making them key tools in multiple fields. Enzymatic sensors offer high selectivity, but they have stability limitations. In this sense, nonenzymatic sensors, based on electrocatalysts such as metals or carbonaceous materials, have gained relevance due to their robustness and lower susceptibility to environmental conditions, although they still present challenges such as selectivity. The development of advanced materials and nanomaterials with controlled physicochemical properties is essential to improve the performance and reliability of these devices, with carbon-based materials being a promising choice in this sense.

## 3. Sol-gel methodology applied to nanomaterials used in sensing

The sol-gel methodology represents a suitable alternative for tailored material synthesis, being an effective option for the preparation of nanomaterials fulfilling the requirements for their application in nonenzymatic electrochemical sensors. **Table 1** shows some advantages of using the sol-gel methodology to synthesize materials compared to other methods such as chemical vapor deposition (CVD), hydrothermal or electrochemical processes, or direct pyrolysis [58, 59]. The sol-gel methodology is a powerful and versatile approach for the synthesis of nanomaterials with precise control of porosity, surface chemistry, and morphology [60], making it ideal for a wide range of applications in electrocatalysis.

### 3.1 Principles of sol-gel synthesis

Over the last decades, the sol-gel method has evolved from its origins in basic chemistry (silica aerogels) to become a fundamental tool for the synthesis of advanced materials (from carbon-based to metal-based aerogels, including hybrids). The first documented study of a sol-gel synthesis is attributed to Jacques-Joseph Ebelmen, who in 1846 obtained silica gels through the hydrolysis of silicon tetrachloride under ambient conditions [61]. This pioneering work allowed the development of silica-based systems, which led to the first industrial application of the method through the creation of glass coatings by the Schott Glass company in the 1930s. Subsequently, in the early 1960s and 1970s, the ceramics industry showed a growing interest in this type of gel, promoting significant activity in the field and the diversification of materials obtained by this route [62].

An important breakthrough in the diversification of the sol-gel method was its application to organic materials. In the 1990s, R.W. Pekala developed the synthesis of carbon aerogels by adapting the sol-gel methodology usually applied to silica and metal oxides to obtain organic polymeric materials with three-dimensional porous structures and high surface areas. Pekala focused on the aqueous polycondensation of resorcinol and formaldehyde (RF) used as precursors to synthesize carbon aerogels through a sol-gel polymerization process, followed by supercritical drying and carbonization post-treatments. These materials exhibited unique properties due to their porous structure of interconnected colloids or nodules in the range of 3–30 nm, resulting in a three-dimensional structure with high surface areas (400–1100 $m^2 g^{-1}$) [63].

| Parameter | Sol-gel methodology | Other techniques (CVD, hydrothermal, pyrolysis...) |
|---|---|---|
| Temperature | Low (< 100°C) | High (up to 1300°C) |
| Structure control | Adjustable porosity and morphology design | Certain limitations |
| Purity | High homogeneity and absence of impurities | Presence of some impurities |
| Versatility | Produces aerogels, films, fibers, and composites | Restricted |
| Cost and sustainability | Low energy consumption | Extreme operating conditions |

**Table 1.**
*Advantages of sol-gel methodology vs. other techniques.*

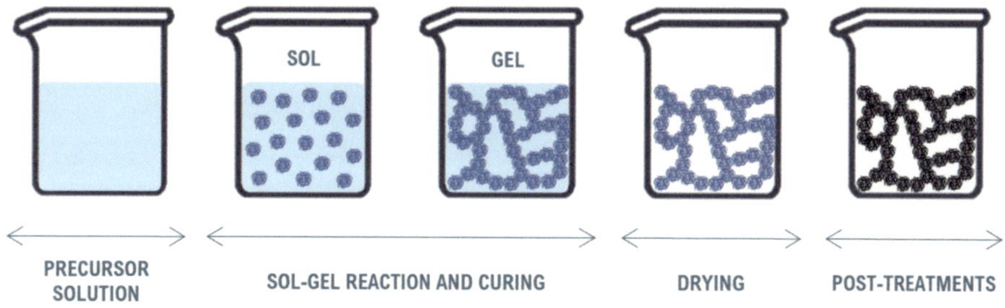


**Figure 2.**
*Visual scheme of the steps involved in the sol-gel methodology.*

**Figure 2** shows the typical sol-gel process involving four steps. Generally, a sol-gel process begins with the dissolution of the precursors in a solvent (water, reducing solutions, etc.) and, occasionally, the addition of a pH regulator that promotes polymerization and cross-linking to form the final structure. For example, in the synthesis of carbon aerogels, the most commonly used precursors are resorcinol and formaldehyde. However, recent research is seeking to replace these types of precursors with more sustainable reactants [64]. In silica aerogels, silicon oxides, such as tetraethylorthosilicate (TEOS) or tetramethoxysilane (TMOS), are typically used [65], while in metal-based aerogels, the precursors are usually metal salts or metal oxides [66]. On the other hand, hybrid systems incorporate simultaneously both inorganic and organic precursors [67].

After applying temperature for a certain time to the precursor solution, the gelation and curing stage occurs, starting with the formation of a colloidal solution (sol), which gradually forms nodules that interact to give rise to a three-dimensional structure (gel) through polycondensation reactions [68]. Then, the solvent must be removed (Drying process) to obtain a porous structure delimited by the nodules formed. There are three solvent removal processes generally applied: evaporation at atmospheric pressure, freeze-drying, or drying under supercritical conditions [69]. Once the material is dried, post-treatments, such as heat treatment, surface modification, or even doping with heteroatoms or metallic precursors, can be applied to modulate the final and desired properties [70].

This methodology allows precise control of the pore size (microporous, mesoporous, and macroporous structures) [60], as well as the surface area and porosity, by modifying the proportion of reactants, varying the amount of solvent or incorporating pH regulators, and even modifying other operating conditions such as temperature or time of reaction. Using the sol-gel methodology, it is possible to synthesize diverse architectures with a homogeneous composition and absence of impurities [63].

Material sol-gel synthesis has undergone a relevant evolution. The reactions can be considerably accelerated by microwave heating without relinquishing control. Although the traditional sol-gel method remains a robust and widely used technique, microwave-assisted technology stands out as a faster and more efficient alternative for the synthesis of materials with tailored properties [71]. Its ability to reduce time and energy costs while keeping the properties controlled makes it a key tool for mass-producing advanced materials for specific applications such as electrocatalysis.

### 3.2 Microwave-assisted sol-gel synthesis

Microwave-assisted sol-gel synthesis represents a relevant progress in the preparation of sol-gel materials, offering notable advantages over conventional

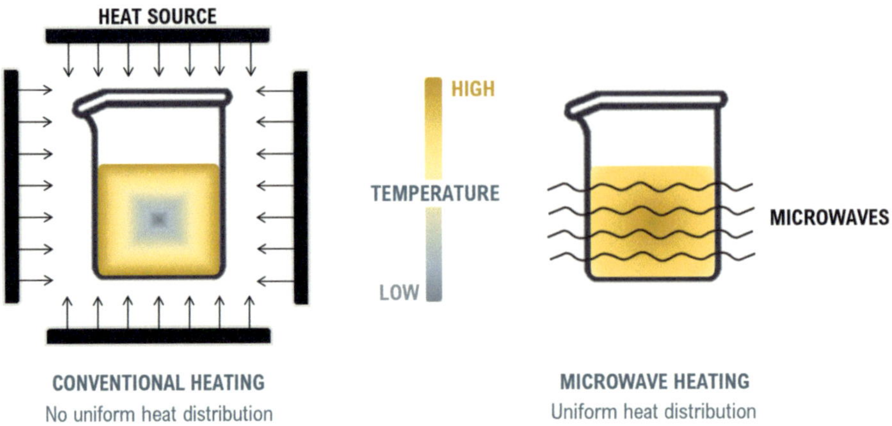


**Figure 3.**
*Comparison between conventional heating vs. microwave heating.*

heating methods [72]. **Figure 3** shows a comparison of how traditional heating takes place compared to the use of microwave technology. Microwave radiation heats the precursor solution uniformly and rapidly, reducing the synthesis time from days to hours [73]. Unlike conventional heating, which relies on conduction and convection heat transfer, microwave heating provides direct energy transfer to the reaction mixture, resulting in more efficient bulk heating. Furthermore, this technology promotes simultaneous and uniformly distributed nucleation points due to the homogeneous heating, minimizing temperature gradients inside the reaction media. As a result, greater control of the properties, such as pore size distribution, is obtained. The accurate determination of the transition point from the sol to the gel state is crucial for controlling the final properties of these materials. Previous works have shown that microwave heating allows the precise identification of this transition and therefore the ability to handle the structure at the desired stage for its further processing [74].

The microwave-assisted process also offers significant advantages in terms of energy efficiency and scalability. Regarding efficiency, this technology reduces energy consumption up to 90% compared to conventional heating methods [75]. Furthermore, this synthesis methodology has been successfully optimized and scaled up, demonstrating its commercial viability [76].

There is a wide variety of materials that can be produced using the sol-gel methodology assisted by microwave technology. Mesoporous carbon materials, graphene aerogels, carbon monoliths, carbon quantum dots, carbon fibers, silica, metals, or hybrid aerogels are some of the most outstanding materials [77]. The possibility to design not only their chemical but also their structural and porous properties makes this methodology a desirable alternative for the tailored synthesis of electrocatalysts for electrochemical sensors.

### 3.3 Porous sol-gel materials as electrocatalysts: Balancing between porosity and electrical conductivity

The performance of sol-gel materials as electrocatalysts in electrochemical sensors depends on achieving an optimal balance between two properties: porosity and electrical conductivity, in addition to surface area optimization.

Specifically, porosity represents a crucial property in the design of sol-gel materials for electrochemical sensors that directly influences their efficiency and sensitivity. Materials with optimized porous structures exhibit larger surface areas, allowing higher densities of active sites and facilitating analyte diffusion, adsorption capacity, and chemical reactivity.

The International Union of Pure and Applied Chemistry (IUPAC) defines porosity as the fraction of the total volume of the material that is occupied by pores, that is, the empty spaces within its structure [76]. This parameter is expressed as a percentage and can include both open and closed pores, which are accessible and not accessible from the outside, respectively. Furthermore, porous materials can be classified according to their pore size into three main categories: micro-, meso-, and macropores, which are shown in **Figure 4**. Micropores contribute to high surface area, mesopores facilitate mass transport and accessibility to the internal surfaces, while macropores serve as transport pathways in the material, reducing diffusion limitations and improving overall accessibility to smaller pores embedded within the structure.

However, the only presence of micropores often suffers from diffusion limitations, while those with only macropores lack sufficient surface area for applications requiring high catalytic activity. Therefore, an interesting approach is to combine all these characteristics to create materials with hierarchical porosity, which integrates pores of different sizes in a single material [69], improving their performance as electrocatalysts. The effect of porosity in materials applied to sensors goes beyond their enhanced electrocatalytic properties, being beneficial also to prevent biofouling [78] and to facilitate the encapsulation of indicator molecules [79, 80].

Porosity and electrical conductivity are inverse properties. An increase in porosity typically compromises the electrical conductivity within the material network. However, both are essential for promoting efficient diffusion of electroactive species and effective charge transfer, which are key parameters in the design of high-performance electrocatalytic materials.

Sol-gel synthesis facilitates the customized modification of both properties at the same time. Thus, for example, the adjustment of the molar ratio of resorcinol to isopropyl alcohol in a synthesis of sol-gel carbon spherogels based on RF enabled precise control of the wall thickness of the hollow carbon nanospheres, allowing simultaneous modulation of their porosity and electrical conductivity. This customization capability proved essential, as the resulting materials combined exceptional conductivity with tunable porosity and an extraordinarily high surface area [81]. In other work, graphene aerogels were developed using a sol-gel process and incorporated nickel into their structure, generating a broader mesopore distribution and

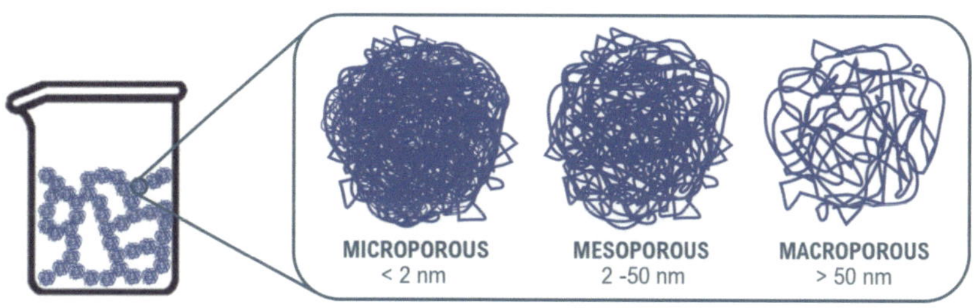


**Figure 4.**
*Types of porosity according to the IUPAC classification.*

much higher electrical conductivity, resulting in a larger electrochemical surface area and improved electrocatalytic performance [82].

In conclusion, the sol-gel methodology represents a versatile technique for designing nanomaterials with tailored porosity, surface area, and morphology. Its ability to operate at low temperatures and its compatibility with microwave-assisted technologies enable a rapid, efficient, and scalable synthesis. The precise control of the porous structure and the incorporation of electrically conductive agents make it possible to optimize both analyte diffusion and charge transfer, key factors for maximizing sensitivity and selectivity.

## 4. Application of sol-gel materials in sensing: Comparison with traditional sensing materials

Despite the extended use of electrochemical sensors, there are still challenges to overcome to improve their analytical performance. Establishing efficient electron transfer and good electrical current density is crucial for achieving a highly sensitive device. In this context, modification of the electrode with organic/inorganic materials has arisen as a solution over the past years, especially with the exponential growth of nanotechnology [83]. Several works have revised the importance of material engineering in the electrochemical biosensing field, highlighting the boost in the use of graphene-based materials and metal nanoparticles, as they increase electron transfer and active surface area and favor the immobilization of recognition elements [84].

The use of sol-gel materials, in particular sol-gel ceramics, was prominent in the early 2000s as entrapment surfaces of bioreceptors [85]. In biosensing, bioreceptors are used to increase selectivity while minimizing interference effects. A proper and oriented immobilization of bioreceptors on the surface of the transducer is mandatory to ensure affinity and stability over time. Several immobilization strategies include covalent linking, absorption, and entrapment. Early works on sol-gel materials combined the high porosity of mainly silica-based bioceramics and doped sol-gel films with bioreceptors (enzymes primarily), added to the mixture solution during the sol-gel synthesis. Thus, siloxane polymers grow around the bioreceptor, leading to the oriented entrapment of the biomolecule in the porous sol-gel structure. As the process is performed at room temperature, the biocatalytic activity of the enzymes is preserved while reducing long-term leakage. Nowadays, other biorecognition elements, as antibodies, DNA, or aptamers, have been used in combination with sol-gel-based materials [86]. Moreover, sol-gel technology has also been applied to the obtaining of synthetic recognition elements as molecularly imprinted polymers, a more stable alternative to traditional bioreceptors. An example of this is the combination of $Ti_3C_2T_x$@$SiO_2$ with TEOS used for the detection of the antibiotic ofloxacin in wastewater samples [87].

The incorporation of novel sol-gel materials, particularly metallic nanoparticles and carbon-based composites such as carbon-based aerogels, xerogels, and conductive hydrogels, has overcome the initial lack of electrical conductivity of silica-based ceramics, broadening the applicability of these materials in electrochemical sensors. In addition, the presence of electrocatalysts that reduce the activation energy is essential for an electrochemical sensor to work efficiently. The use of sol-gel materials has evolved from solely being applied as immobilization surfaces to also enhancing sensitivity through their electrocatalytic properties [6].

### 4.1 Metal-based materials

The use of both noble metals and transition metal oxides is extended as electrocatalysts in sensor devices. The first are incorporated due to their excellent conductivity and catalytic properties, with gold being the predominant choice. Their catalytic effect against reactions occurring at neutral pH has been exploited in biosensing both as electrode modifiers and as labels [88]. However, the combination of these metallic materials with sol-gel synthesized substrates is gaining great attention. A composite of porous carbon black with nanostructured gold nanoceramic conductive film was obtained via sol-gel methodology using methyltrimethoxysilane as alkoxide [89]. The solution was used for sensing dopamine in human body fluids with a limit of detection (LOD) of 0.75 nM. Gold has also been combined with tin dioxide sol-gel technology and used for the modification of indium tin oxide (ITO) electrodes. The nanocomposite combines the excellent semiconducting and biocompatibility properties of tin dioxide with the electrocatalytic and selectivity of gold against vitamin B12 [90].

Transition metal oxide nanoparticles are applied in biosensing due to their large surface area and electron transfer compared to their bulk counterparts. Sol-gel synthesis has also been applied to transition metal oxide nanoparticles, producing high-quality materials in a faster and more economic way without compromising homogeneity, batch variability, and flexibility [91]. In electrochemical sensors, sol-gel synthesized metal oxide nanoparticles have been used as electrocatalysts for several biologically relevant analytes. Zinc oxide nanoparticles were used as electrode modifiers in a competitive label-free immunosensor for the detection of clenbuterol in animal feed samples, showing better performance than previously reported nanomaterial-based sensors [92]. The same type of particles has also been used in combination with copper for the direct electrocatalytic detection of myoglobin [93] and glucose [94]. Other novel materials, such as metal-organic frameworks with promising applicability in sensor devices, have also been synthesized through sol-gel processing, providing highly ordered porous architectures not obtained with traditional synthetic techniques [95]. Although their use so far is predominantly as immobilization surfaces and to increase the electroactive area of the electrode [96], they are also suitable candidates as electrocatalysts.

### 4.2 Carbon-based materials

Carbon is the predominant material used as a working electrode in electrochemical devices, as it is low cost and easily accessible. Carbon-based materials have been more recently introduced in electrochemical sensors as electrode modifiers, with graphene, porous carbon, nanotubes, and quantum dots standing out. Apart from their wide operational potentials and high chemical stability, they provide a large surface area and high electrical conductivity.

Carbon nanotubes are most actively used in enzymatic and nonenzymatic biosensors, owing to their high porosity, electrocatalytic properties, and fast electron transfer. However, they tend to precipitate over time and are difficult to uniformly disperse. Although they can be used alone as electrode modifiers, they are frequently combined with other electrocatalytic materials. This approach has been followed for the nonenzymatic detection of glucose, combining cobalt-based ferrites with carbon nanotubes and reporting a reduction in the charge transfer resistance by the combination of both materials [97]. A similar approach has been followed for the detection of bisphenol A, combining multi-walled carbon nanotubes with copper ferrite nanocomposites obtained through the citrate sol-gel method [98]. In this case, the combination of both materials increased sensitivity by reducing particle size and hence increasing the amount of exposed active sites. Bisphenol A was detected with an LOD of 3.2 nM without noticeable interference in real samples.

A promising sol-gel type of material with wide applicability in biosensing is carbon aerogel. Their highly porous structure and high surface area facilitate immobilization of biomolecules while preventing fouling and increasing electron transfer, which enhances the specificity and sensitivity of sensing devices [99]. The high surface area of these materials also promotes their electrocatalytic activity. Moreover, the flexibility of aerogels, such as graphene-based aerogels, can also be applied to wearable electronics. Although its use has been mainly applied to piezoelectric sensors [100], preliminary works have been conducted for the application of these aerogels in health diagnostics, such as pH monitoring in chronic wounds utilizing a 3D carbon fiber aerogel [101]. In summary, sol-gel carbon materials have revolutionized the field of electrochemical sensors by offering stable, biocompatible, and highly porous structures that improve sensitivity and reduce detection limits.

Sol-gel materials have evolved from simple bioreceptor immobilization matrixes to active components that improve the sensitivity and stability of electrochemical sensors. Thus, this methodology allows us to combine metallic species and carbon-based materials with tailored porosity, surface area, and morphology without compromising electrical conductivity, enabling the development of more efficient and versatile sensors. These innovations have significantly expanded sensor applications, particularly in the biomedical and environmental fields, thanks to their high porosity, surface area, and electrocatalytic capacity.

## 5. General conclusions and future vision

Sol-gel materials are revolutionizing electrochemical sensors by offering unprecedented control over porosity, morphology, and surface chemistry. Hierarchical porosity (micro-meso-macropores) can be precisely tailored by tuning synthesis parameters, thereby optimizing both diffusion kinetics and the active surface area. This, coupled with the ability to tune electrical conductivity through the incorporation of metals or heteroatoms, enables the creation of electrocatalysts with superior performance, creating matrixes that offer rapid electron transfer and detection limits in the nanomolar range for target analytes.

Sol-gel materials represent a paradigm shift in electrochemical sensors technology, surpassing, for example, traditional approaches based on enzymes. In healthcare, environmental monitoring, and food safety fields, sensors created with sol-gel materials demonstrate remarkable stability to variations in pH, temperature, and humidity by eliminating fragile biomolecules and replacing them with these robust electrocatalytic materials. Moreover, microwave-assisted synthesis has reduced production times from days to hours in a sol-gel process while simultaneously improving homogeneity in the final material obtained and energy efficiency for their production. These advances are critical for scaling laboratory prototypes to industrial manufacturing without sacrificing nanoscale precision. Therefore, sol-gel materials have evolved from their initial role as simple immobilization matrixes to become active electrocatalysts and key components of advanced sensing platforms.

Significant challenges in sol-gel synthesis remain despite the important progress made during the last decades: (i) more sustainable sol-gel synthetic routes to strengthen the industry commitment to sustainability and cost reduction without compromising the performance, (ii) maintaining nanostructured properties during large-scale synthesis, and (iii) developing advanced synthesis techniques that ensure industrial-scale reproducibility and operational stability.

In the sensor field, advances will focus on the integration of microfluidic platforms, 3D printing, and sol-gel composite materials to develop multifunctional wearable devices with specific geometries. These developments seek to improve the detection of target analytes, especially through nonenzymatic sensors, which still have the challenge of achieving more selective detection in complex matrixes. In this sense, the combination of metallic structures with porous sol-gel platforms offers new possibilities for improving the sensor performance. Thus, the next generation of implantable and flexible sensors will also focus on biocompatibility and the simultaneous monitoring of multiple parameters in medical diagnostics, environmental monitoring, and quality control in the food industry.

In summary, porous sol-gel materials will play a central role in the advancement of electrochemical sensing, thanks to their tunable properties, scalability, and adaptability to diverse applications. Continued innovation in device synthesis, functionalization, and integration will further improve their performance and open new avenues in material and analytical sciences.

## Acknowledgements

The authors thank the financial support received from Fundación BBVA through the Beca Leonardo 2023 and Principado de Asturias and SEKUENS for the project IDE/2024/000749 co-funded by the EU. NRR is grateful to the RyC2021-031456-I grant funded by MCIN/AEI/10.13039/501100011033 and the European Union NexGenerationEU/PRTR. MSS is grateful to the PRE2021-098471 PhD grant funded by MCIN/AEI/10.13039/501100011033 and FSE+.

## Author details

Mario Sánchez Suárez[1], Celia Toyos Rodríguez[2], Ana Arenillas de la Puente[1], Natalia Rey Raap[1] and Ignacio Cameán Martínez[1*]

1 Carbon Science and Technology Institute (INCAR-CSIC), Oviedo, Spain

2 University of Antwerp, Antwerp, Belgium

*Address all correspondence to: icamean@incar.csic.es

## References


[1] Vigneshvar S, Sudhakumari CC, Senthilkumaran B, Prakash H. Recent advances in biosensor technology for potential applications – An overview. Frontiers in Bioengineering and Biotechnology. 2016;**4**:11. DOI: 10.3389/FBIOE.2016.00011

[2] Wang J. Analytical Electrochemistry, 3rd ed. 2006:1-250. DOI:10.1002/0471790303.

[3] Hassan MH, Vyas C, Grieve B, Bartolo P. Recent advances in enzymatic and non-enzymatic electrochemical glucose sensing. Sensors (Basel). 2021;**21**:4672. DOI: 10.3390/S21144672

[4] Kimmel DW, Leblanc G, Meschievitz ME, Cliffel DE. Electrochemical sensors and biosensors. Analytical Chemistry. 2012;**84**:685-707. DOI: 10.1021/AC202878Q/ASSET/AC202878Q.FP.PNG_V03

[5] Chiu WT, Chang TFM, Sone M, Hosoda H, Tixier-Mita A, Toshiyoshi H. Developments of the electroactive materials for non-enzymatic glucose sensing and their mechanisms. Electrochem. 2021;**2**:347-389. DOI: 10.3390/ELECTROCHEM2020025

[6] Myasoedova TN, Kalusulingam R, Mikhailova TS. Sol-gel materials for electrochemical applications: Recent advances. Coatings. 2022;**12**(11):1625. DOI: 10.3390/COATINGS12111625

[7] Khanna V. Nanosensors: Physical, Chemical, and Biological. Boca Raton, Florida, USA: CRC Press; 2011

[8] Hulanicki A, Glab S, Ingman F. Chemical sensors definitions and classification. Pure and Applied Chemistry. 1991;**63**:1247-1250. DOI: 10.1351/PAC199163091247/MACHINEREADABLECITATION/RIS

[9] Barhoum A, Hamimed S, Slimi H, Othmani A, Abdel-Haleem F, Bechelany M, et al. Modern Designs of Electrochemical Sensor Platforms for Environmental Analyses: Principles, Nanofabrication Opportunities, and Challenges. 2023. DOI:10.1016/j.teac.2023.e00199ï.

[10] Grieshaber D, MacKenzie R, Vörös J, Reimhult E. Electrochemical biosensors - Sensor principles and architectures. Sensors (Basel). 2008;**8**:1400. DOI: 10.3390/S80314000

[11] Singh R, Gupta R, Bansal D, Bhateria R, Sharma M. A review on recent trends and future developments in electrochemical sensing. ACS Omega. 2024;**9**:7336. DOI: 10.1021/ACSOMEGA.3C08060

[12] Baranwal J, Barse B, Gatto G, Broncova G, Kumar A. Electrochemical sensors and their applications: A review. Chem. 2022;**10**:363. DOI: 10.3390/CHEMOSENSORS10090363

[13] McGrath MJ, Scanaill CN. Sensing and sensor fundamentals. Sensor Technologies. 2013;**Chapter 2**:15-50. DOI: 10.1007/978-1-4302-6014-1_2

[14] Stradiotto NR, Yamanaka H, Zanoni MVB. Electrochemical sensors: A powerful tool in analytical chemistry. Journal of the Brazilian Chemical Society. 2003;**14**:159-173. DOI: 10.1590/S0103-50532003000200003

[15] Walker NL, Roshkolaeva AB, Chapoval AI, Dick JE. Recent advances in potentiometric biosensing. Current Opinion in Electrochemistry.

2021;**28**:100735. DOI: 10.1016/J. COELEC.2021.100735

[16] Villagrasa JP, Colomer-Farrarons J, Miribel PL, Villagrasa JP, Colomer-Farrarons J, Miribel PL. Bioelectronics for amperometric biosensors. State of the Art in Biosensors - General Aspects. 2013:241-274. Chapter 10. DOI: 10.5772/52248

[17] Liu Y, Lv J, Yang L, Cui Y. Electrochemical uric acid sensors: Fundamentals and commercial status. Journal of the Electrochemical Society. 2022;**169**:127520. DOI: 10.1149/1945-7111/ACAB37

[18] Song C, Guo J, Wang Y, Xiang H, Yang Y. Electrochemical glucose sensors: Classification, catalyst innovation, and sampling mode evolution. Biotechnology Journal. 2024;**19**:e202400349. DOI: 10.1002/BIOT.202400349

[19] Saputra HA, Karim MM. Enzymatic and enzyme-free electrochemical lactate sensors: A review of the recent developments. Electrochemical Science Advances. 2025;**5**:e202400021. DOI: 10.1002/ELSA.202400021

[20] Fredj Z, Singh B, Bahri M, Qin P, Sawan M. Enzymatic electrochemical biosensors for neurotransmitters detection: Recent achievements and trends. Chem. 2023;**11**:388. DOI: 10.3390/CHEMOSENSORS11070388

[21] Singh S, Pavithra N, Kaur H, Varshney R, Khan NA, Kumar R, et al. Enzyme-based sensor for the real-time detection of atrazine: Evidence from electrochemical and docking studies. Scientific Reports. 2024;**14**:1-13. DOI: 10.1038/s41598-024-65801-y

[22] Mohammadpour-Haratbar A, Mohammadpour-Haratbar S, Zare Y, Rhee KY, Park SJ. A review on non-enzymatic electrochemical biosensors of glucose using carbon nanofiber nanocomposites. Biosensors (Basel). 2022;**12**:1004. DOI: 10.3390/BIOS12111004

[23] Ding Y. Developments of materials for nonenzymatic biosensing applications. Highlights in Science, Engineering and Technology. 2023;**73**:484-493. DOI: 10.54097/HSET.V73I.14061

[24] Li R, Deng X, Xia L. Non-enzymatic sensor for determination of glucose based on PtNi nanoparticles decorated graphene. Scientific Reports. 2020;**10**:1-8. DOI: 10.1038/s41598-020-73567-2

[25] Magar HS, Hassan RYA, Abbas MN. Non-enzymatic disposable electrochemical sensors based on CuO/Co3O4@MWCNTs nanocomposite modified screen-printed electrode for the direct determination of urea. Scientific Reports. 2023;**13**:1-16. DOI: 10.1038/s41598-023-28930-4

[26] Zhou C, Feng J, Tian Y, Wu Y, He Q, Li G, et al. Non-enzymatic electrochemical sensors based on nanomaterials for detection of organophosphorus pesticide residues. Environmental Science: Advances. 2023;**2**:933-956. DOI: 10.1039/D3VA00045A

[27] Rohilla D, Chaudhary S, Umar A. An overview of advanced nanomaterials for sensor applications. Engineering Sciences. 2021;**16**:47-70. DOI: 10.30919/ES8D552

[28] Li H, Xiao N, Jiang M, Long J, Li Z, Zhu Z. Advances of transition metal-based electrochemical non-enzymatic glucose sensors for glucose analysis: A review. Critical Reviews in Analytical Chemistry. 2024;**54**(7):1987-2006. DOI: 10.1080/10408347.2024.2339955

[29] Qurat-ul-Ain MKRR, Pervaiz M, Saeed Z, Majid A, Liaqat M. Graphene-based enzymatic and non-enzymatic electrochemical glucose sensors: Review of current research and advances in nanotechnology. ChemistrySelect. 2023;**8**(46):e202303952. DOI: 10.1002/SLCT.202303952

[30] Annalakshmi M, Kumaravel S, Chen TW, Chen SM, Lou BS. 3D flower-like NiCo layered double hydroxides: An efficient Electrocatalyst for non-enzymatic electrochemical biosensing of hydrogen peroxide in live cells and glucose in biofluids. ACS Applied Bio Materials. 2021;**4**:3203-3213. DOI: 10.1021/ACSABM.0C01600

[31] Yan R, Zhou Q, Xie H, Lei C. Au nanoparticle-embellished UiO-66 on reduced graphene oxide as a non-enzymatic electrocatalyst at a remarkably low oxidation potential for glucose oxidation and sensing. Analyst. 2024;**149**:761-767. DOI: 10.1039/D3AN02127K

[32] Osuna V, Aparicio Martínez EP, Dominguez RB, Vega RA. A review on the advances in nanomaterials for electrochemical non-enzymatic glucose sensors working in physiological conditions. Chem. 2024;**12**(8):159. DOI: 10.3390/CHEMOSENSORS12080159

[33] Ahmad R, Khan M, Tripathy N, Khan MIR, Khosla A. Hydrothermally synthesized nickel oxide nanosheets for non-enzymatic electrochemical glucose detection. Journal of the Electrochemical Society. 2020;**167**:107504. DOI: 10.1149/1945-7111/AB9757

[34] Wu D, Xu Z, Zhang T, Shao Y, Xi P, Li H, et al. Cu2O/CuO@rGO heterostructure derived from metal–organic-frameworks as an advanced electrocatalyst for non-enzymatic electrochemical H2O2 sensor. RSC Advances. 2016;**6**:103116-103123. DOI: 10.1039/C6RA23551D

[35] Wang F, Li Y, Yan C, Ma Q, Yang X, Peng H, et al. Bismuth-decorated honeycomb-like carbon nanofibers: An active electrocatalyst for the construction of a sensitive nitrite sensor. Molecules. 2023;**28**:3881. DOI: 10.3390/MOLECULES28093881/S1

[36] Wang MQ, Ye C, Bao SJ, Zhang Y, Xu MW, Li Z. Bimetal-organic-frameworks-derived yolk-shell-structured porous Co2P/ZnO@PC/CNTs hybrids for highly sensitive non-enzymatic detection of superoxide anion released from living cells. Chemical Communications (Cambridge). 2016;**52**:12442-12445. DOI: 10.1039/C6CC06974F

[37] Xie Y, Tu X, Ma X, Fang Q, Lu L, Yu Y, et al. High-performance voltammetric sensor for dichlorophenol based on β-cyclodextrin functionalized boron-doped graphene composite aerogels. Nanotechnology. 2019;**30**(18):185502. DOI: 10.1088/1361-6528/AB0092

[38] Khatami MJ, Mazloum-Ardakani M, Mohammadian-Sarcheshmeh H, Vafazadeh R. Highly stable and selective determination of glucose by a modified glassy carbon electrode based on micro-rods [Ni(HL)NCS] complex as a novel modifier. Journal of Electrochemical Science and Engineering. 2024;**14**(1):185-202. DOI: 10.5599/JESE.2368

[39] Simsek M, Wongkaew N. Carbon nanomaterial hybrids via laser writing for high-performance non-enzymatic electrochemical sensors: A critical review. Analytical and Bioanalytical Chemistry. 2021;**413**:6079. DOI: 10.1007/S00216-021-03382-9

[40] Lavanya AL, Bala Kumari KG, Prasad KRS, Brahman PK. Fabrication

of electrochemical sensor based on electrochemically co-deposited Ru-Co bimetallic nanoparticles on glassy carbon electrode: An analytical measurement tool for monitoring of hydrazine in water samples. International Journal of Environmental Analytical Chemistry. 2020;**102**:720-735. DOI: 10.1080/03067319.2020.1726333

[41] Nawaz M, Shaikh H, Buledi JA, Solangi AR, Karaman C, Erk N, et al. Fabrication of ZnO-doped reduce graphene oxide-based electrochemical sensor for the determination of 2,4,6-trichlorophenol from aqueous environment. Carbon Letters. 2023;**34**:201-214. DOI: 10.1007/S42823-023-00562-8

[42] Rajakumaran R, Ramki S, Chen SM, Chen TW, Veerasankar S, Tseng TW, et al. Rose-petal-like morphology of yttrium molybdate nanosheets (YMoO4) anchored on functionalized carbon nanofibers: An efficient electrocatalyst for the electrochemical sensing of bisphenol-A. International Journal of Electrochemical Science. 2019;**14**:6571-6585. DOI: 10.20964/2019.07.22

[43] Teradale AB, Unki SN, Ganesh PS, Das KK, Das SN. Development of a diethylcarbamazine citrate-based electrochemical sensor for quick and affordable detection of sulfadiazine and uric acid in environmental monitoring. ChemistrySelect. 2024;**9**(22):e202401047. DOI: 10.1002/SLCT.202401047

[44] Liu X, Li Y, Qiao W, Chang M, Li Y. A non-enzymatic electrochemical sensor based on nitrogen-doped mesoporous carbon@hydroxyl-functionalized ionic liquid composites modified electrode for the detection of fenitrothion. RSC Advances. 2023;**13**:13030. DOI: 10.1039/D3RA01011B

[45] Mohammad A, Zamzami MA. Construction of carbon cloth modified-Al2O3-g-C3N4 sensor for non-enzymatic electrochemical detection of hydrogen peroxide. Diamond and Related Materials. 2022;**132**:109600. DOI: 10.1016/J.DIAMOND.2022.109600

[46] Dey B, Ahmad MW, Kim BH, Kamal T, Yang DJ, Patra CN, et al. Manganese cobalt-MOF@carbon nanofiber-based non-enzymatic histamine sensor for the determination of food freshness. Analytical and Bioanalytical Chemistry. 2023;**415**:3487-3501. DOI: 10.1007/S00216-023-04737-0

[47] Jamal M. (Digital Presentation) Non-enzymatic lactose sensor based on nickel oxide nanoparticles modified carbon fiber cloth electrode. ECS Meeting Abstracts. 2022;**MA2022-02**:2631. DOI: 10.1149/MA2022-02632631MTGABS

[48] Suropati RTL, Wibowo R, Nurani DA, Krisnandi YK. Modified flower-like nickel oxide on carbon paste electrode (CPE) for analysis cholesterol based on non-enzymatic sensor. IOP Conference Series: Materials Science and Engineering. 2020;**763**(1):012002. DOI: 10.1088/1757-899X/763/1/012002

[49] Song H, Huo M, Zhou M, Chang H, Li J, Zhang Q, et al. Carbon nanomaterials-based electrochemical sensors for heavy metal detection. Critical Reviews in Analytical Chemistry. 2024;**54**(7):1987-2006. DOI: 10.1080/10408347.2022.2151832

[50] Hernández-Ramírez D, Franco-Guzmán M, Ibarra-Ortega IS, Álvarez-RomeroGA, Rebolledo-PeralesLE. Review—Trends on the development of non-enzymatic electrochemical sensors modified with molecularly imprinted polymers for the quantification of glucose. Journal of the Electrochemical Society. 2024;**171**:077506. DOI: 10.1149/1945-7111/AD5D1F

[51] Selvi Gopal T, James JT, Gunaseelan B, Ramesh K, Raghavan V, Malathi ACJ, et al. MXene-embedded porous carbon-based Cu2O nanocomposites for non-enzymatic glucose sensors. ACS Omega. 2024;**9**:8448. DOI: 10.1021/ACSOMEGA.3C09659

[52] Jazi HK, Sarafbidabad M, Ben HM, Ahmadipour M. The effect of laser surface texturing on ZnO/MWCNT nanocomposite modified screen-printed carbon electrode for non-enzymatic glucose biosensor. Diamond and Related Materials. 2025;**151**:111845. DOI: 10.1016/J.DIAMOND.2024.111845

[53] Sivakumar M, Vivekanandan AK, Panomsuwan G, Veeramani V, Chen SH, Jiang Z, et al. Flower-like NiCo2O4 nanoflake surface covered on carbon nanolayer for high-performance electro-oxidation of non-enzymatic glucose biosensor. Materials Today Chemistry. 2022;**26**:101156. DOI: 10.1016/J.MTCHEM.2022.101156

[54] Abrori SA, Septiani NLW, Hakim FN, Maulana A, Suyatman N, et al. Non-enzymatic electrochemical detection for uric acid based on a glassy carbon electrode modified with MOF-71. IEEE Sensors Journal. 2021;**21**:170-177. DOI: 10.1109/JSEN.2020.3014298

[55] de Souza OPL, Tiba DY, Ferreira JHA, Lieb LC, Canevari TC. Non-enzymatic biosensor based on F,S-doped carbon dots/copper nanoarchitecture applied in the simultaneous electrochemical determination of NADH, dopamine, and uric acid in plasma. Analyst. 2024;**149**:2728-2737. DOI: 10.1039/D3AN02239K

[56] Li Y, Zhai Z, Du X, Liao R, Wang Y, Yang Y, et al. Carbon-based-polypyrrole composite nanowires for portable and remote trace sweat lactate detection. Advances in Engineering Technology Research. 2024;**10**:97. DOI: 10.56028/AETR.10.1.97.2024

[57] Abou-Elyazed A, Li S, Mohamed GG, Li X, Meng J, El-Sanafery S. Graphitic carbon nitride/MOFs hybrid composite as highly selective and sensitive electrodes for calcium ion detection. Molecules. 2023;**28**:8149. DOI: 10.3390/MOLECULES28248149/S1

[58] Bokov D, Turki Jalil A, Chupradit S, Suksatan W, Javed Ansari M, Shewael IH, et al. Nanomaterial by sol-gel method: Synthesis and application. Advances in Materials Science and Engineering. 2021;**2021**:5102014. DOI: 10.1155/2021/5102014

[59] Ndlwana L, Raleie N, Dimpe KM, Ogutu HF, Oseghe EO, Motsa MM, et al. Sustainable hydrothermal and solvothermal synthesis of advanced carbon materials in multidimensional applications: A review. Materials. 2021;**14**:5094. DOI: 10.3390/MA14175094

[60] Wei Z, Yanfei Z, Jiao W. Mesoporous carbon materials: Synthesis methods, properties, and advanced applications. Frontiers in Materials. 2025;**12**:1548671. DOI: 10.3389/FMATS.2025.1548671/PDF

[61] Ben Arbia M, Helal H, Comini E. Recent advances in low-dimensional metal oxides via sol-gel method for gas detection. Nanomaterials. 2024;**14**:359. DOI: 10.3390/NANO14040359

[62] Esposito S. "Traditional" sol-gel chemistry as a powerful tool for the preparation of supported metal and metal oxide catalysts. Materials. 2019;**12**:668. DOI: 10.3390/MA12040668

[63] Pekala RW, Alviso CT. Carbon aerogels and xerogels. MRS Proceedings. 1992;**270**:3-14. DOI: 10.1557/PROC-270-3/METRICS

[64] Maetz A, Delmotte L, Moussa G, Dentzer J, Knopf S, Ghimbeu CM. Facile and sustainable synthesis of nitrogen-doped polymer and carbon porous spheres. Green Chemistry. 2017;**19**:2266-2274. DOI: 10.1039/C7GC00684E

[65] Emrie DB. Sol-gel synthesis of nanostructured mesoporous silica powder and thin films. Journal of Nanomaterials. 2024;**2024**:6109770. DOI: 10.1155/2024/6109770

[66] Liu W, Herrmann AK, Bigall NC, Rodriguez P, Wen D, Oezaslan M, et al. Noble metal aerogels-synthesis, characterization, and application as electrocatalysts. Accounts of Chemical Research. 2015;**48**:154-162. DOI: 10.1021/AR500237C/SUPPL_FILE/AR500237C_SI_001.PDF

[67] Geppi M, Mollica G, Borsacchi S, Marini M, Toselli M, Pilati F. Solid-state nuclear magnetic resonance characterization of PE–PEG/silica hybrid materials prepared by microwave-assisted sol-gel process. Journal of Materials Research. 2007;**22**:3516-3525. DOI: 10.1557/JMR.2007.0434

[68] Pekala RW. Organic aerogels from the polycondensation of resorcinol with formaldehyde. Journal of Materials Science. 1989;**24**:3221-3227. DOI: 10.1007/BF01139044

[69] Feinle A, Elsaesser MS, Hüsing N. Sol-gel synthesis of monolithic materials with hierarchical porosity. Chemical Society Reviews. 2016;**45**:3377-3399. DOI: 10.1039/C5CS00710K

[70] Pekala RW, Kong FM. A synthetic route to organic aerogels - mechanism, structure, and properties. Le Journal de Physique Colloques. 1989;**24**:C4-33-C4-40. DOI: 10.1051/JPHYSCOL:1989406

[71] Predoană L, Attila Karajz D, Otieno Odhiambo V, Stanciu I, Szilágyi IM, Pokol G, et al. Influence of the microwaves on the sol-gel syntheses and on the properties of the resulting oxide nanostructures. Microwave Heating - Electromagnetic Fields Causing Thermal and Non-Thermal Effects. 2020:87-112. Chapter 5. DOI: 10.5772/INTECHOPEN.94931

[72] Menéndez JA, Arenillas A, Fidalgo B, Fernández Y, Zubizarreta L, Calvo EG, et al. Microwave heating processes involving carbon materials. Fuel Processing Technology. 2010;**91**:1-8. DOI: 10.1016/J.FUPROC.2009.08.021

[73] Siddiqui MTH, Chan FL, Nizamuddin S, Baloch HA, Kundu S, Czajka M, et al. Comparative study of microwave and conventional solvothermal synthesis for magnetic carbon nanocomposites and bio-oil from rice husk. Journal of Environmental Chemical Engineering. 2019;**7**:103266. DOI: 10.1016/J.JECE.2019.103266

[74] Juárez-Pérez EJ, Calvo EG, Arenillas A, Menéndez JA. Precise determination of the point of sol–gel transition in carbon gel synthesis using a microwave heating method. Carbon N Y. 2010;**48**:3305-3308. DOI: 10.1016/J.CARBON.2010.05.013

[75] Lobato B, Flores SL, dos Santos-Gómez L, García AB, Pernía AM, Prieto MJ, et al. Designing and optimizing electrode materials for energy harvesting in CAPMIX cells. Nanomaterials. 2024;**14**:2031. DOI: 10.3390/NANO14242031

[76] The IUPAC Compendium of Chemical Terminology. The IUPAC Compendium of Chemical Terminology2019. DOI: 10.1351/GOLDBOOK

[77] Cai X, Tan G, Deng Z, Liu J, Gui D. Preparation of hierarchical porous

carbon aerogels by microwave assisted sol-gel process for supercapacitors. Polymers (Basel). 2019;**11**:429. DOI: 10.3390/POLYM11030429

[78] Miecznikowski K, Cox JA. Electroanalysis based on stand-alone matrices and electrode-modifying films with silica sol-gel frameworks: A review. Journal of Solid State Electrochemistry. 2020;**24**:2617-2631. DOI: 10.1007/S10008-020-04697-W/METRICS

[79] Gan W, Li Y, Liu T, Yang Y, Song B, Dai S, et al. Rapid and sensitive detection of ammonia in water by a long period fiber grating sensor coated with sol-gel silica. Optics Express. 2022;**30**:33817. DOI: 10.1364/OE.472205

[80] Stocker MK, Sanson ML, Bernardes AA, Netto AM, Brambilla R. Acid–base sensor based on sol–gel encapsulation of bromothymol blue in silica: Application for milk spoilage detection. Journal of Sol-Gel Science and Technology. 2021;**98**:568-579. DOI: 10.1007/S10971-021-05529-7

[81] Salihovic M, Pameté E, Arnold S, Sulejmani I, Bartschmid T, Hüsing N, et al. Black goes green: Single-step solvent exchange for sol-gel synthesis of carbon spherogels as high-performance supercapacitor electrodes. Energy Advances. 2024;**3**:482. DOI: 10.1039/D3YA00480E

[82] González-Barriuso M, Sánchez-Suárez M, González-Lavín J, Arenillas A, Rey-Raap N. Synthesis of Ni-doped graphene aerogels for electrochemical applications. Gels. 2024;**10**:180. DOI: 10.3390/GELS10030180/S1

[83] Beaver K, Dantanarayana A, Minteer SD. Materials approaches for improving electrochemical sensor performance. Journal of Physical Chemistry B. 2021;**125**:11820-11834. DOI: 10.1021/acs.jpcb.1c07063

[84] Fontana-Escartín A, Bertran O, Alemán C. Materials engineering in electrochemical biosensors: A review of cost-effective approaches to efficient biodetection. Materials Today Communications. 2024;**41**:111030. DOI: 10.1016/j.mtcomm.2024.111030

[85] Wang J. Sol–gel materials for electrochemical biosensors. Analytica Chimica Acta. 1999;**399**:21-27. DOI: 10.1016/S0003-2670(99)00572-3

[86] Almeida CMR, Merillas B, Pontinha ADR. Trends on aerogel-based biosensors for medical applications: An overview. International Journal of Molecular Sciences. 2024;**25**(2):1309. DOI: 10.3390/ijms25021309

[87] Yang H, Wang W, Zeng Y, Tang R, Yan H, Yang Y, et al. A novel composite of aptamer-based Ti3C2Tx and molecularly imprinted polymer with double recognition property for sensitive electrochemical detection of ofloxacin. Journal of Analysis and Testing. 2024;**8**(1):335-347. DOI: 10.1007/s41664-024-00335-w

[88] Valero-Calvo D, García-Alonso FJ, de la Escosura-Muñiz A. Bimetallic nanoparticles as electrochemical labels in immunosensors for the detection of biomarkers of clinical interest. Analysis and Sensing. 2024;**4**(1):e202400017. DOI: 10.1002/anse.202400017

[89] Alanazi AZ, Alhazzani K, El-Wekil MM, Ali AMBH, Darweesh M, Ibrahim H. Nanostructured gold-embedded porous carbon black nanoceramic film coating on pencil graphite rods: An innovative disposable ultrasensitive sensor for monitoring dopamine. Microchemical Journal. 2024;**206**:111454. DOI: 10.1016/j.microc.2024.111454

[90] Sharma A, Arya S, Chauhan D, Solanki PR, Khajuria S, Khosla A. Synthesis of Au–SnO2 nanoparticles for electrochemical determination of vitamin B12. Journal of Materials Research and Technology. 2020;**9**:14321-14337. DOI: 10.1016/j.jmrt.2020.10.024

[91] Parashar M, Shukla VK, Singh R. Metal oxides nanoparticles via sol–gel method: A review on synthesis, characterization and applications. Journal of Materials Science: Materials in Electronics. 2020;**31**:3729-3749. DOI: 10.1007/s10854-020-02994-8

[92] Zhan B, Zhang Y, Zhao X. High sensitive sol-gel based electrochemical immunosensor for clenbuterol determination. International Journal of Electrochemical Science. 2021;**16**:1-11. DOI: 10.20964/2021.11.23

[93] Haque M, Fouad H, Seo HK, Alothman OY, Ansari ZA. Cu-doped ZnO nanoparticles as an electrochemical sensing electrode for cardiac biomarker myoglobin detection. IEEE Sensors Journal. 2020;**20**:8820-8832. DOI: 10.1109/JSEN.2020.2982713

[94] Mahmoud A, Echabaane M, Omri K, Boudon J, Saviot L, Millot N, et al. Cu-doped ZnO nanoparticles for non-enzymatic glucose sensing. Molecules. 2021;**26**(4):929. DOI: 10.3390/molecules26040929

[95] Sumida K, Liang K, Reboul J, Ibarra IA, Furukawa S, Falcaro P. Sol-gel processing of metal-organic frameworks. Chemistry of Materials. 2017;**29**:2626-2645. DOI: 10.1021/acs.chemmater.6b03934

[96] Chakraborty T, Das M, Kao CH. Copper-loaded ZIF-8-based immunosensor for early diagnosis of chronic kidney disease by detecting neutrophil gelatinase-associated lipocalin (NGAL) biomarker. Materials Chemistry and Physics. 2024;**326**:129845. DOI: 10.1016/j.matchemphys.2024.129845

[97] Ali A, Shoukat R, Ashraf AR, Rasheed Z, Muqaddas S, Iqbal M, et al. Cobalt-based ferrite modified carbon nanotubes fibers for flexible and disposable microelectrode toward electrochemical glucose sensing. Analytical Science Advances. 2024;**5**(11 12):e202400032. DOI: 10.1002/ansa.202400032

[98] Baghayeri M, Amiri A, Fayazi M, Nodehi M, Esmaeelnia A. Electrochemical detection of bisphenol a on a MWCNTs/CuFe2O4 nanocomposite modified glassy carbon electrode. Materials Chemistry and Physics. 2021;**261**:124247. DOI: 10.1016/j.matchemphys.2021.124247

[99] Mahmoudpour M, Dolatabadi JEN, Hasanzadeh M, Soleymani J. Carbon-based aerogels for biomedical sensing: Advances toward designing the ideal sensor. Advances in Colloid and Interface Science. 2021;**298**:102550. DOI: 10.1016/j.cis.2021.102550

[100] Zhang W, Liu S, Liang X, He J, Lu Y, Wu Q. Super-elastic and multifunctional graphene aerogels with multilayer cross-linked pore structure for dynamic force sensing arrays. Carbon NY. 2025;**236**:120105. DOI: 10.1016/j.carbon.2025.120105

[101] Yang M, Choy K, leong. A nature-derived, flexible and three dimensional (3D) nano-composite for chronic wounds pH monitoring. Materials Letters. 2021;**288**:129335. DOI: 10.1016/j.matlet.2021.129335

Chapter 11

# A State-of-the-Art of Understanding of the Separation of Microplastics from Groundwater: Development through Mathematical Modeling

*Sobana Subramani, Rames C. Panda, Sundararaman T.R., Murugesan S.N., Venkatachalam Varadharajan and Balasubramanian Karuppiah*

## Abstract

Microplastics (MP) are acknowledged as a worldwide pollutant of concern, and there has been a steady escalation in the amount of public attention and academic research on these materials. The movement of MPs toward the depth of the soil horizons is due to the added effect of adsorption and blocking processes in the unsaturated zone. Porous media, specifically, plastic size, shape, polymer type, solution chemistry, input concentration, and grain size, are the variables and parameters that influence the MPs' mass transport in sediments, sea, river, groundwater bodies with different concentrations, soil, and human beings. The plastic density has been conferred as an important manipulated variable that controls the final settling velocities of the MP. The final velocity of MP increases with increasing particle density. This chapter reviews MP mass transport through its mathematical modeling by gathering more than 59 works of literature.

**Keywords:** modeling, microplastic movement, adsorption, ground water, separation

## 1. Introduction to microplastics

Microplastics originate from plastics and cause pollution in the ground and water [1]. Microplastics (MP) are smaller-sized particles (<5 mm) of plastics composed of complex chemicals, generally toxic and non-biodegradable, which are scattered (almost 5.25 trillion particles weighing 268,940 tons approximately) around soil and aqueous environment and are considered pollutants resulting in contamination through ingestion by animals, and entrapment by marine fauna, zooplankton, cetaceans, and aquatic lives. These particles arrive in the environment due to the degradation of microbeads, plastic and polymeric substances, tires, synthetic cloth, and packaging plastic materials. It has been reported that a one-liter water bottle contains

325 pieces of MPs. Human blood contains 4 μg of MP per ml. Out of 13 types of microplastics, polyethylene, polypropylene, and polystyrene are important. It has been reported that microplastics are found in drinking water (0.001 to $10^6$ particles/$m^3$) and other aquatic sources, causing health issues for human beings. It is necessary to assess the distribution of MPs over the sources and devise their separation or treatment methods. It has been observed that the microplastics thrown in aquatic sources (lake, river, and ocean) float for a few days and then get submersed on the bottom of aquatic sources, while the fine particles diffuse and are consumed by the aquatic culture, resulting in ecological damage [1]. The transport of MPs is limited to marine and terrestrial environments that only include coastal areas and surface water as suggested in Ref. [1,2] Research on the movement of MPs in soil has indicated that biological activity, human activity, soil features (pores and cracks), and precipitation may all contribute to the potential influx of MPs into groundwater through the soil [3, 4]. The study examined modeling the physical system that simulated the transmission of microplastics and a steady current, reporting microplastic occurrences in soil and groundwater. Furthermore, the challenges and constraints associated with studying microplastics in soil and groundwater are examined in Refs. [5, 6]. By creating governing equations with an advection and dispersion model, the movement of MPs in a saturated and unsaturated porous medium was modeled with the finite element method was used to simulate and investigate how the governing equations' parameters affected MP transport and to limit the number of MPs that leak into groundwater habitats by combining low diffusivity, fast water flow velocity and high soil porosity were described in [7]. With time, the MPs start deforming, degrading, and with an aqueous medium, they form colloids. The motivation for writing this review chapter is to cover the mathematical modeling for removing microplastics from aquatic environments. These MPs should be removed from the environment as it is causes harm to animals, human beings, etc. A large portion of microplastics in wastewater is removed through primary, secondary, and tertiary treatment. Approximately 35–59% of the microplastics are removed during the primary treatment, and 50–98% are removed during the secondary treatment in wastewater treatment plants [8]. It has also been reported that overall microplastic elimination can be achieved up to 88% in the absence of tertiary treatment and up to 97% with tertiary treatment [8]. However, sequences of too many unit operations make the integrated process costlier. Microplastics on the road have to adopt automated sampling devices, and a continuous monitoring unit could provide real-time data, capturing progressive variations in microplastic concentrations [9]. This explains that there is a dire need for process modeling for scaling up and optimization studies for the separation of MPs using the mechanical filtration process. Different types of separation techniques have been suggested for this. However, the separation techniques must be scientifically adjudged based on the properties and characteristics of MPs. The subject is under research and needs a review for technical feasibility and implementation on commercial up-scales.

## 2. Occurrence

There are almost 13 types of polymers from which MPs are generated and enter the environment. However, out of them, polyethylene, polyvinyl chloride, polypropylene, polystyrene, polyester, polyethylene terephthalate, polyamide, and polyurethane are important. Researchers estimate that 4–12 million tons of plastic are flowing into the oceans each year. By 2050, that will affect the fish and other aquatic populations.

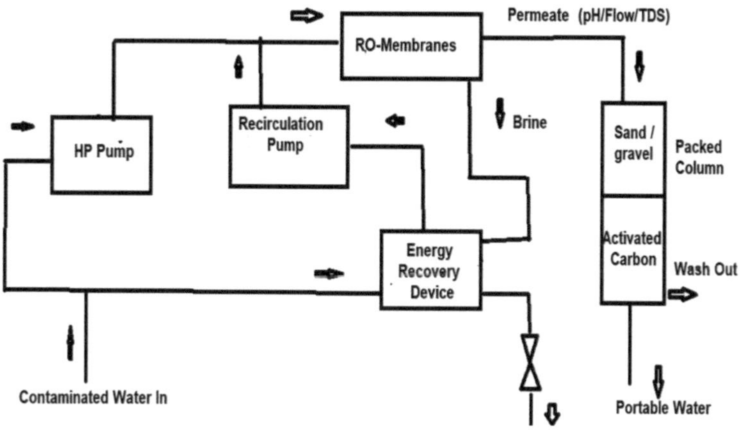


**Figure 1.**
*Block diagram of RO and packed column to remove microplastics from drinking water.*

The availability of polymers in the form of MPs is polyethylene (PE, 76.8%), polyethylene terephthalate (PET, 6.9%), polypropylene (PP, 6%), and polystyrene (PS, 4%). The compositions and physicochemical properties of the particles are described by many authors. MPs are tiny polymeric solid particles, light in weight, chemically stable, insoluble, and nondegradable. The particles can be easily molded into different sizes/shapes, are good insulators, have low thermal conductivity, and do not rust [2]. These MPs can adsorb contaminants from the environment and can act as vectors for toxins and pathogens. However, the properties change with runoff, tides, and natural calamities. The breakdown of bigger plastic materials like toys, bags, and beverage bottles *via* mechanical or photooxidative mechanisms like sun, heat, wind, and waves causes the secondary MPs that may never completely disappear and the rate of improperly managed plastic debris entering the aquatic environment and potentially forming secondary MPs is increasing day by day in nearly every nation.

These microplastics are generated from broken plastic bottles, bags, synthetic textiles, tires, city dust, road-marking, personal care products, etc. The used solids, when thrown away into the environment, are carried away with domestic and municipal sewage and industrial effluents to rivers, lakes and seas [3]. However, the potential question is: 'How to remove it from aqueous media using feasible technology.' Removal—The contamination of MPs and their effects on physiological metabolism are not yet properly known. However, the effect on human health is very serious. Generally, the exposure comes through ingestion, inhalation, dermal contact, and many other means. There are many techniques to remove microplastics from water. Generally, the techniques are divided into physical, chemical, and biological methods. **Figure 1** shows the block diagram used to remove MP from the drinking water. The inlet water consists of microplastics and passes through the HP pump and the RO membrane module. The outlet of RO permeate water is allowed to pass through the packed sand column to remove microplastics from the portable drinking water.

## 3. Modeling

There are many techniques to remove microplastics from water. Generally, the techniques are divided into physical, chemical, and biological methods. MPs can be either in ground or in water, where they slowly form colloids. The modeling of colloidal particles was considered in this section, which will lead to the transport of the microplastics in groundwater [10–13]. By creating governing equations using the advection and dispersion model, the movement of MPs in a saturated and unsaturated porous medium was modeled with the finite element method. This model was used to simulate and investigate the parameters that affect MP transport and to limit the number of MPs that leak into groundwater habitats by combining the effect of low diffusivity, fast water flow velocity and high soil porosity which was described in Ref. [14]. Researcher [15] studied the transport of smaller glass particles in a horizontal narrow path under laminar flow and derived formulations for Brownian diffusivity through material transfer, where it was observed that the colloidal particles maintained a constant deposition rate through gravity that was confirmed through image processing. Therefore, the deposition flux and thickness of the boundary layer are controlled by gravity. A mathematical model was constructed to demonstrate the orientation of non-Brownian motion of particles based on the shape that percolates through porous media with the rectangular grain, where the solution trajectory was found to be dependent on flow rate and flow anisotropy [16].

Researcher [17] investigated the contribution of acid-base (AB) (polar) interactions to the total interaction energy between membranes and colloids with the help of the van der Waals (LW) give the polar interaction energies between various surfaces from measurements of contact angles of different probe liquids on these surfaces and the surface potentials of the membranes and colloids were determined using electrokinetic measurements, yielding the electrostatic (EL) interaction. They presented the total interaction energy, or energy of adhesion, between a colloid and a membrane as the sum of LW and EL interaction energies according to

$$U_{mic}^{DLVo} = U_{mic}^{LW} + U_{mic}^{EL} \tag{1}$$

where $U_{mic}^{\mathrm{DLVO}}$—total interaction energy between the membrane and colloid immersed in water. $U_{mic}^{\mathrm{LW}}$—LW interaction term. $U_{mic}^{\mathrm{EL}}$—EL interaction term.

The subscripts m, l, and c correspond to the membrane, bulk feed solution, and colloid, respectively.

The addition of the AB interaction term results in the prolonged Derjaguin-Landau-Verwey-Overbeek (DLVO) method, which may be written as

$$U_{mic}^{XDLVo} = U_{mic}^{LW} + U_{mic}^{EL} + U_{mic}^{AB} \tag{2}$$

where

$U_{mic}^{\mathrm{XDLVO}}$—total interaction energy between the membrane and colloid immersed in water.

$U_{mic}^{\mathrm{AB}}$—AB interaction term.

According to van der Waals' acid-base approach, the total surface tension of a pure substance is the added form of an LW and an AB component as

$$\gamma^{TOT} = \gamma^{LW} + \gamma^{AB} \tag{3}$$

where

$\gamma^{\mathrm{TOT}}$—total surface tension.

$\gamma^{\mathrm{LW}}$ and $\gamma^{\mathrm{AB}}$—LW and AB components of surface tension.

The polar AB constituent of a material's surface energy is given by

*DOI: http://dx.doi.org/10.5772/intechopen.1011483*

$$\gamma^{AB} = \sqrt{\gamma^{+} + \gamma^{-}} \tag{4}$$

where

$\gamma^{+}$—electron acceptor parameter.

$\gamma^{-}$—electron donor parameter.

The extended Young's equivalence relates the contact angle of a liquid on a solid surface to the surface tension parameters of both the solid and the liquid as

$$(1 = cos\theta)\gamma_1^{TOT} = 2\left(\sqrt{\gamma_s^{LW}\gamma_1^{LW}} + \sqrt{\gamma_s^{+}\gamma_1^{-}} + \sqrt{\gamma_s^{-}\gamma_1^{+}}\right) \tag{5}$$

where

$\theta$—The angle of contact between the liquid and solid.

The subscripts s and 1 correspond to the solid surface and the liquid, respectively.

The extended Young's model is described as an equilibrium force balance. The left-hand side of Eq. (5) represents the free energy of interconnection per unit area of the liquid (l).

The right-hand side of Eq. (5) represents the free energy of adhesion per unit area between the liquid (l) and the solid (s).

Surface tensions for membranes and colloids were calculated using the acid-base approach, which can be used to evaluate the free energies of adhesion per unit area $\Delta G_{y0}{}^{LW}$ and $\Delta G_{y0}{}^{AB}$, between the membrane and colloid surfaces.

Expressions for the LW and AB adhesion energies per unit area are,

$$\Delta G_{y0}^{LW} = 2\left(\sqrt{\gamma_I^{LW}} - \sqrt{\gamma_m^{LW}}\right)\left(\sqrt{\gamma_c^{LW}} - \sqrt{\gamma_I^{LW}}\right) \tag{6}$$

and

$$\Delta G_{y0}^{AB} = 2\sqrt{\gamma_I^{+}}\left(\sqrt{\gamma_m^{-}} + \sqrt{\gamma_c^{-}} - \sqrt{\gamma_I^{-}} + 2\sqrt{\gamma_1^{-}}\right) + 2\sqrt{\gamma_I^{-}}\left(\sqrt{\gamma_m^{+}} + \sqrt{\gamma_C^{+}} - \sqrt{\gamma_I^{+}}\right) - 2\left(\sqrt{\gamma_m^{+}\gamma_C^{-}} + \sqrt{\gamma_m^{-}\gamma_C^{+}}\right) \tag{7}$$

The free energy of adhesion per unit area indicates the interaction energy per unit area between two planar surfaces, such as a membrane and a colloid, when they are brought into contact. The LW interaction energy per unit area decays with the inverse square of the distance between two infinite planar surfaces, like a membrane and a colloid:

$$\Delta G_{(Plate-Plate)}^{LW} = \frac{A}{12\pi y^2} \tag{8}$$

where

$A$—Hamaker constant.

$y$—the separation distance between the two interacting planar surfaces, such as a membrane and a colloid.

The Hamaker constant can be calculated from Eq. (8) as

$$A = -12\pi u_0^2 \Delta G_{y0}^{LW} \tag{9}$$

The LW interaction energy between an infinite flat plate and a spherical colloid is

$$U^{LW}_{(Plate-Sphere)} = -\frac{Aa_c}{6h} \quad (10)$$

where $a_c$ is the radius of the spherical colloid and $h$ is the surface-to-surface separation distance between the flat plate (membrane) and the sphere (colloid).

The LW interaction energy between a membrane and a colloid in an aqueous system is given as

$$U^{LW}_{mic}(h) = 2\pi\Delta^{LW}_{y0}\frac{y_0^2 a_c}{h} \quad (11)$$

The AB interaction energy between two infinite planar surfaces is given as

$$\Delta G^{AB}_{(Plate-Plate)} = \Delta G^{AB}_{Yo}\ exp\left[\frac{Y_0 - Y}{\lambda}\right] \quad (12)$$

where

$\lambda$—characteristic decay length of AB interactions in water, whose value is between 0.2 and 1.0 nm. A commonly used value of $\lambda$ for an aqueous environment is 0.6 nm. By applying Derjaguin's procedure to Eq. (12), the decay behaviors of AB interactions between a membrane and a colloid in an aqueous situation are

$$U^{AB}_{mlc} = 2\pi a_c\ \lambda\Delta G^{AB}_{Yo}\ exp\left[\frac{Y_0 - Y}{\lambda}\right] \quad (13)$$

The EL contact energy per unit area between two infinite planar shells with a separation distance is given as

$$\Delta G^{EL}_{(Plate-Plate)} = \frac{\varepsilon_0\varepsilon_r k}{2}\left(\zeta_m^2 + \zeta_C^2\right)(1 - coth(k - y) + \frac{2\zeta_m\zeta_c}{\left(\zeta_m^2 + \zeta_c^2\right)}csch(ky)) \quad (14)$$

where

$\varepsilon_0\varepsilon_r$—dielectric permittivity of the suspending fluid.

$\kappa$—the inverse Debye screening length.

$\zeta_m$—surface potentials of the membrane.

$\zeta_c$—surface potentials of the colloid.

Once again, applying Derjaguin's technique to Eq. (14), the decay performance of the EL interaction energy between a membrane and a colloid in an aqueous environment is given as

$$U^{EC}_{mlc}(h) = \pi\varepsilon_r\varepsilon_0 a_c\left(2\zeta_c\zeta_m\, ln\left(\frac{1 + e^{-kh}}{1 - e^{-kh}}\right) + \left(\zeta_c^2 + \zeta_m^2\right) ln\left(1 - e^{-2kh}\right)\right) \quad (15)$$

With time, microplastics become colloids as they are rejected in the aqueous environment. The mathematical modeling of the colloid transport experiments was carried out using conventional colloid attachment theory to emphasize variations in colloid attachment and straining behavior and to determine parameter ranges relevant to attachment [10]. MP particles adhere to certain supports through the attachment mechanism and later get dragged away from them due to the application of an external force. First-order rate expressions are used to depict colloid attachment and detachment, with an irreversible first-order straining part that depends on depth used to explain straining. To model exclusion, transport parameters for colloid-accessible pore space are changed [11]. The transport or percolation through ground soil can be compared to transport through porous media. Using direct observations at the pore scale and a realistic micromodel of porous media, they present experimental evidence of the effect of colloid exclusion from areas of small aperture sizes and discovered that larger particles travel along with multiple streamlines, which results in a slightly lower velocity than the small colloids, while smaller particles appear to travel along faster streamlines in pore throats [12]. Column-scale experiments connect explanations at the pore scale to a larger scale for the effect of transport velocity and colloid size on the early stage of free-moving colloids. It was discovered that, in contrast to the tracer, the colloids increase with colloid size and water velocity [13]. Different groupings of a cylindrical sand lens entrenched in the center of a larger cylinder of matrix sand and porous media (Ottawa quartz sands) that are negatively charged, stable monodispersed colloids (carboxyl latex microspheres) were presented in the saturated water, physically heterogeneous arrangements. Physical heterogeneity and colloid size were discovered to have a significant influence on colloid migration, when the attachment weightage has increased and the colloid size was less than the sand pore size. Spatial differences in the pore water velocity had an impact on colloid retention in the heterogeneous systems, and when the ratio of the median to colloid particle sizes was less than 0.5 percent, while straining occurred [14]. To distinguish the straining performance of negatively charged latex colloids (1.1 and 3.0 mm) at textural interfaces, a saturated column experiment was conducted. The effluent concentration curves method, final three-dimensional disseminations in the columns, mass balance information, microscopic investigation of deposition behaviors in micromodel experiments, and numerical modeling were used to conclude mechanisms of colloid transport and retention behaviors. Compared to the sand surface, less deposition happened at the textural interfaces inside the column because advection, dispersion, and size exclusion tend to confine colloid transport to the larger pore networks [18]. Five sizes of carboxylate-modified microspheres (0.1 to 2.0 μm) were tested for their deposition and re-entrainment behaviors in impinging jet systems and porous media at pore fluid velocities (2–8 m $day^{-1}$) that are relevant to the environment of with and without an energy barrier to deposition and observed that in the present of energy barrier, pore geometry plays a crucial role in controlling colloid deposition with the excess deposition that occurs is due to retention in flow stagnation zones [19]. The conclusion has been obtained that straining can be a significant factor in colloid deposition in unfavorable attachment circumstances based on analysis of pore size distribution, size exclusion, time- and concentration-dependent deposition behaviors, colloid size distribution, hyper-exponential deposition profiles, batch release rates, micromodel observations, and deposition at textural interfaces and reported many of the limitations of filtration theory also [20]. In a laboratory investigation, saturated porous media (gravel and glass beads) were subjected to steady-state flow to ascertain the transport and deposition rate of suspended particles. The analytical solution of a convection-dispersion equation with a first-order deposition rate skillfully represented the breakthrough curves, with the primary variables kept in this investigation to examine the mechanisms influencing the transport and deposition kinetics were the suspended particle size distribution, the porous media, and the flow rates themselves [21]. Non-perturbed deposition and re-entrainment dynamics of biological and non-biological colloids in porous media in the presence of an energy barrier to deposition at the grain surface were examined and the profiles of retained colloids and numerical simulation of breakthrough-elution behaviors were used to calculate the re-entrainment rate coefficients and deposition rate [22]. Experimental analysis using packed columns and basic mathematical modeling were used to investigate the effects of grain size, pore water ionic strength, and water capacity on the movement of latex microspheres (1.1 microns) in porous media. Negatively charged and hydrophilic colloids were used in the chemically aggressive experiments condition to determine the colloid attachment to solid-water interfaces (SWI) and air-water interfaces (AWI), with the help of the solution chemistry being adjusted to pH 10 using a bicarbonate buffer. Complementary batch experiments and interaction energy calculations are used to show the segmentation of colloids to the SWI and AWI was negligible over the range of ionic strengths taken into consideration. Each experiment yielded measurements of the breakthrough curve and final deposition profile, which showed that the sand grain size, water content, and suspension ionic strength all had a significant impact on colloid retention with the majority of colloids discovered to be deposited at the column intake, defying standard filtering theory, and hyper-exponential deposition profiles [23]. Rapid microbial biofilm development on the plastic was accompanied by notable alterations in its physicochemical characteristics by the experiment, the author found that the submerged plastic grew less hydrophobic and more neutrally buoyant and was easily colonized by bacteria, but there was no sign of the presence of microorganisms that break down plastic and pays to improved understanding of the fate of plastic debris in the marine environment [24]. Studies using glass beads and sand-filled saturated columns are used to measure the impact of surface roughness on the release and deposition of micro- and nano-sized colloids at various solution ionic strengths (e.g., >0.01 M) and tuned the Derjaguin approximation method and computed interaction energies, which show that by lowering the energy barrier, the sharp asperities on sand surfaces can promote colloid deposition in primary minima [25]. Using a parallel plate flow chamber system, the impact of gravitational force on the deposition of bacteria *Burkholderia cepacia* G4g and carboxylate-modified polystyrene latex (CML) microspheres at 0.5, 1.1, and 1.8 mum has been assessed and suggested that gravity was a significant driving force for the initial stages of deposition of particles that were larger than 1 mm size confirmed by a particle trajectory model [26]. However, in the absence of energy barriers, they carried out unit model simulations and experimental observations of the retention of colloids of varying sizes and densities in porous media under both up-flow and downflow circumstances and suggested significantly different deposition under both flow conditions (e.g., for large or dense colloids with gravity number NG > 0.01 at pore water velocity of 4 m/day), which reflect underlying gravity and flow influences on simulated colloid trajectories that produced significantly different attached colloidal distributions over the model surfaces [27]. They predicted a model for colloid transport in dual-permeability media that takes into account first-order exchange between the two areas of aqueous phases, as well as the reversible and irreversible retention of colloids and colloid concentrations in aqueous and solid phases, have analytical solutions by using matrix decomposition and Laplace transformation worked well for evaluating how model parameters affected the colloid distribution [28]. The mechanistic model that forecasts colloid attachment in unfavorable circumstances also predicts colloid release when subjected to elevated fluid velocity and decreased ionic strength (IS), which are thought to be common circumstances for the mobilization of environmental colloids. This model replaces mean-field colloid-collector interaction with a discrete representation of surface heterogeneity, which takes into account both repulsive and attractive interactions at the same time and leads to a distribution of attraction strengths in an attached colloid population that is in primary minimum contact with the surface [29]. Colloid transport under the same flow field circumstances is not well described by dispersion coefficients measured by fitting advection-dispersion transport models to nonreactive tracer breakthrough curves, by laboratory and field experiments. In columns filled with glass beads, 48 colloid transport experiments were carried out under circumstances that hindered the chemical attachment of colloid particles, and nine distinct colloid sizes and a range of flow rates were investigated. A mathematical model that describes the colloid transport in homogenous, water-saturated porous media by the breakthrough curves was effectively recreated [30]. The influence of gravity on the transport of NP in porous media was thoroughly assessed by using carbon tubes combined with graphene oxide (carbon-based NP) along with $TiO_2$ and ZnO metal-oxide NP. Packed column experiments were performed on the two types of carbon-based NP, utilizing both upward (gravity-negative) and downward (gravity-positive) flow orientations under unfavorable conditions. The experiment involving the other two types of metal-oxide NP was carried out in both upward and downward flow orientations under both favorable and unfavorable conditions. Gravity had little impact on the transport and retention of carbon-based NPs under unfavorable conditions, as shown by the breakthrough curves and retained profiles of two carbon-based NPs in up-flow orientation equal to those in downflow orientation [31]. Transport of microplastics *via* phytoplankton aggregates, which make up the majority of the sinking flux, from the ocean's surface to its deeper layers was studied and a novel apparatus, the flow-through roller tank, which simulates the behaviors of laboratory-made aggregates sinking through a thick layer of microplastics, interactions between microplastics and aggregates were investigated. The diatom Chaetoceros neo gracile, the cryptophyte *Rhodomonas salina*, and a grouping of the two created three different forms of collections that were utilized as models. The microplastic sinking rates increased significantly from tenths to hundreds of meters per day as a result of the incorporation and concentration of microplastics in all model aggregates [32].

Researcher [33] investigated the aggregation and sedimentation of carboxyl-functionalized CB-NPs according to the changes in the solution chemistry (0.1–10 mM NaCl and 0.01–1 mM $CaCl_2$) in the presence/absence of natural organic matter (1 and 5 mg $L^{-1}$) humic acid and the humic acid was found to have the highest effect on the stability of CB-NPs under the aquatic circumstances. The sedimentation caused by CB-NP accumulation was more actively observed in NaCl than in $CaCl_2$. With the increase in the ionic strength of the NaCl solution, the accumulation rate of CB-NPs also increased, whereas in the $CaCl_2$ solution, the CB-NPs holdup stability was almost insensitive to the ionic strength. Using a pulse injection method, three polydisperse suspensions of slit particles, with median sizes of 3.5, 9.5 and 18.3 μm, were inspected under high ionic strength conditions, ranging from 0 to 600 mM (NaCl), with two high flowrates of 0.15 and 0.30 cm/s. These conditions were verified in a column system subjected to a continuous flow rate of 0.05 cm/s over 24-hr duration, using the same saline water as during the injection test. The experimental setup closely mirrored the one described in Ref. [34]. Prior to the injection of slit particles, the porous media underwent saline water flow at the specified velocity for a duration of 5 pore volumes, achieving equilibrium when the pH and electrical conductivity of the effluent matched those of the influent. The total interaction energy, $\Phi$, for the system of particles and collectors was defined as the combination of electrostatic ($\Phi EDL$), van der Waals (ΦVDW), and born (ΦBORN) repulsion interaction energy, represented by the equation.

$$\Phi\ (\delta) = \Phi \mathrm{EDL}\ (\delta) + \Phi \mathrm{VDW}\ (\delta) + \Phi \mathrm{BORN}\ (\delta) \qquad (16)$$

The overall interaction energy $\Phi$ was computed by modeling the particle-sand grain system as a sphere-plate interaction. The repulsive electrostatic double-layer interaction energies were determined using the following formula

$$\varnothing_{EOL}^{(\delta)} = \varepsilon_0 \varepsilon_\tau \pi \frac{d_p}{2} \left( 2\varepsilon_g \varepsilon_p \, ln \left( \frac{1 + e^{-k_d \delta}}{1 - e^{-}k_d \delta} \right) = \left( \varepsilon_g^2 + \varepsilon_p^2 \right) ln \left( 1 - e^{-2k_d \delta} \right) \right) \qquad (17)$$

where $\varepsilon_0$ represents the dielectric permittivity, valued at 8.854 × 10–12 F/m, while $\varepsilon_r$ denotes the relative dielectric permittivity of water (equal to 78.4 for water at 20°C), The parameter $\delta$ indicates the distance between the surface of the SP and the sand grain, $d_p$ refers to the diameter of the SP and $\kappa_d$ is identified as the Debye length with $K_d^{-1}$ representing the characteristic thickness of the diffuse charged layer. The Extended DLVO (XDLVO) interaction energy profiles for the SP were determined using the measured zeta potentials at various ionic strengths. The calculation of drag forces (*F*D), adhesive forces (*F*A), repulsive forces (*F*R), and gravity forces (*F*G) were performed according to the established formulas, and one-dimensional transport experiments (short-pulse injection) were tested in a saturated sand column with constant flow velocities and SP. Concentration breakthrough data obtained at the effluent of the packed columns were inspected by the convection-dispersion equation (CDE) including a term of a first-order deposition kinetics as

$$\frac{\partial C}{\partial t} = D_L \frac{\partial_C^2}{\partial_x^2} - u \frac{\partial C}{\partial x} - K_{dep} C \qquad (18)$$

where

$C$—the particle concentration in solution ($ML^{-3}$),

$t$—the elapsed time (T),

$x$—the travel distance (L),

$D_L$—the longitudinal dispersion coefficient ($L^2 T^{-1}$),

$u$—the average velocity of particles ($LT^{-1}$],

$K_{dep}$the SP deposition rate coefficient ($T^{-1}$).

The analytical solution for the above equation no (18) is as follows

$$C(t,x) = \frac{mx}{Q\sqrt{4\pi D_L t^3}} exp\left(-K_{dep} t\right) exp\left( -\frac{(x - ut)^2}{4D_L t} \right) \qquad (19)$$

$m$—the mass of particles injected (M).

The regression parabolic solutions used in Eq. (19) are used to determine the SP transport parameters ($D_L$, $K_{dep}$, and the residence time $t$).

This study showed that the extent of deposition increased with higher ionic strength, reduced flow velocity, and larger ratios of the median diameter of the suspended particles to the median diameter of the sand grains. Additionally, the kinetics of suspended particle recovery and deposition were significantly influenced by the chemistry of the solution, the hydrodynamics, and the size of the suspended particles [35].

A theoretical model that can replicate how biofouling affects the destiny of microplastics. The model is based on contours of ocean depth for light, water density, temperature, salinity, viscosity, settling, and biofilm development. Maximum concentration is reached at middle depths due to the size-dependent upward migration of microplastic particles as predicted, and comparatively low levels of tiny particle abundances are anticipated at the ocean's surface. The settling of plastic particles over depth was modeled as a function of the settling velocity $V_s$ ($ms^{-1}$)

$$\frac{dz}{dt} = V_s(Z,t) \tag{20}$$

$V_s$ was calculated using an alteration of the Stokes equation, provided by [36]. $V_s$ can be used to calculate the equivalent spherical diameter of both spherical and non-spherical particles. This equation was validated for microplastics using the settling data for spherical and pure microplastic particles in Ref. [37] as

$$V_s(Z,t) = -\left(\frac{\rho_{tot} - \rho_{sw,z}}{\rho_{sw,z}} g\omega v_{s\omega,z}\right)^{\frac{1}{3}} \tag{21}$$

where

$\rho_{tot}$—the density of the plastic particle with biofilm (kg $m^{-3}$),
$\rho_{sw,z}$—sea water density at depth z (kg $m^{-3}$),
$v_{sw,z}$—is the kinematic viscosity of the seawater ($m^2$ $s^{-1}$),
g—the gravitational acceleration (m $s^{-2}$).
The settling velocity ω was calculated as

$$\omega = 1.74 * 10^{-4} D^2 \; for \; D < 0.05 \tag{22}$$

$$log\,(\omega) = -3.76715 + 1.92944\,log\,(D) - 0.09815(logD)^2 - 0.00575(logD)3 + 0.00056(logD)4$$

For 0.05

$$\geq D \geq 5 * 10^9 \tag{23}$$

The dimensionless particle diameter (D) was calculated from the equivalent spherical diameter D in (m) as

$$D = \frac{(\rho_{tot} - \rho_{sw,z}) g D_n^3}{\rho_{s\omega,z} v_{sw,z}^2} \tag{24}$$

Almost all the plastic particle sizes were assumed to be spherical, and the biofilm was homogeneously distributed over the particles. Then, the density of plastic particles together with biofilm will be calculated as

$$\rho_{tot}(Z,t) = \frac{r_{pl}^3 \rho_{pl} + \left[(r_{pl} + t_{bf})^3 - r_{pl}^3\right] \rho_{bf}}{(r_{pl} + t_{bf})^3} \tag{25}$$

in which $\rho_{tot}$ the total density (kg $m^{-3}$) was calculated from the particle radius $r_{pl}$ (m), the biofilm thickness $t_{bf}$ (m), and the biofilm and plastic density, $\rho_{bf}$ and $\rho_{pl}$, respectively (kg $m^{-3}$). Biofilm volume was calculated using the volume of algae cells, $V_A$ in ($m^3$), the number of attached algae, A ($m^{-2}$), and the plastic surface area

$$V_{bf} = V_A A \theta_{pl} \tag{26}$$

The attached algal growth, $dA/\mathrm{dt}$, was dynamically modeled as:

$$\frac{dA}{dt} = \frac{\beta_A A_A}{\theta_{pl}} + \mu_A(T, l)A - m_A A - Q_{10}^{\frac{T-20}{10}} R_{20} A \tag{27}$$

In the right side of the first term of the models (27) signifies the fouling through the collision of algae with the plastic particles, the second term calculates light intensity and the temperature-limited growth of the algae, the third term calculates for (grazing) mortality, and the last term computes for respiration of algal, respectively. When it is believed that the algae on the microplastics have access to enough nutrients. Then, the modification of a Gaussian model to the vertical chlorophyll profile concentration model is an accepted method in oceanography as

$$Chl(Z) = \frac{chl\ a(z)}{c^{hla}_{\mathrm{zbase}}} = C_b - sz + C_{max} e^{-\left(\frac{z-z_{max}}{\Delta z}\right)^2} \tag{28}$$

in which Chl (z) is the normalized chlorophyll concentration at depth, z, $C_b$ is the normalized surface concentration, s is the normalized slope, $C_{max}$ is the normalized maximum concentration, $Z_{max}$ the depth at which the maximum concentration can be found, and Δz (m) the width of the peak. $c^{hia}_{zbaze}$ is the average chlorophyll-a concentration of the vertical profile (mg $m^{-3}$). For the Brownian encounter kernel rate, the diffusivity of the plastic was calculated as

$$D_{pl} = \frac{k(T + 273.16)}{6\pi\mu_{sw} r_{tot}} \tag{29}$$

where k is the Boltzmann constant, μ is the dynamic water viscosity, and $r_{tot}$ is the sum of the shear rate of the plastics and the biofilm thickness taken from Ref. [38].

### 3.1 Convective-dispersive of MP model

Using the convective–dispersive equation (CDE), and the column pore water velocity and dispersity, the transport of MPs was simulated for one-site kinetic retention with total mass balance by Dong et al. [39] as

$$\frac{\partial c}{\partial t} + \frac{\rho}{\theta}\frac{\partial s}{\partial t} = D\frac{\partial^2_C}{\partial x^2} - v\frac{\partial C}{\partial x} \tag{30}$$

and

$$\frac{\rho}{\theta}\frac{\partial s}{\partial t} = K\psi C \tag{31}$$

where c ($M/L^3$) is the concentration of the MPs in the aqueous suspension, s (M/M) are the concentrations of the MPs on the sand surface, $t$ is time (T), $x$ (L) is the distance from the column inlet, **ρ** ($M/L^3$) is the dry bulk density of the packed column, θ is the porosity of the packed column, $D$ ($L^2/T$) is the hydrodynamic dispersion coefficient, $v$ (L/T) is the pore water velocity, $K$ is the first-order attachment coefficient ($T^{-1}$), and ψ is the blocking factor.

$\psi$ a blocking factor is calculated as follows

$$\Psi = \left(1 - \frac{S}{S_{max}}\right)\left(\frac{d_c + x}{d_c}\right)^{-\beta} \tag{32}$$

Or

$$\psi = max\,(1, S^{s_{max}}), \tag{33}$$

$S^{smax}$ (M/M) – the maximum solid phase concentration of the MPs,$d_c$ (L)—the median diameter of the sand grains, and β (0.432)—an empirical parameter that controls the shape of the colloid spatial distribution. The overall MP-MP interaction energies were determined using the classical Derjaguin-Landau-Verwey-Overbeek (DLVO) theory in order to better understand the stability of MPs in saltwater [40, 41] as given in Ref. [39].

Freshwater and oceanic sediments had the greatest diversity in polymer composition. Thus, it is contended that a significant portion of plastic debris gathers in this area. This demonstrates that plastics' density, surface area, and size had the greatest effects on the patterns of plastic trash transit and accumulation [42]. The most significant impacts on the movement and accumulation patterns of plastic waste were assessed. They studied how soil colloids and surface chemistry jointly influenced the movement of tetracycline and ciprofloxacine through sand under steady-state saturated flow conditions. In comparison with sand (5.11 mg/kg for ciprofloxacin and 2.80 mg/kg for tetracycline), both antibiotics showed a much higher adsorption capacity on soil colloids (17,500 mg/kg for ciprofloxacin and 8600 mg/kg for tetracycline). The adsorption of ciprofloxacin increased to 8.91 mg/kg when treated with humic acid. Tetracyclin adsorption rose to 7.99 mg/kg after the sand was coated with iron oxide and to 8.35 mg/kg after being coated with humic acid. The strong adsorption capacity of ciprofloxacin resulted in a recovery rate of less than 4% in the effluents from the columns containing 0, 20, and 50% of iron oxide/humic acid-coated sand. The surface coating decreased the recovery rates of tetracycline from 35.4% (in uncoated sand) to 12.0% (in a column containing 50% iron oxide-coated sand) and 0.010% (in a column containing 50% humic acid-coated sand) [43].

### 3.2 Adsorption-desorption mechanism

The kinetic adsorption model of antibiotics, fitted on sand and colloids in the form of the pseudo-second-order kinetics formula, is given in Eq. (34)

$$q(t) = \frac{Kq_e^2 t}{1 + Kq_e t} \tag{34}$$

where K is the rate constant of adsorption (g/mg/h), $q_e$ is the amount of antibiotics adsorbed at equilibrium (mg/g), q(t) is the amount of antibiotics adsorbed at time t (mg/g), and $kq_e^2$ is the initial rate of change of adsorption (mg/g/h). A one-dimensional form of the convection-dispersion two kinetic retention sites was used to describe the transport behaviors of antibiotics and colloids [44].

$$\frac{\partial \theta_c}{\partial t} = \rho_b \frac{\partial (s_1)}{\partial t} + \rho_b \frac{\partial (s_2)}{\partial t} = \frac{\partial}{\partial x}\left(\theta D \frac{\partial C}{\partial x}\right) + \frac{\partial q_c}{\partial x} \tag{35}$$

where θ is the volumetric water content (dimensionless), c is the colloid/antibiotic concentration (mg/L), t is time (h), $\rho_b$ is the bulk density of the porous media (mg/cm$^3$), x is the travel distance (cm), D is the hydrodynamic dispersion coefficient, and q is the Darcy velocity (cm/h). $S_1$ and $S_2$ are the solid phase concentrations associated with retention Type 1 and 2 sites, respectively.

This two-site model assumes that the total number of adsorption sites consists of two types of sites. On site 1 (Type 1 site, [Eq. (36)]), the adsorption is assumed to be instantaneous and reversible, while on site 2 (Type 2 site, [Eq. (37)]), the adsorption is kinetically controlled and irreversible.

$$\frac{\partial(S_1)}{\partial t} = \frac{\theta}{\rho_b}\varphi_t K_{att1} C - K_{det1} S_1 \tag{36}$$

$$\frac{\partial(S_2)}{\partial t} = \frac{\theta}{\rho_b}\varphi_x K_{att2} C \tag{37}$$

where $K_{att1}$ is the first-order attachment rate of site 1, $K_{det1}$ is the first-order detachment rate of site 1, $K_{att2}$ is the first-order attachment rate of site 2, and $\psi_t$ and $\psi_x$ are dimensionless functions that account for time-dependent retention and depth-dependent retention, respectively.

$$\varphi_t = 1 - \frac{S_1}{S_{max\,1}} \tag{38}$$

$$\varphi_2 = \left(\frac{d_c + x}{d_c}\right)^{-\beta} \tag{39}$$

The values of these functions are given as ([45, 46]) where.

$S_{max1}$—the parameter in the blocking function for the first adsorption site, β—an empirical factor controlling the shape of the spatial distribution, and $d_c$—the median diameter of the sand packed in the column.

Eqs. (35)–(37) are solved by Hydrus-1D software in Ref. [47].

Results from Grand Canonical Monte Carlo (GCMC) and molecular dynamics (MD) simulation suggested that the primary mechanism for pesticide adsorption on MP was surface adsorption. The final configurations absorption showed that intricate interactions existed between the pesticide molecules and the carbon and hydrogen atoms in polyethylene (PE), indicating they could serve as potential carriers for pesticide in aquatic environments. Microplastics' adsorption capability was computed as

$$Q_t = \frac{(C_0 - C_t)V}{m} \tag{40}$$

where $Q_t$ represents the adsorption capacity of microplastics (mg/g), $C_0$ denotes the initial pesticide concentration (mg/L), $C_t$ reflects the residual pesticide concentration at any given time (mg/L), V signifies the volume of the pesticide solution (L), and m indicates the mass of microplastics (g). The adsorption mechanism of pesticides onto microplastics was analyzed using both pseudo-first-order kinetics and pseudo-second-order kinetics.

$$\frac{t}{Q_t} = \frac{1}{K_2 Q_e^2} + \frac{t}{Q_e} \tag{41}$$

where $Q_e$ represents the equilibrium unit adsorption capacity of microplastics (mg/g), while $Q_t$ indicates the unit adsorption capacity of microplastics at any given moment (mg/g). $K_1$ denotes the rate constant for pseudo-first-order kinetics (1/h), $K_2$ signifies the rate constant for pseudo-second-order kinetics (g/(mg.h)), and t refers to the duration of adsorption (h).

The Langmuir adsorption isotherms model assumes that there are no interactions between the adsorbents and that of the adsorbent surface, which is uniform. Adsorption merely takes place on the adsorbent's outside, or it is single-layer adsorption.

$$\frac{C_e}{Q_e} = \frac{1}{Q_m K_L} + \frac{C_e}{Q_m} \tag{42}$$

where $Q_m$ represent the monolayer saturated adsorption capacity (mg/g), $C_e$ signify the equilibrium mass concentration of pesticides (mg/L), and $K_L$ indicates the Langmuir adsorption equilibrium constant (L/mg). The adsorption mechanism of an uneven surface could be described by the Freundlich empirical model for adsorption isotherms, which is suitable to low concentration adsorption.

$$ln\, Q_e = ln\, K_f + \frac{1}{n} ln\ C_e \tag{43}$$

$K_f$ represents the Freundlich adsorption equilibrium constant, while n denotes the concentration index, indicating both the adsorption capacity and the uniformity of the adsorbent's surface. When n falls between 2 and 10, the process is considered favorable for adsorption. The adsorption thermodynamic calculation was performed by using the Van't Hoff equation as follows:

$$\partial G = -RTlnK \tag{44}$$

$$\partial G = \partial H - T\partial S \tag{45}$$

$$ln\ K = \frac{\partial H}{RT} + \frac{\partial S}{R} \tag{46}$$

where K is the equilibrium constant, R is the gas constant, 8.314 J/ (mol/K), T is the temperature (K), δG is the Gibbs-free energy change (J/mol), δH is the enthalpy change (J/mol), and δS is the entropychange (J/(molK)) [48]. Using column experiments, this work compared the transport differences between degradable and nondegradable MPs under various conditions (flow rates, ionic strengths (ISs), pH, and coexisting cations). Ultraviolet (UV) irradiation was utilized to further replicate the effects of aging on different types of microplastics (MPs). For this comparison, polylactic acid (PLA) and polyvinyl chloride (PVC) were selected as the polymers. The advection-dispersion equation (ADE) was used to model the transport data of the MPs. The interaction energy between MPs and quartz sand was precisely predicted by the Derjaguin-Landau-Verwey-Overbeek (DLVO) theory, which describes the forces between charged surfaces interacting within a liquid medium and offers a mathematical framework for understanding the aggregation and kinetic stability of aqueous dispersions. It combines the electrostatic repulsion caused by the so-called double layer of counterions with the van der Waals attraction:

$$\beta U(\gamma) = Z^2 \lambda_B \left(\frac{e^{Ka}}{1+Ka}\right)^2 \frac{e^{-KT}}{T} \tag{47}$$

$\lambda_B$ is the Bierrum length, U is the potential energy, e = 2.71929, Euler's number, K is the inverse of the Debye-Huckel screening length ($\lambda_D$), $K^2 = 4\,T\,\lambda_B\,n$, $\beta^{-1} = K_B$ thermal energy scale at absolute temperature [49].

An investigation into the horizontal movement of macro- and microplastics ranging in size from 50 to 0.25 mm on the soil surface caused by rain-induced runoff was conducted using a surface runoff plot experiment and the influences of vegetation cover, the characteristics of plastics, and rainfall scenarios were investigated and showed that the presence of vegetation significantly enhance the retention of plastics by about 20% and the herb plant (Photinia×fraseri Dress) showed a better interception efficiency on plastics than the shrub plant (*Ophiopogon japonicus).* Increase rainfall amount from 20 to 60 mm boosted the transport of plastics and controlling the soil erosion could also be used to prevent plastics in soil to enter into aquatic environment [50]. Particle tracking and hydrodynamic modeling were combined to create a method for examining the movement and spread of microplastic particles in riverine environments [51]. The fate and transport behaviors of both pollutants in porous media are transformed when MPs and PFOA (an emerging surfactant pollutant) coexist in the same environment and this interaction is highly interrelated with the amount of PFOA adsorbed onto MPs and the MPs' preliminary surface properties [52]. Colloid transport study in the column experiments was simulated using a one-dimensional discrete random-walk particle-tracking model, which was created at PNNL. This model applies to advection and dispersion, as outlined in Ref. [53]. Deposition of the suspended colloids model is governed by a relationship of the form

$$\frac{\partial C}{\partial t} = -K_f C \tag{48}$$

C refers to the concentration of suspended colloids, which is defined as the number of particles per unit volume of fluid, while $K_{ds}$ denotes the deposition rate coefficient. This coefficient is represented in a discrete manner by assigning a probability of deposition to each numerical particle during each time step. A similar approach is employed to model the re-entrainment process through a re-entrainment rate coefficient ($K_f$) that is applied to the population of colloids that have been reversibly deposited. The particle model also facilitates the simulation of partially irreversible deposition through a fraction of the deposited particles that are irrevocably attached ($f_{ir}$). The mean deposition rate coefficient and the standard deviation of the natural logarithm of the deposition rate coefficient are specified as in Ref. [54].

$$\mu_{kf} = \theta_{kf} e^{-\left(\frac{\sigma_{lnk^2}}{2}\right)} \tag{49}$$

$$\sigma_{kf}^{2=\mu_{kf}^2}\left(exp\left(\sigma_{lnkf}^2\right)-1\right) \tag{50}$$

$$\mu_{lnkf=ln\,(\mu kf)+\left(\frac{\sigma_{lnkf}^2}{2}\right)} \tag{51}$$

where $\Theta$ is the median of the distribution. Note that $\ln\Theta_{kf} = \Theta_{lnkf}$ because the natural log transformation is monotonic, and that $\Theta_{lnkf} = \mu_{lnkf}$ because the normal distribution is symmetric. It follows that $\mu_{lnkf} = \ln(\mu_{kf}) + (\sigma l_{nkf}^{\,2}/2)$. The aforementioned models yield the mean and standard deviations of $K_f$, given the input data [55].

They developed governing equations to represent the movement of MPs within a saturated porous medium. The simulations indicate that to effectively manage this environmental threat, it is essential to decrease the diffusivity of MPs while enhancing water flow velocity, porosity, and the first-order attachment coefficient. The mass balance equations for MPs are as follows:

$$\frac{\partial C_t}{\partial t} = -\nabla . f \tag{52}$$

$$C_t = \theta C + \rho S \tag{53}$$

$$f = -\theta D \nabla C + v\theta C \tag{54}$$

where $C_t$ represents the total concentration of microplastics ($N/m^3$), f denotes the total flux of microplastics ($N/m^2s$), C is the concentration of microplastics in the aqueous phase ($N/m^3$), S indicates the concentration of microplastics in the solid phase (N/kg), θ stands for the volumetric water content (−), **ρ** refers to soil bulk density ($kg/m^3$), D is the diffusivity ($m^2/sec$), and v represents the average velocity of pore water (m/s). By substituting Eqs. (53) and (54) into Eq. (52), the one-dimensional flow can be expressed as follows.

$$\theta \frac{\partial C}{\partial t} + \rho \frac{\partial S}{\partial t} = \theta D \frac{\partial_C^2}{\partial x^2} - v\theta \frac{\partial C}{\partial x} \tag{55}$$

MPs in the solid phase can be defined using the following attachment-detachment model for colloids by Lee et al. [56]

$$\rho \frac{\partial S}{\partial t} = K_a \theta C - K_d \rho S \tag{56}$$

$k_a$ is the first-order attachment coefficient (1/s)

$$K_a = \frac{3(1-\theta)}{2d_{50}} \eta \alpha v \tag{57}$$

$$\rho \frac{\partial S}{\partial t} = K_a \theta C \tag{58}$$

Galerkin's residual scheme is used to solve the important partial differential equations of (55)-(58).

Equation (55) described a one-dimensional advection-dispersion model (ADE) for tracer transport as given in [57]

$$\frac{\partial C}{\partial t} = -v \frac{\partial C}{\partial x} + D_l \frac{\partial_C^2}{\partial_x^2} \tag{59}$$

where *C*—the tracer concentration in the liquid phase [$ML^{-3}$],
*t* is the time [T],
*v*—the pore water velocity [$LT^{-1}$] $v = q/n$ (*q* is Darcy velocity),
$q = Q/A$ (A is the cross area),
*n*—the effective porosity,
x—the flow length [L],
$D_L$—the longitudinal dispersion coefficient [$L^2T^{-1}$]

$$D_L = \alpha_L \theta \tag{60}$$

where $\alpha_L$ is the dispersity [L]. The effect of the molecular diffusion is measured as negligible [58] and a steady-state flow condition with constant water content and water flux was assumed by Wang et al. [59]. So

$$C(t) = \frac{M}{Qt_0} \frac{1}{\sqrt{4\pi P\left(\frac{t}{t_0}\right)^3}} exp\left[-\frac{\left(1-\frac{t}{t_0}\right)^2}{4P\frac{t}{t_0}}\right] \tag{61}$$

where $t_0$ is the mean transit time, $t_0$ = x/v, **ρ** is the dispersion parameter [$ML^{-3}$]

$$\rho = \frac{D_{L\ t_0}}{x^2} \tag{62}$$

$M$ is the mass of injected tracer in SI unit as [M], $Q$ is the flow rate in [$L^{-3}\ T^{-1}$]. By minimizing the root mean square error (RMSE), the dispersity is inversely calculated using the restrained tracer breakthrough curves from the column experiments as

$$RMSE = \sqrt{\frac{\sum_{i=1}^{n}(P_i - O_i)^2}{n}} \tag{63}$$

$P_i$ and $O_i$ are the predicted and observed tracer concentrations, and n is the number of observations.

## 4. Results and discussion

It is necessary to identify model parameters for prediction and control. In this study, some models are considered to simulate response on exit MP concentration. Those experimental data are used to identify parameters using least square ARX-based model's structure.

Parameter Identification:

Wang et al. [47] presented a MP transport model through porous media as explained below.

Microplastic transport model [60]:

From the attachment, detachment, and depth-dependent retention model, the following equation has been derived,

$$\theta\frac{\partial C}{\partial t} + P_b\frac{\partial S_1}{\partial t} + P_b\frac{\partial S_2}{\partial t} = \theta\left(D_L\frac{\partial_C^2}{\partial x^2} - q\frac{\partial C}{\partial x}\right) \tag{64}$$

After rearranging above Eq. (64) for site1 and site2

$$\frac{\partial C}{\partial t} = D_L\frac{\partial_C^2}{\partial x^2} - q\frac{\partial C}{\partial x} - \frac{P_b}{\theta}\frac{\partial S_1}{\partial t} - \frac{P_b}{\theta}\frac{\partial S_2}{\partial t} \tag{65}$$

For site 1

$$P_b\frac{\partial S_1}{\partial t} = \theta K_{aat1}C - P_b K_{det1}S_1 \tag{66}$$

For site2

$$P_b \frac{\partial S_2}{\partial t} = \theta K_{aat2} \varphi C - P_b K_{det2} S_2 \tag{67}$$

By substituting Eqs. (66) and (67) into Eq. (65), the model for the removal of MP has been derived:

$$\frac{\partial C}{\partial t} = D_L \frac{\partial_C^2}{\partial x^2} - q \frac{\partial C}{\partial x} - \frac{1}{\theta}[\theta Kaat_1 C - P_b Kdet_1 S_1] - \frac{1}{0}[\theta Kaat_2 \varphi C - P_h Kdet_2 S_2] \tag{68}$$

$$\frac{\partial C}{\partial t} = D_L \frac{\partial_C^2}{\partial x^2} - q \frac{\partial C}{\partial x} - Katt_1 C + \frac{P_b}{\theta} Kdet_1 S_1 - Katt_2 C \left[\frac{d_{50} + x}{d_{50}}\right]^{-\beta} + \frac{P_b}{\theta} Kdet_2 S_2 \tag{69}$$

$$\frac{\partial C}{\partial t} = \frac{D_L}{\rho_{MP}} \frac{\partial_C^2}{\partial x^2} - q \frac{\partial C}{\partial x} - C\left\{Katt_1 + Katt_2 \left[\frac{d_{50} + x}{d_{50}}\right]^{-\beta}\right\} + \frac{P_b}{\theta}\{Kdet_1 S_1 + Kdet_2 S_2\} \tag{70}$$

Equation (70) is solved using the MP transport model's identified parameter values that are presented in **Table 1**.

| Serial NO | Term | Unit | Value | Reference Paper |
|---|---|---|---|---|
| 1. | C-Deuterated water tracer concentration. | mg/ liter = ppm | 49.29 | Wang et al. [47] |
| 2. | $D_L$—Longitudinal dispersion coefficient of tracer. | $cm^2/sec$ | $0.18*10^{-3}$ | |
| 3. | $\alpha_L$—Dispersity of the tracer[L]. | cm | 0.22 | |
| 4. | v-Tracer velocity. | cm/sec | 0.00085 | |
| 5. | Darcy's law and data from a constant head experiment, the hydraulic conductivity. | cm/min | 0.051 | |
| 6. | $D_L$—Dispersion coefficient of microplastic. | $m^2/s$ (Given) | 0.016 cm2/sec | Ma et al. [60] |
| 7. | $\rho_{MP}$—Density of microplastic. | $g/cm^3$ | 1.05 $g/cm^3$ | Ma et al. [60] |
| 8. | C-MPs concentration in the effluent. | $kg/m^3$ (Given) | MP concentration in inlet =20 ppm = 0.02 mg/ml | Effluent concentration 0.001 mg/ml |
| 9. | x-Distance from column inlet. | m (Given) | Total length of column = 32 cm | 300 nm, Ma et al. [60] |
| 10. | q-Pore water velocity. | m/s (Given) | 0.086 cm/sec | Ma et al. [60] |
| 11. | $K_{att1}$—First-order attachment coefficient for the site1. | 1/min(Given) | 9.67*exp.(−3) | |

| Serial NO | Term | Unit | Value | Reference Paper |
|---|---|---|---|---|
| 12. | $K_{att2}$—First-order attachment coefficient for the site-2. | 1/min (Given) | 1.39*exp.(−2) | |
| 13. | $D_{50}$—Median diameter of sediments. | m (Given) | The grain size distribution of sand is 630–800 μm with a median diameter of 715 μm and a uniformity coefficient of 1.13. | Ma et al. [60] |
| 14. | β—The empirical factor that controls the shape of RPs. | dimensionless (Given) | 0.432 | |
| 15. | $P_b$-Bulk density of packed columns. | $kg/m^3$ | 1.58 $g/cm^3$ | Ma et al. [60] |
| 16. | Θ—Porosity. | $m^3/m^3$ (Given) | 0.393 (±0.005) | Ma et al. [60] |
| 17. | $K_{det1}$—First-order detachment coefficient for site1. | 1/min (Given) | 6.56*exp.$^{(-2)}$ | |
| 18. | S1—MPs concentration in the solid phase 1. | kg/kg (Given) | S1 > S2 | 0.5 mg/ ml = (500 ppm) |
| 19. | $K_{det2}$—First-order detachment coefficient for site 2. | 1/min (Given) | 3.45*exp.$^{(-4)}$ | |
| 20. | S2—MPs concentration in the solid phase 2. | kg/kg (Given) | | 0.001 mg/ml (100 ppm) |

**Table 1.**
*Important model parameters identified in tracer models by the authors.*

After substituting the values given in **Table 1**, the MP transport model has been derived as given in Eq. (71). Using the PDEPE command in MATLAB, the equation is simulated in three-dimensional (3D) and two-dimensional (2D) views, and shown in **Figure 2a** and **b**, respectively. As time increases, the microplastic concentration

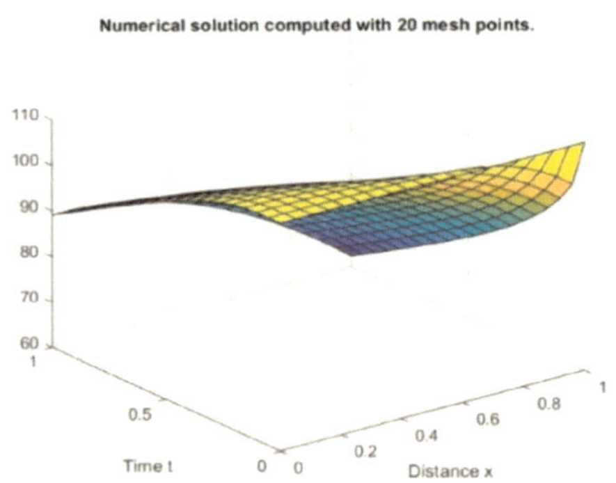


(a)

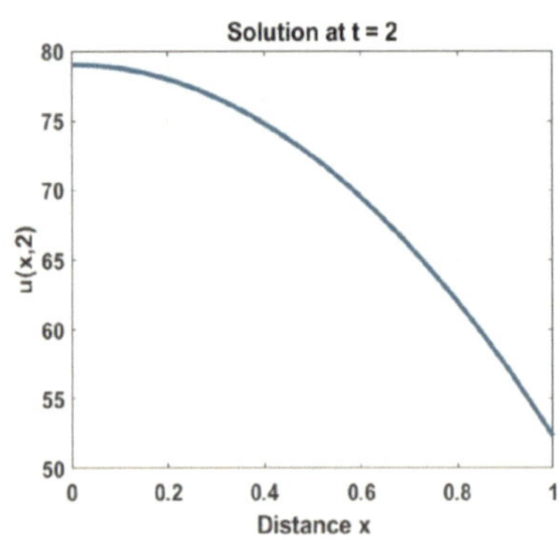


(b)

**Figure 2.**
*a and b show the response of microplastic concentration in 3D and 2D models.*

decreases in 2D figure. With 20 mesh points in a 3D plot, the microplastic concentration decreases over time as well as with distance.

$$\frac{\partial C}{\partial t} = 16\frac{\partial_C^2}{\partial x^2} - 0.086\frac{\partial C}{\partial x} - 28.7711 \quad (71)$$

These profiles are important as it hints for the design of the packed bed.

## 5. Future direction of research

Contamination of MPs in groundwater has wide implications on human physiology when it is used for drinking purposes. However, the awareness and studies are still in the dark. The mechanism of migration of MPs through soil needs to be more transparent as other pollutants also migrate along with it. Tracking the movement with an isotope study may be beneficial in finding traces of MPs in different layers of soil. The structure of soil is more complicated than that of water. Soil is an uneven mixture of clay, sand, plant residues, straw, impurities, and asphalt. Therefore, percolation of MPs through soil is different than diffusion through water. After pretreatment the average microplastic concentration was $0.33 \pm 0.34$ particles/$m^3$ as suggested in [61]. Thus, practical approaches to the separation technologies of MPs need more studies and research. There are enriched reports of soil pollution with heavy metals in the past; however, when MPs mix with these heavy metals, the contaminated groundwater quality and separation techniques need modifications and further studies with risk assessment models. Pollution control and mitigation studies can be supportive in formulating policy decisions. Also, it is more dependent on terrestrial conditions.

## 6. Conclusion

The systematic review using existing research bases reveals that there is a lack of analytical techniques, contamination of MPs through drinking water, separation methods, and operational strategies. The study consisted of finding the formation, characteristics of MPs, and their occurrences in aquatic bodies, and the associated mathematical models have been reviewed in this study. The pre-filtration treatment followed by filtration in a packed tower is the prime important technique for the purification of potable water. The migration of MPs through the ground soil has been explained in many research articles.

## Acknowledgements

The authors would like to thank Rajalakshmi Engineering College, Rajalakshmi Nagar, Thandalam, Chennai-602105, and Erode Sengunthar Engineering College, Perundurai, Erode-638057, TamilNadu, India, for their support in carrying out the study.

## Conflict of interest

The authors declare no conflict of interest.

**Author details**

Sobana Subramani[1]*, Rames C. Panda[2], Sundararaman T.R.[2], Murugesan S.N.[2], Venkatachalam Varadharajan[1] and Balasubramanian Karuppiah[1]

1 Department of EIE, CSE and Biomedical Engineering, Erode Sengunthar Engineering College, Erode, Tamil Nadu, India

2 Department of Chemical Engineering, Rajalakshmi Engineering College, Thandalam, Chennai, India

*Address all correspondence to: sobbana.subramani@gmail.com

## References

[1] Frank YA, Vorobiev ED, Vorobiev DS, Trifonov AA, Antisferov DV, Hunter TS, et al. Preliminary screening for microplastic concentrations in the surface water of the Ob and Tom Rivers in Siberia, Russia. Sustainability. 2021;**13**: 80. DOI: 10.3390/su13010080

[2] Eo S, Hong SH, Song YK, Lee J, Lee J, Shim WJ. Abundance, composition, and distribution of microplastics larger than 20 μm in sand beaches of South Korea. Environmental Pollution. 2018;**238**:894-902. DOI: 10.1016/j.envpol.2018.03.096

[3] Rillig MC, Ingraffia R, de Souza Machado AA. Microplastic incorporation into soil in agroecosystems. Frontiers in Plant Science. 2017;**8**:1805. DOI: 10.3389/fpls.2017.01805

[4] Rillig MC, Ziersch L, Hempel S. Microplastic transport in soil by earthworms. Scientific Reports. 2017;**7**: 1362. DOI: 10.1038/s41598-017-01594-7

[5] Youn-Tae K, Han WS, Yoon H-O. Mobility of microplastics in subsurface environments: Current knowledge and perspectives. Journal of soil Groundwater Environment. 2019;**24**:1-12. DOI: 10.7857/JSGE.2019.24.3.001

[6] Hyeok KJ, Myeong H, Kwan K. Application of quartz crystal microbalance to understanding the transport of microplastics in soil and groundwater. Korean Journal of Mineralogy and Petrology. 2020;**33**:463-475. DOI: 10.22807/KJMP.2020.33.4.463

[7] Ryu H-S, Moon J, Kim H, Lee J-Y. Modeling and parametric simulation of microplastic transport in groundwater environments. Applied Science. 2021;**11**: 7189. DOI: 10.3390/app11167189

[8] Sun J, Dai X, Wang Q, van Loosdrecht MCM, Ni B-J. Microplastics in wastewater treatment plants: Detection, occurrence and removal - A review. Water Research. 2019;**152**(1 April):21-37. DOI: 10.1016/j.watres.2018.12.050

[9] Choran N, Ormeci B. Micro-flow imaging for in-situ and real-time enumeration and identification of microplastics in water. Frontiers in Water. Sec. Environmental Water Quality. 2023;**5**(May):1148379. DOI: 10.3389/frwa.2023.1148379

[10] Bradford SA, Yates SR, Bettahar M, Simunek J, Brown GE. Physical factors affecting the transport and fate of colloids in saturated porous media. Water Resources Research. 2002;**38**(12): 63-61. DOI: 10.1029/2002WR001340

[11] Bradford SA, Simunek J, Bettahar M, Van Genuchten MT, Yates SR. Modeling colloid attachment, straining, and exclusion in saturated porous media. Environmental Science & Technology. 2003;**37**(10):2242-2250. DOI: 10.1021/es025899u

[12] Sirivithayapakorn S, Keller AA. Transport of colloids in saturated porous media: A pore-scale observation of the size exclusion effect and colloid acceleration. Water Resources Research. 2003;**39**:4. DOI: 10.1029/2002WR001583

[13] Keller AA, Sirivithayapakorn S, Chrysikopoulas CV. Early breakthrough of colloids and bacteriophage MS2 in a water-saturated sand column. Water Resources Research. 2004;**40**:8. DOI: 10.1029/2003WR002676

[14] Bradford SA, Bettahar M, Simunek J, Th M, Genuchten V. Straining and attachment of colloids in physically heterogeneous porous media. Vadose

Zone Journal. 2004;**3**(2):384-394. DOI: 10.2136/vzj2004.0384

[15] Yiantsios S, Karabelas A. The effect of gravity on the deposition of micronsized particles on smooth surfaces. International Journal of Multiphase Flow. 1998;**24**(2):283-293. DOI: 10.1016/S0301-9322(97)00062-1

[16] Frey JM, Schmitz P, Dufreche J, Gohr Pinheiro I. Particle deposition in porous media: Analysis of hydrodynamic and weak inertial effects. Transport in Porous Media. 1999;**37**(1):25-54. DOI: 10.1023/A:1006546717409

[17] Brant JA, Childress AE. Assessing short-range membrane–colloid interactions using surface energetics. Journal of Membrane Science. 2002;**203**(1–2):257-273. DOI: 10.1016/S0376-7388(02)00014-5

[18] Bradford SA, Jiri JS, Bettahar M. Straining of colloids at textural interfaces. Water Resources Research. 2005;**41**:10. DOI: 10.1029/2004WR003675

[19] Tong M, Joainson WP. Excess colloid retention in porous media as a function of colloid size, fluid velocity, and grain angularity. Environmental Science & Technology. 2006;**40**(24):7725-7731. DOI: 10.1021/es061201r

[20] Bradford SA, Simunek J, Bettahar M, Van Genuchten MT, Yates SR. Significance of straining in colloid deposition: Evidence and implications. Water Resources Research. 2006;**42**:12. DOI: 10.1029/2005WR004791

[21] Ahfir N-D, Wang HQ, Benamar A, Alem A, Massei N, Dupont J-P. Transport and deposition of suspended particles in saturated porous media: Hydrodynamic effect. Hydrogeology Journal. 2007;**15**(4):659-668. DOI: 10.1007/s10040-006-0131-3

[22] Johnson WP, Li X, Assemi S. Deposition and re-entrainment dynamics of microbes and non-biological colloids during non-perturbed transport in porous media in the presence of an energy barrier to deposition. Biological Processes in Porous Media: From the Pore Scale to the Field. 2007;**30**(6):1432-1454. DOI: 10.1016/j.advwatres.2006.05.020

[23] Tarkzaban S, Bradford SA, Martinus T, Genuchten V, Walker SL. Colloid transport in unsaturated porous media: The role of water content and ionic strength on particle straining. Journal of Contaminant Hydrology. 2008;**96**(1):113-127. DOI: 10.1016/j.jconhyd.2007.10.006

[24] Shen C, Li B, Wang C, Huang Y, Jin Y. Surface roughness effect on deposition of nano- and micro-sized colloids in saturated columns at different solution ionic strengths. Vadose Zone Journal. 2011;**10**(3):1071-1081. DOI: 10.2136/vzj2011.0011

[25] Chen G, Hong Y, Walker SL. Colloidal and bacterial deposition: Role of gravity. Langmuir : The ACS Journal of Surfaces and Colloids. 2010;**26**(1):314-319. DOI: 10.1021/la903089x

[26] Lobelle D, Cunliffe M. Early microbial biofilm formation on marine plastic debris. Marine Pollution Bulletin. 2011;**62**(1):197-200. DOI: 10.1016/j.marpolbul.2010.10.013

[27] Ma H, Pazmino EF, Johnson WP. Gravitational settling effects on unit cell predictions of colloidal retention in porous Media in the Absence of energy barriers. Environmental Science & Technology. 2011;**45**(19):8306-8312. DOI: 10.1021/es200696x

[28] Leij FJ, Bradford SA. Colloid transport in dual-permeability media. Journal of Contaminant Hydrology. 2013;**150**:65-76. DOI: 10.1016/j.jconhyd.2013.03.010

[29] Pazmino E, Trauscht J, Johnson WP. Release of colloids from primary minimum contact under unfavorable conditions by perturbations in ionic strength and flow rate. Environmental Science & Technology. 2014;**48**(16):9227-9235. DOI: 10.1021/es502503y

[30] Chrysikopoulos CV, Katzourakis VE. Colloid particle size-dependent dispersivity. Water Resources Research. 2015;**51**(6):4668-4683. DOI: 10.1002/2014WR016094

[31] Cai L, Zhu J, Hou Y, Tong M, Kim H. Influence of gravity on transport and retention of representative engineered nanoparticles in quartz sand. Fate and transport of bio colloids and nanoparticles in soil and groundwater systems. Journal of Contaminant Hydrology. 2015;**181**:153-160. DOI: 10.1016/j.jconhyd.2015.02.005

[32] Long M, Mariceau B, Gallinari M, Lambert C, Huvet A, Raffray J, et al. Interactions between microplastics and phytoplankton aggregates: Impact on their respective fates. Particles in Aquatic Environments: From Invisible Exopolymers to Sinking, Marine Chemistry. 2015;**175**:39-46. DOI: 10.1016/j. marchem.2015.04.003

[33] Han Y, Hwang G, Park S, Gomez-Flores A, Jo E, Eom I-C, et al. Stability of carboxyl-functionalized carbon black nanoparticles: The role of solution chemistry and humic acid. Journal of Environmental Science Nano. 2017;**4**(4):800-810

[34] Bennacer L, Ahfir N-D, Bouanani A, Alem A, Wang H. Suspended particles transport and deposition in saturated granular porous medium: Particle size effects. Transport in Porous Media. 2013;**100**:377-392

[35] Bennacer L, Ahfir N-D, Alem A, Wang H. Coupled effects of ionic strength, particle size, and flow velocity on transport and deposition of suspended particles in saturated porous media. Transport in Porous Media. 2017;**118**(2):251-269. DOI: 10.1007/s11242-017-0856-6

[36] Dietrich WE. Settling velocity of natural particles. Water Resources Research. 1982;**18**(6):1615-1626. DOI: 10.1029/WR018i006p01615

[37] Kowalski N, Reichardt AM, Waniek JJ. Sinking rates of microplastics and potential implications of their alteration by physical, biological, and chemical factors. Marine Pollution Bulletin. 2016;**109**(1):310-319. DOI: 10.1016/j.marpolbul.2016.05.064

[38] Kooi M, Van Nes EH, Scheffer M, Koelmans AA. Ups and downs in the ocean: Effects of biofouling on the vertical transport of microplastics. Environmental Science & Technology. 2017;**51**:14

[39] Dong Z, Qiu Y, Zhang W, Yong Z, Wei L. Size-dependent transport and retention of micron-sized plastic spheres in natural sand saturated with seawater. Water Research. 2018;**143**:518-526. DOI: 10.1016/j.watres.2018.07.007

[40] Xia T, Fortner JD, Zhu D, Qi Z, Chen W. Transport of sulfide-reduced graphene oxide in saturated quartz sand: Cation-dependent retention mechanisms. Environmental Science & Technology. 2015;**49**(19):11468-11475

[41] Hoek EMV, Agarwal GK. Extended DLVO interactions between spherical

particles and rough surfaces. Journal of Colloid and Interface Science. 2006;**298**(1):50-58. DOI: 10.1016/j.jcis.2005.12.031

[42] Schwarz AE, Ligthart TN, Boukris E, Van Harmelen T. Sources, transport, and accumulation of different types of plastic litter in aquatic environments: A review study. Marine Pollution Bulletin. 2019;**143**:92-100. DOI: 10.1016/j.marpolbul.2019.04.029

[43] Xing Y, Chen X, Wagner RE, Zhuang J, Chen X. Coupled effect of colloids and surface chemical heterogeneity on the transport of antibiotics in porous media. Science of The Total Environment. 2020;**713**:136644. DOI: 10.1016/j.scitotenv.2020.136644

[44] Schijven JF, Šimůnek Jiří. Kinetic modelling of virus transport at the field scale. Journal of Contaminant Hydrology. 2002;**55**:113-135. DOI: 10.1016/S0169-7722(01)00188-7

[45] Adamczyk Z, Siwek B, Zembala M, Belouschek P. Kinetics of localized adsorption of colloid particles. Advances in Colloid and Interface Science. 1994;**48**:151-280. DOI: 10.1016/0001-8686(94)80008-1

[46] Zhou J, Jia R, Brown RW, Yang Y, Zeng Z, Jones DL, et al. The long-term uncertainty of biodegradable mulch film residues and associated microplastics pollution on plant-soil health. Journal of Hazardous Materials. 2023;**442**:130055. DOI: 10.1016/j.jhazmat.2022.130055

[47] Li W, Brunetti G, Zafiu C, Kunaschk M, Debreczeby M, Stumpp C. Experimental and simulated microplastics transport in saturated natural sediments: Impact of grain size and particle size. Journal of Hazardous Materials. 2024;**468**:133772. DOI: 10.1016/j.jhazmat.2024.133772

[48] Li H, Wang F, Li J, Deng S, Zhang S. Adsorption of three pesticides on polyethylene microplastics in aqueous solutions: Kinetics, isotherms, thermodynamics, and molecular dynamics simulation. Chemosphere. 2021;**264**:128556. DOI: 10.1016/j.chemosphere.2020.128556

[49] Fei J, Xie H, Zhao Y, Zhou X, Sun H, Wang N, et al. Transport of degradable/nondegradable and aged microplastics in porous media: Effects of physicochemical factors. Science of The Total Environment. 2022;**851**(1):158099. DOI: 10.1016/j.scitotenv.2022.158099

[50] Han N, Zhao Q, Ao H, Hu H, Wu C. Horizontal transport of macro- and microplastics on soil surface by rainfall induced surface runoff as affected by vegetations. Science of The Total Environment. 2022;**831**:154989. DOI: 10.1016/j.scitotenv.2022.154989

[51] Xiaorong L, Wang X, Liu X, Singh VP. Dispersal and transport of microplastic particles under different flow conditions in riverine ecosystem. Journal of Hazardous Materials. 2023;**442**:130033. DOI: 10.1016/j.jhazmat.2022.130033

[52] Rong H, Qin J, He L, Tong M. Cotransport of different electrically charged microplastics with PFOA in saturated porous media. Environmental Pollution. 2023;**331**:121862. DOI: 10.1016/j.envpol.2023.121862

[53] Scheibe TD, Wood BD. Water Resources Research. 2003;**39**:4. DOI: 10.1029/2001WR001223

[54] Li X, Scheibe TD, Johnson WP. Apparent decreases in colloid deposition rate coefficients with distance of

transport under unfavorable deposition conditions: A general phenomenon. Environmental Science & Technology. 2004;**38**(21):5616-5625. DOI: 10.1021/es049154v

[55] Rodriguez-Seijo A, Lourenco J, Rocha-Santos TAP, da Costra J, Duarte AC, Vala H, et al. Histopathological and molecular effects of microplastics in Eisenia and rei Bouché. Environmental Pollution. 2017;**220**(Pt. A):495-503. DOI: 10.1016/j.envpol.2016.09.092. Epub 2016 Oct 13

[56] Lee S-Y, Baik M-H, Song J-K. Removal characteristics of dissolved uranium by Shewanella p. and application to radioactive waste disposal. Economic and Environmental Geology. 2009;**42**(5):471-477

[57] Li W, Brunetti G, Bolshakova A, Stumpp C. Effect of particle density on microplastics transport in artificial and natural porous media. Science of the Total Environment. 2024;**935**:173429. DOI: 10.1016/j.scitotenv.2024.173429

[58] Gramling CM, Harvey CF, Meigs LC. Reactive transport in porous media: A comparison of model prediction with laboratory visualization. Environmental Science & Technology. 2002;**36**(11): 2508-2514. DOI: 10.1021/es0157144

[59] Wang H, Lacroix M, Masséi N, Dupont J-P. Transport des particules en milieu poreux: Transport des particles en milieu poreux : détermination des paramètres hydrodispersifs et du coefficient de dépôt. Comptes Rendus de l'Académie des Sciences - Series IIA - Earth and Planetary Science. 2000;**331** (2):97-104. DOI: 10.1016/S1251-8050(00)01388-4

[60] Ma J, Zhao J, Zhu Z, Li L, Yu F. Effect of microplastic size on the adsorption behavior and mechanism of triclosan on polyvinyl chloride. Environmental Pollution. 2019;**254**: 113104. DOI: 10.1016/j.envpol.2019.113104

[61] Zhang W, Zhang S, Wang J, Wang Y, Jingli M, Wang P, et al. Microplastic pollution in the surface water of the Bohai Sea, China. Environmental Pollution. 2017;**231**(Pt.-1):541-542. DOI: 10.1016/j.envpol.2017.08.058. Epub 2017 Aug 29